Servicing Zenith Televisions

Also by Bob Rose and Prompt® Publications

Servicing RCA/GE Televisions

Servicing Zenith Televisions

By

Bob Rose

©2000 Sams Technical Publishing

PROMPT© Publications is an imprint of Sams Technical Publishing located at 5436 W. 78th Street, Indianapolis, IN 46268

All rights reserved. No part of this book shall be reproduced, stored in a retrieval system, or transmitted by any means, electronic, mechanical, photocopying, recording, or otherwise, without written permission from the publisher. No patent liability is assumed with respect to the use of the information contained herein. While every precaution has been taken in the preparation of this book, the author, the publisher or seller assumes no responsibility for errors or omissions. Neither is any liability assumed for damages resulting from the use of information contained herein.

International Standard Book Number: 0-7906-1216-X
Library of Congress Catalog Card Number: 00-103515

Acquisitions Editor: Alice Tripp
Editor: Cricket A. Franklin
Typesetting: Cricket A. Franklin
Proofreader: Kim Heusel
Cover Design: Christy Pierce
Graphics Conversion: Philip Velikan, Christy Pierce
Illustrations and Other Materials: Courtesy of Zenith Electronics and PHOTOFACT

Trademark Acknowledgments:
All product illustrations, product names and logos are trademarks of their respective manufacturers. All terms in this book that are known or suspected to be trademarks or services have been appropriately capitalized. PROMPT® Publications and Sams Technical Publishing cannot attest to the accuracy of this information. Use of an illustration, term or logo in this book should not be regarded as affecting the validity of any trademark or service mark.

PRINTED IN THE UNITED STATES OF AMERICA

9 8 7 6 5 4 3 2 1

Dedication

Servicing Zenith Televisions is dedicated to those who have taught me more than I have taught them: our children – Mark, Jennifer, and David, and our grandchildren – Wesley Ryan and Zackary Lawrence.

An Expression of Gratitude

A project like this is never the product of just one person, rather a collaboration among many. Therefore, I want to add to the acknowledgments a very special "thank you" to those who have supported and sustained me in this effort:

Ms. Alice Tripp of Sams Technical Publishing who always had something encouraging to say and never failed to answer my questions promptly and with common sense,

My partner, Robert Graves, who often carried more than his share of the work load at the shop to give me more time to work on this project,

Brian Graves who also shared the work load,

Mr. Conrad Persson of Electronic Servicing And Technology who never failed to tell me that I could do it if I wanted to,

And my wife, Vicki, who always supports me unconditionally even in those times I am difficult to live with.

Acknowledgements

The author gratefully acknowledges the following, without whom this project could not have been completed:

Ms. Alice Tripp of Sams Technical Publishing for her editorial expertise and words of advice at several points during the development of the manuscript.

Sams Technical Publishing for permission to use material from PHOTOFACTS®.

Zenith Electronics for permission to use excerpts and illustrations from factory service literature and technical training materials for each of the chassis mentioned in the thirteen chapters of this book, specifically Mr. Jeff Massie, Mr. Phil Brigman, and Mr. Stacy McAdams.

Finally, Mr. Conrad Persson, editor of Electronic Servicing And Technology, for permission to refer to his magazine at several places in the book.

Contents

Introduction 1

CHAPTER 1 - THE C-2 CHASSIS 7
Available Literature 8
Service Menu 11
Power Supply 11
System Control 12
Sweep Circuits 17
Scan-Derived Voltages 23
High-Voltage Shutdown 23
Video Processing 23
Video Output Circuit 24
Audio Processing 24
Repair History 24

CHAPTER 2 - THE C-3 CHASSIS 31
Features 32
Available Literature 32
Factory Service Menu 32
Power Supply 38
Standby Voltages 40
Power-On Sequence 44
Sweep (Deflection) Circuits 44
Pincushion Circuit 47
System Control 48
Video Processing 48
    Jack Pack 50
    Picture-In-Picture Module 54
35-Inch CRT and Dynamic Focus 54
Audio System 58
    Surround Sound 58
    SEQ 61
Troubleshooting Tips 61
Repair History 64

CHAPTER 3 - THE C-5 CHASSIS 67
Chassis Familiarization and Available Literature 68
Customer Menu 68
Service Menu 68
Power Supply 72

Standby Mode	72
Full Power Mode	72
System Control	72
Video Processing System	78
Deflection Circuits	79
XRP Circuit	84
Audio Circuit	84
Closed Captions	85
Repair History	89

CHAPTER 4 - THE C-6 CHASSIS 91

Available Literature	92
Service Menu	92
Power Supply	94
DC/DC Converter	99
System Control	100
Video Processor and Related Circuitry	100
Tuner	101
Video Output Printed Circuit Board	101
Jack Pack	104
Deflection Circuits	104
Audio Circuit	104
Repair History	107

CHAPTER 5 - THE C-7 CHASSIS 109

Available Literature	112
Service Menu	112
Service Adjustments	113 X
Power Supply	113
System Control and Tuner Functions	118
Deflection Circuits	118
Video Circuit	119
Audio Circuit	119
Repair History	119

CHAPTER 6 - THE C-8 CHASSIS 127

Available Literature	127
Troubleshooting Techniques	129
C-8 Modules	129
Customer Menu	130
Service Menu	130
Service Adjustments	132
Power Supply	135

System Control 139
Troubleshooting the Microprocessor 139
Tuner 139
Video Processing Circuit 146
    Video Processor 146
    IF Module 147
    Source Switch 147
    Picture-In-Picture Assembly 147
    Video Output Module 152
    Sweep Circuits 152
Audio Circuits 157
Repair History 157

CHAPTER 7 - THE C-10 CHASSIS 161

Available Literature 161
Customer Menu 163
Service Menu 163
Power Supply 164
System Control 166
Tuner 172
Deflection Circuits 172
Video Processor 173
Video Output Module 173
Operating Signals 178
Audio System 178
Jack Pack 179
Repair History 179

CHAPTER 8 - THE C-11 CHASSIS 181

Chassis Service Controls 181
Available Literature 182
Menus 182
Power Supply 182
Microprocessor 186
Deflection Circuits 186
Video Circuit 186
Operating Signals 187
Video Output Module 187
Audio Circuits 191
Jack Pack 191
Repair History 191

CHAPTER 9 - THE C-12 CHASSIS 197

Available Literature 199

Service Menu And Chassis Adjustments ... 199
Power Supply ... 199
System Control ... 202
Deflection Circuits ... 205
Video Circuit ... 205
Audio Circuit ... 206
Audio/Video Input Schematic ... 207
Repair History ... 212

CHAPTER 10 - THE Y-LINE ... 213
Overview of GX Chassis ... 214
Available Literature ... 215
Customer Menu ... 215
Service Menu ... 217
Power Supply ... 220
Troubleshooting Primer ... 224
System Control ... 225
Tuner ... 227
Deflection Circuits ... 231
Video Circuit ... 236
    Vertical and Horizontal Drive ... 236
    Video Muting ... 236
    Luma and Chroma Processing ... 236
Audio Signal ... 236
    Key Voltages and Waveforms ... 236
Video Output Module ... 237
AFC Search Problems ... 238
Picture-In-Picture Module ... 239
Audio Circuit ... 240
    Monophonic Audio ... 240
    MTS Stereo ... 241
    Non-MTS Stereo Audio ... 242
Repair History ... 246

CHAPTER 11 - THE Z-LINE ... 247
Available Literature ... 249

GA Chassis ... 249
Service Menu ... 252
Power Supply ... 255
    19/20" Power Supply ... 255
    25/27" Power Supply ... 257
System Control ... 258
Tuner ... 267
Deflection Circuits ... 267
Pincushion Correction ... 273

XRP Shutdown	273
Video Processor and Associated Circuits	273
Video Output Module	276
Audio Circuits	276
Monophonic Audio	276
Non-MTS Stereo Audio	277
MTS Stereo	277
Troubleshooting the Audio Circuits	277
Jack Pack	277

GH Chassis 280

Customer Menu	280
Service Menu	281
Alignment Procedures	281
Video Detector	283
AGC Delay	283
Audio Detector	283
Stereo Level Adjust	283
Gray Scale Adjustment	284
Power Supply	284
System Control	284
Deflection Circuits	288
Video Processing Circuit	292
The Video Output Module	292
The Video Processing IC	292
PIP Circuits	293
Audio Circuit	295

GX Chassis 302

Service Menu	302
Service Procedures	303
Power Supply	305
System Control	311
Deflection Circuits	311
Video Processing Circuits	315
Servicing the Video Processor	315
Video Output Circuit Board	315
PIP Circuit	319
Audio Circuit	320
Repair History	322

CHAPTER 12 - THE A-LINE CHASSIS 325

Available Literature	325

GA Chassis 326

Customer Menu	327
Service Menu	328

Service Adjustments	329
Mechanical Adjustments	330
RGB Cutoff Adjustment	330
Servicing the GA Chassis	330
Power Supplies	334
System Control	338
Tuner	338
Keyboard	338
Deflection Circuits	339
Video Circuits	349
Video Output Modules	355
Audio Circuit	355
Monophonic audio	355
Non-MTS Stereo	356
MTS Stereo	356
Auxiliary Video Switch	356
GB Chassis	**376**
Customer Menu	377
Service Menu	377
Servicing the GB Chassis	378
Power Supply	382
System Control	383
Deflection Circuits	383
Video Circuits	383
Video Output Module	388
PIP and Video Switching Circuits	388
Audio Circuits	388
Repair History	389

CHAPTER 13 - THE B-LINE CHASSIS 401

Available Literature	404
Customer Menu	404
Service Menu	408
Servicing Notes	408
Power Supply	412
Degauss Cycle	413
System Control	418
Deflection Circuits	418
CRT Protection Circuit	419
Video Processing System	422
Main Tuner IF	422
Jack and Switch Circuit	422
Digital Comb Filter	422

Video Processor 423
    Y Signal 423
    Chroma Signal 423
    RGB Processing 423
    Automatic Brightness Limiter (ABL) 423
    White and Black Balance 423
Gemstar And PIP Connections 432
Video Output Module 432
The Audio Circuit 432
Repair History 437

Appendix 447

    Brief Explanation of a Few Audio Terms 447
    Technical Training Literature 448
    A Sample Order Form 450

Introduction

Zenith is one of the oldest and most venerated names in the consumer electronics business. The name has – for about eighty years – stood for quality and innovation, and is a name that ranks among the best in a field that includes true giants in the electronics industry. I checked their web site (www.Zenith.com) and came across a quotation with which I heartily agree: "A pioneer in electronics technology, Zenith has invented countless industry-leading developments, including the first portable and push-button radios, the first wireless TV remote controls, and the first HDTV system using digital technology."

I.

I really didn't know how Zenith got its start or what the name stood for until I read a blurb found on the web site. I was delighted to learn that Zenith Electronics Corporation got its start in 1918 when two "wireless-radio enthusiasts" (HAM radio operators), turned a kitchen in Chicago into a factory and began making radio equipment for other radio amateurs or HAMS. I said that I was delighted to make the discovery because I have been a HAM radio operator for more years now than I care to remember. It has been a hobby that I have been able to share with my two sons. By the early 1920's, the radio factory that began on a kitchen table had grown, at least in part because the infant radio business had also begun to grow. Our two HAM radio operators had gotten in "on the ground floor," a knack that Zenith still has.

But what about the name? I used to think that it meant something like the traditional definition of zenith: a culminating point or the highest possible point. But, I soon found out that it came from one of the founders' amateur radio station call sign, 9ZN. The original radios were sold under the name Z-Nth which became the trademark Zenith. The name was formalized in 1923 when Zenith Radio Corporation was incorporated.

The young company's list of achievements were phenomenal, including the world's very first portable radio (1924), the first home radio receiver to operate on ordinary household electricity as opposed to a battery (1926), and the first push-button tuning radio (1927). It was also about this time that Zenith's famous slogan appeared, "*The Quality Goes In Before The Name Goes On*."

Zenith soon became a leader in other consumer electronics developments such as the first all-electronic TV station (1939), the first FM radio station in the Midwest (1940), and the world's very first subscription TV system (1947). Zenith even pioneered AM and FM radio broadcasting, including the invention of the stereo FM radio broadcast system authorized by the FCC in 1961 and still used world-wide today. It also played a key role in the development of broadcast standards for black-and- white and color television.

Just as they were leaders in the development of early radios and radio technology, Zenith was at the forefront in the development of televisions and television technology. Early developments included some of the first prototype television receivers in the 1930's and experimental TV broadcasts that began in 1939 and continued during World War II. They continued the innovate pace by introducing their first line of black-and-white TV sets in 1948, the industry's first 21-inch, three-gun rectangular color picture tube (1954), and the first practical wireless remote control that revolutionized TV tuning worldwide (1956). The system was marketed under the now famous rubric "Space Command."

Mounting competitive pressure led Zenith to use its broad engineering and marketing expertise to enter the cable television products and original equipment manufacturer components business in the late 1970's They also acquired the Heath Company in 1979 and capitalized on Heath's entry into personal computers to form Zenith Data Systems in 1980. The company's management built the computer business into a billion-plus dollar operation by the late 1980's and sold it in 1989.

The company marketed its last radio in 1982 and changed its name from "Zenith Radio Corporation" to "Zenith Electronics Corporation" in 1984. However, it remained committed to audio engineering as it related to television and co-developed the multichannel television sound (MTS) transmission system adopted by the industry for TV broadcasts (1984) and received an Emmy for that pioneering work (1986).

Zenith kept up its innovative pace by patenting "flat tension mask" technology for high-resolution color video displays with perfectly flat screens. The company also produced TV's with "Sound by Bose" (1988) and "Dolby Surround Sound" (1988), became the first to manufacture televisions with built-in closed captions decoders (1991), produced the first TV's with an electronic program guide (1994), and was the first to feature a set with a track-ball operated remote control (1995). Of course, Zenith has been among the leaders of the industry in the development of digital (HDTV) television.

But the dream story came to an end in the 1990's. I guess a better way to put it is to say that Zenith Corporation has gone through drastic changes in the last ten years. Due to almost cut-throat competition, severe price erosion, and aging manufacturing plants, Zenith's last full year of profits was 1988. Facing the necessity to compete more effectively against giant, foreign-owned firms, the company sought a global partner with complementary skills and looked to L G. Electronics, Inc. (LGE) who acquired a majority interest (55%) in the company in the early 1990's.

Other changes included a major brand revitalization program in 1997 by updating its lightning bolt logo and creating a new marketing campaign designed to introduce the company and its products to a new generation of consumers. It also began to ship digital set-top boxes to major telecommunications companies. Their efforts were rewarded because Zenith was recognized in 1998 as the top consumer electronics brand in a massive customer satisfaction survey in Fortune magazine.

Despite the strengthened product line and the promise of new digital technology, Zenith's financial condition worsened. The corporation asked for and received a restructuring plan designed to return it to financial health. Part of the story ended in 1999 when L. G. Electronics became the sole owner of this proud American company and opened a new chapter to the venture that began in 1918. Good luck to you, Zenith, as the year 2000 begins. May you find the new millennium full of prosperity. I believe you deserve it.

II.

I have devoted just a few pages to the history of the "Zenith Electronics Corporation," and I hope I have given you a feel for what they have stood for and accomplished since their beginning in 1918. Now, let's turn our attention to servicing the televisions they have manufactured.

The first question I faced was, "Where do I begin?" Implicit in the answer is, "What do leave out?" Let's face it, Zenith has manufactured many, many televisions over the years, and lots of them have been in service for twenty to twenty-five years and in some instances even longer. After more than a little thought and on the advice from my editor, I decided to limit my discussion to those sets that have been manufactured in the last ten years, give or take a year or two.

The next question was, "How am I going to organize the material?" Given the multitude of chassis manufactured and sold over the last ten years, how can I choose which ones to include, and how can I organize the included material in a comprehensive way? Zenith organizes its televisions two ways, by chassis type and by model year or "line." The situation becomes complicated because a certain chassis type may belong to several model years or "lines." I decided to use the chassis type of classification to discuss the televisions manufactured up to the time Zenith changed its way of doing service by switching from module support to repair-to-the-component level support that began with the Y-Line. So, I use chassis types in Chapters 1 through 9 and model years, or "lines," as the basic organizing principle in Chapters 10 through 13.

The third question I faced was, "Where do I spend the bulk of my time?" I couldn't spend equal amounts of time on each chassis. Neither could I treat each and every chassis manufactured. I decided to cover the early chassis as comprehensively as possible within certain limits and devote the bulk of my time to the newer products. Chapters 1 through 9 make up about half of the book while Chapters 10 through 13 make up the other half. I decided, in other words, to spend most of my allotted space on the new products, those that require "repair-to-the-component" level attention.

Projection televisions are a world unto themselves which led me to exclude them altogether. Maybe the time will come when I can devote a book just to projection televisions, but that time is not now. Moreover, I decided not to include certain direct view sets, the really high end products, simply because I didn't have time or space to include them. I opted for the most popular chassis on the market, and I think I have been able to treat almost all of them.

III.

There are a number of service aids that make servicing Zenith televisions less a chore and more of a pleasure. May I go on to say that the successful servicer won't neglect resources like these?

1) Web Sites and Addresses (Please keep in mind that these addresses are subject to change without notice. If you work with the internet, you have a feel for how quickly web sites and addresses change! As far as I am able to determine, the addresses are current as of this writing.)

parts.sales@zenith.com - to place orders for Zenith replacement parts and accessories

customer.service@zenith.com - to check pricing, availability, and to get answers to questions about available orders

parts.research@zenith.com - to request research for parts for which you don't have part numbers

parts.credit@zenith.com - to inquire about account status or request copies of invoices or credit memos

www.zenithservice.com - the basic web site that offers a lot of useful information and which provides links to other sites

Questions about any of these sites and their uses may be addressed to:

Webmaster
Zenith Electronics Corporation,
201 James Record Road, Building 3
Huntsville, AL 35824

2) Service Literature – Zenith offers a variety of technical publications. See the appendix for a fairly complete list. You may start a literature subscription or order individual service manuals either from Zenith parts or by contacting:

Zenith Electronics Corporation
Service Literature Distribution
1000 Milwaukee Avenue
Chicago, Illinois 60025

If you prefer, place your order by phone, 847-391-8941, or by fax at 847-391-8726.

I have mixed feelings about some of the early literature (literature published prior to the Y-Line) produced, a fact you will discover as you read through my book. For example, the service data didn't give the detailed information publications like Sams PHOTOFACT offered and required you to use a microfiche to find a part number. Zenith operated with the philosophy, "Trace the problem to the module only, and then replace it. Don't even attempt to fix the module." Such an approach was workable, but it didn't satisfy the dedicated servicer and left those who bought the publications feeling just a little disappointed. Thankfully, the system changed with the introduction of the Y-Line and a "repair to the component level" philosophy. Their service manuals now rank, at least in my opinion, among the best in the industry. I urge you who see lots of Zenith television to consider seriously investing in the factory service literature.

Sams – who has been supplying technical literature for years – still publishes PHOTOFACTS which are some of the most helpful publications for people in the service industry. Address questions about their monthly subscription program to:

Sams Technical Publishing
Attention: Customer Service
5436 W. 78th Street
Indianapolis, Indiana 46268

You can also call them at 1-800-428-SAMS (7267).

Sams now offers a program called E-Facts, a feature that permits you to get a PHOTOFACT via e-mail on the same day you order it. Simply call them up at 800-428-SAMS (7267), tell them what you want, give them a credit card number, and download the information a few hours later.

I believe you will find other publications helpful too. I have no hesitation about recommending the periodical *Electronic Servicing And Technology* to which you may subscribe by writing to:

Mr. Conrad Persson, Editor
Electronic Servicing And Technology
Post Office Box 12486
Overland Park, Kansas 66212

Don't forget ProService published by NESDA, one of the leaders in the field of consumer electronics. The publication is a small one, just a few pages long, but it has the benefit of keeping you current with respect to industry trends. Contact them by writing to:

NESDA
2708 West Berry Street
Fort Worth, Texas 76109

Or, if you prefer, call them at 817-921-9061 or fax them at 817-921-3741.

3) Training Material – Zenith also offers a variety of training material, including paper manuals and video tapes. Zenith is one of the few remaining companies that offers regular service meetings at strategic locations in the country. I can't give you information about the meetings, but I do ask that you see the appendix for the publications that are available and how to order them.

4) Z-Tips is a database sold on floppy disk and contains thousands of easy to follow snippets that let you stay on top of your Zenith repair work. They are reasonably priced and will pay for themselves every time you use them. Z-Cat is a companion program that contains parts pricing and a parts subscription program.

IV.

Consumer electronic products are changing at least every six months. It seems that when you get accustomed to a particular product, the product changes and leaves you to find out how it changed and the best way to effect a repair. The biggest changes seem to be occurring in the fields of miniaturization and microcomputer technology. That is, product size is shrinking because the components that make them are getting smaller and smaller. And the more sophisticated microprocessors become the more functions they are able to control, eliminating extra components and the hardware necessary to fit them into a circuit.

These trends have a profound impact on the servicer, the one who is called upon to fix the product when it breaks. For example, a Zenith TV manufactured several years ago may have as many as dozen modules in it, particularly if it is one of the "digital" chassis. We who have worked on Zenith products for years learned how to navigate the maze of wires and modules to find the problem and seldom needed more than a piece of paper and a good voltmeter. Pop the back off one of today's sets, and what do you find? Usually, a single circuit board that is loaded on both sides with components and, has far more features than its "dozen module" cousin of a few years ago. And, forget about finding the problem with just a piece of paper and a voltmeter. Now you need – at the very least – a good scope and a service manual that may have as many as 100 pages in it. Things certainly have changed!

These changes place certain demands upon the servicer. First, the changes require us to learn all we can. I spend on the average at least an hour a day reading about new products, perusing a new CD-ROM, or watching a training video tape, and I and still can't keep up. The desk in my study at home is piled with manuals, magazines, and tapes that I haven't had time to examine but will get around to soon. Do I have a choice? Not if I plan to stay in business. And neither do you.

Let me give you an example of the way circuits are now being configured and the need to learn about them by pointing out how the new vertical circuits work. The other day I turned a fairly new TV on and listened as it promptly shut down. A few years ago, I would have begun my investigation by checking the power supply voltages first and the XRP circuit second. This time, however, I began by focusing my attention on the vertical circuit which, as it turned out, was the cause for the shut down. That's right. When the vertical deflection fails, the microprocessor responds by turning the TV off.

A second example that comes to mind is operating some of the newer TV's with the video output module removed from the picture tube. If you do, you will more than likely cause irreparable damage to the picture tube and create a situation where you might receive a severe electric shock.

Change is the name of the game, and you and I have "to go with the flow" or be left behind.

Second, the changes require us to learn new service techniques. Some of us made the transition from vacuum tubes to transistors, from transistors to integrated circuits, and now must make the transition from conventional solid state devices to surface mount technology. This means we have to shift our thinking, acquire the skills necessary to cope with the new miniature jungle, and buy the tools that we need to work with the surface mount devices. Griping about it won't make it go away. Learning how to work with it is the only viable option open to us.

Third, the changes necessitate the development of skills that include using the newer electronic test equipment. You really need, for example, a very good signal generator and a very good scope and the ability to use both. You simply will not find the solution to some of the problems the new equipment causes without them. So, forget about popping the back off a modern TV and finding the problem just by using your trusty Simpson 260.

Well, enough of this. Let's see what we can find out about repairing Zenith televisions.

CHAPTER 1
THE C-2 CHASSIS

When it was introduced, the C-2 chassis was assigned to the SENTRY 2 series. Externally, selected models had the then new Euro-style cabinet. Internally, the chassis (or module) was constructed on a one-piece printed circuit board (figure 1-1) and came in a mono and a stereo version.

The following table lists modules that belong to the C-2 family of televisions:

Mono		**Stereo**	
9-913-01	27"	9-896-01	27"
9-914-01	25"	9-911-01	25"
9-915-01	20"	9-912-01	20"

Figure 1-2 gives you an idea about how the components are mounted on the one-piece circuit board. It also serves as a location guide to help you find key voltage and waveform test points. Figures 1-3 and 1-4 provide additional information with respect to cable hookups and the difference between the mono and stereo versions. Figure 1-5 is included to help you locate key adjustment points in the event that you have to replace the module or need to tweak one or more of the so-called "service adjustments."

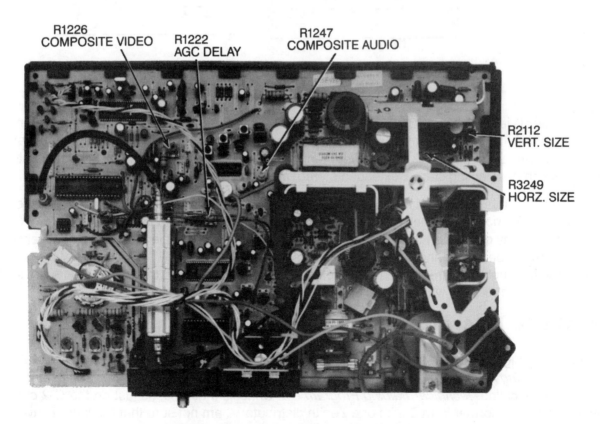

Figure 1-1 One-Piece Printed Circuit Board

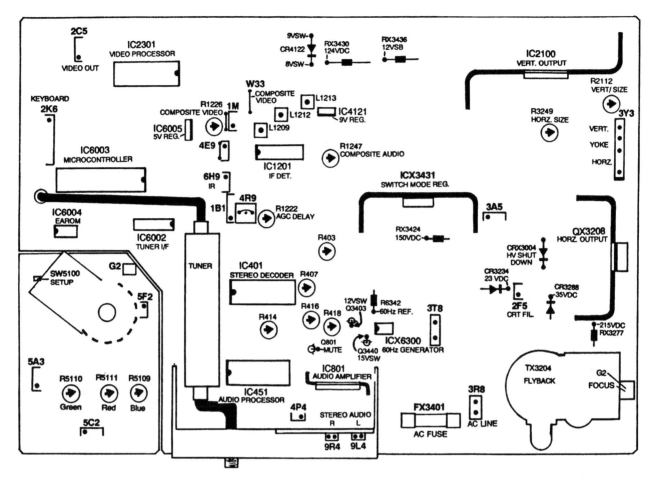

Figure 1-2 Serves as a Location Guide for Components

The C-2 chassis is available in 20", 25," and 27" versions. It features a 178-channel tuning system, full on-screen menu, the now-familiar service menu, and Zenith's auto channel search. The stereo features include a stereo audio jack pack, and a two-channel, one-watt audio output amplifier. The customer has the option of selecting stereo, extended stereo, mono, or SAP from the audio set-up menu.

Let me conclude this brief introduction by noting that the C-2 chassis are aging rapidly. Given the cost of repair and the price of replacement sets, I am seeing fewer and fewer in the shop. Therefore, I have decided to limit my discussion to those areas that are prone to give problems and are relatively easy to fix. You might be wondering why I even discuss these chassis in the first place. My rationale is simple. Even though they are aging and sometimes too expensive to fix, there are still lots of them in use, and some people still want them fixed.

Available Literature

Technical training literature is skimpy. I have been able to locate two booklets that might be of help. *Technical Training Program TP45* covers the C-2 chassis along with several others. Zenith also has a booklet simply called *Technical Training Program C-2 Chassis* that focuses just on the C-2 chassis. You may order the former from Zenith or a Zenith distributor. I am not sure that the latter is available

The C-2 Chassis

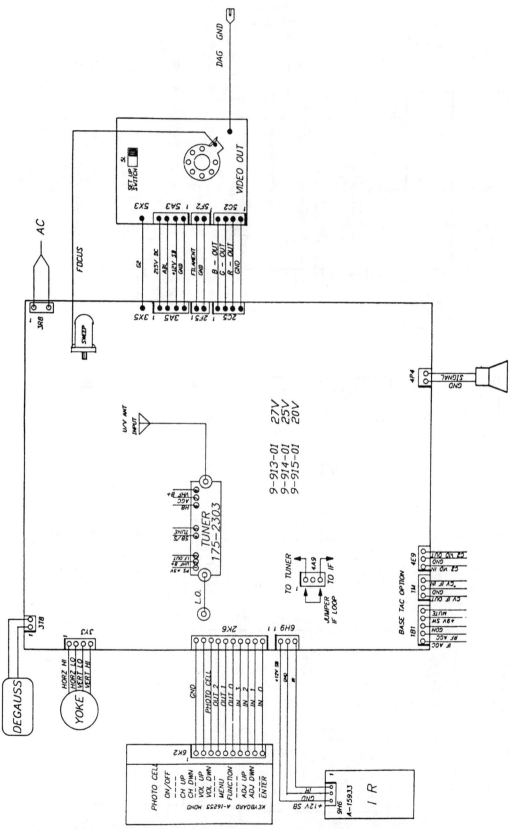

Figure 1-3 Interconnect Diagram for Mono C-2 Chassis

Servicing Zenith Televisions

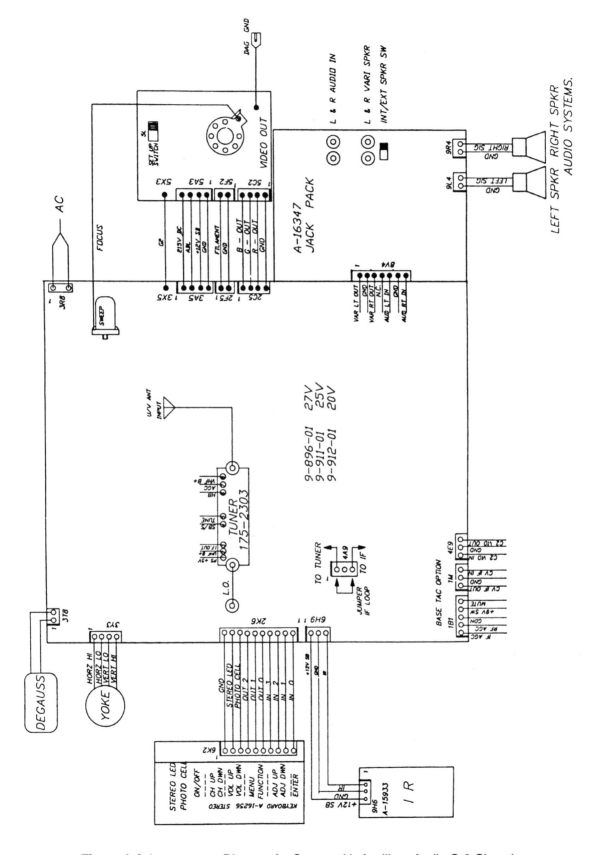

Figure 1-4 Interconnect Diagram for Stereo with Auxiliary Audio C-2 Chassis

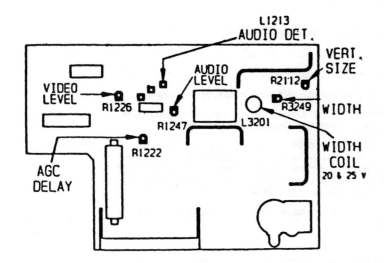

Figure 1-5 Key Adjustment Points for Service Adjustments

because I got my copy at a Zenith service meeting. Since both books contain essentially the same information, I suggest you secure the first one. Besides, it's far easier to read. Because Zenith's literature for the C-2 chassis doesn't always provide the servicer with all the necessary information, I usually use Sams PHOTOFACT 2863 (for the 9-896) or 2872 (for the 9-911) when I need schematics.

Service Menu

The servicer accesses the service menu in one of two ways. First, using the controls on the front of the TV, press and hold at the same time the menu, channel down, and volume down keys or press and hold at the same time the select, adjust left, and adjust right keys. Or second, using the remote control, press and hold the menu button until the customer menu disappears from the picture. Release the menu button and press 9-8-7-6-enter. Page one pops up on the screen.

Page one (service menu #1) contains the video menu and is pink in color. Page two (service menu #2) contains the service features and is green in color. If you want an explanation for each of the adjustments, consult figure 1-6 which is a flyleaf that is distributed with a new or rebuilt module.

Incorrect settings in the service menu cause problems even in televisions as "old" as the C2 chassis. For instance, suppose the on-screen clock runs too fast after you replace the module. How do you correct the problem? Go to the service menu and check to see if the replacement module has been set for 50 hertz instead of 60 hertz!

When you encounter any unusual problem, check the settings in the service menu. Rule out a software problem before you think about a hardware failure.

Power Supply

Zenith calls the power supply "a flyback type of switchmode power supply." Figure 1-7, taken from PHOTOFACT 2872, will be my reference as I discuss it. The power supply comes on in standby mode when the set is plugged into an AC outlet, and it begins to generate not only +124 volts for horizontal deflection but also most of the rest of the voltages necessary to operate the television set. Voltage source 3 (the +12-volt standby) is a major player, serving as a source for eight other voltages. Voltage source 4 provides B+ to the audio output amplifier. The switched voltages become available when the microprocessor initiates the power-on sequence by turning on Q3402.

DESCRIPTION OF FACTORY MENU ITEMS

FACTORY VIDEO MENU:

- BRIGHTNESS — Adjusts 16 levels video brightness.
- VERTICAL SYNC — Adjusts between standard and non-standard vertical mode.
- STORE CUSTOMER CONTROLS — Stores Picture, Black Level, Color Level, Tint, Sharpness, Balance, Bass, Treble, and Volume in the EAROM.
- SENTRY COLOR — Adds a positive color offset to the customer color level. This is intended to compensate for a shift in color level when the photo diode is activated. Now that the photo diode is a feature by itself (Light Sentry), no color change is desired when togging it ON and OFF. Eleven levels of adjustment.
- SENTRY TINT — Same as Sentry Color except with the tint setting. Sixteen levels of adjustment.

FACTORY FEATURES MENU:

- AC CYCLE RATE — Toggles between 50 and 60 hertz systems. This is used for the real time clock so that it keeps time to either 50 or 60 cycles per second.
- VOLUME LIMIT — Limits the maximum volume. Adjust Key will toggle this feature between non-active and active with active assuming the current volume level. Adjusting the volume while in this menu will automatically set this feature in the ACTIVE mode.
- START CHANNEL — Forces the TV to tune to a particular channel upon turn on. Adjust Keys will toggle this feature between NOT ACTIVE and ACTIVE using the currently tuned channel. Tuning to a particular channel while in this menu will automatically set this feature in the ACTIVE mode for the channel which has been tuned.
- CHANNEL LOCK — Allows the TV to tune only one channel. Adjust keys will toggle this feature between NOT ACTIVE and ACTIVE with active assuming the currently tuned channel. Tuning to a particular channel while in this menu will automatically set this feature in the ACTIVE mode for the channel which has been tuned.
- ENVIRONMENT — Toggles between STEREO and MONO chassis. When in MONO mode there is no customer stereo menu since volume is the only audio adjustment which can be made. This menu is also used to permit the factory IR code by adjusting until an "F" is seen in the upper right corner of the display.

All factory adjustments are automatically stored in the EAROM as each adjustment is made.

Figure 1-6 Flyleaf Distributed with a New or Rebuilt Module

Zenith has used this power supply for several years. If you have been keeping up with their products, you know that they still use a modified version of it in the newer sets. Zenith's engineers have changed a component or two to accommodate the needs of the new TV's, but the basic design is still there. Because it is so popular, I have chosen to defer discussion of its operation until Chapter Ten, at which point I intend discuss it in depth and also talk about how to repair it.

System Control

The "brains" of system control is IC6003, an eight-bit device very similar to the microprocessor used in some early microcomputers. It communicates with the EAROM, the memory chip (IC6004), and the tuner control chip (IC6002) via the IM or "InterMetal" bus, a bus that utilizes clock data, and enables lines to facilitate inter-chip communication. Of course, the newer televisions gravitate toward the I2C

The C-2 Chassis

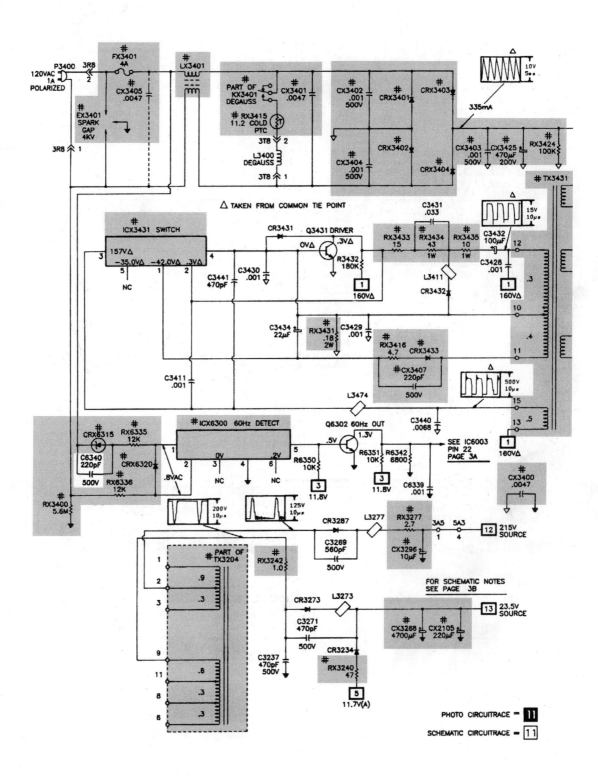

Figure 1-7 A Flyback Type of Switchmode Power Supply

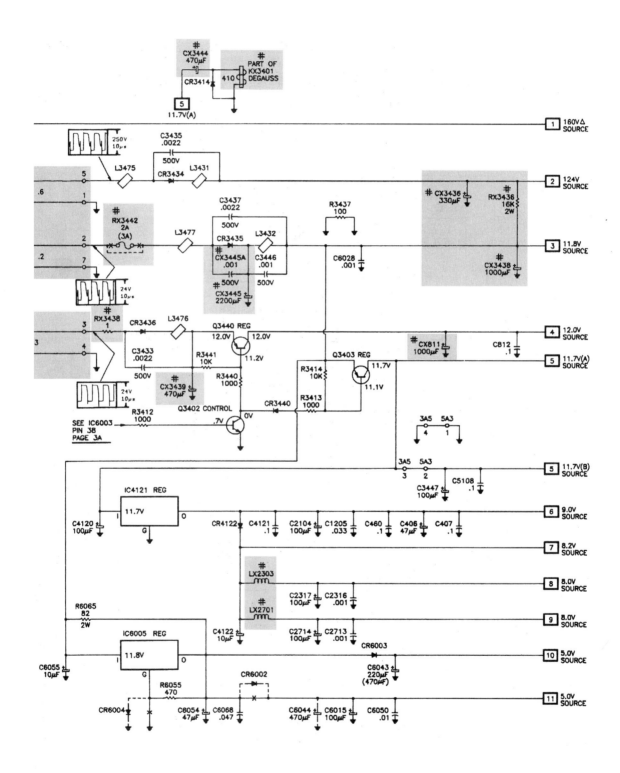

Figure 1-7 A Flyback Type of Switchmode Power Supply (Continued)

The C-2 Chassis

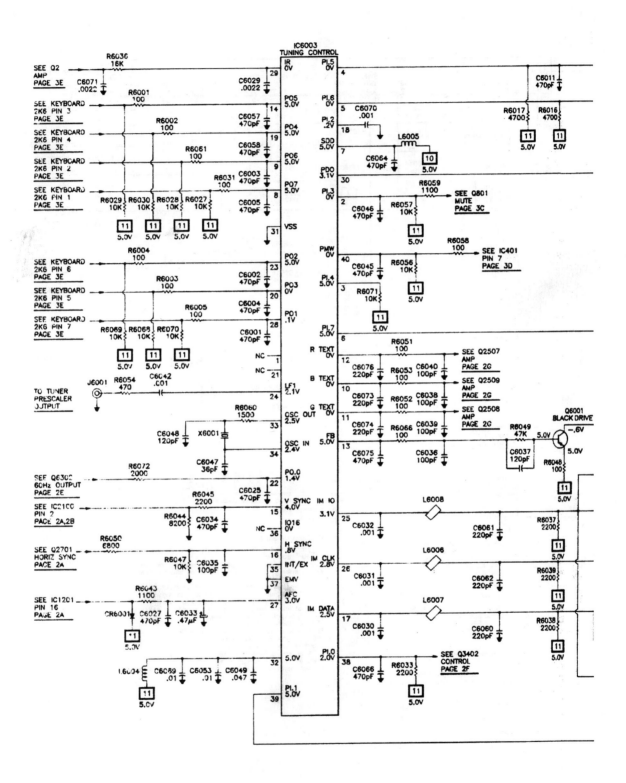

Figure 1-8 Tuner Control Schematic

Servicing Zenith Televisions

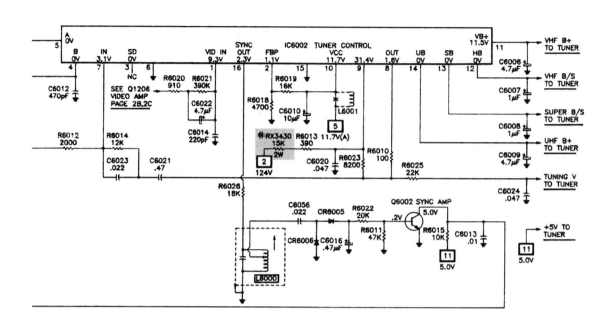

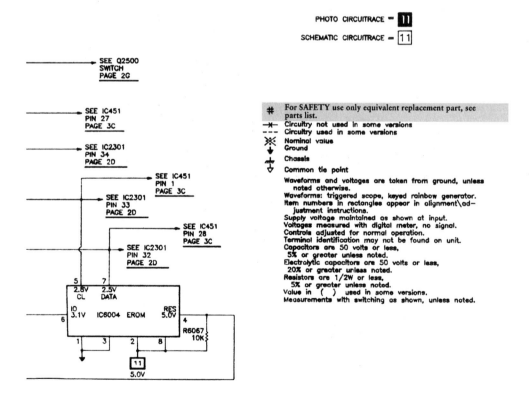

Figure 1-8 Tuner Control Schematic (Continued)

("I-Squared-C") bus Philips engineers developed. It uses just clock and data lines to do the job that three lines used to do. Figure 1-8, from the PHOTOFACT, should give you an idea about how system control is configured and how it works.

Figure 1-8 also lays out the basics of the tuning system with the exception of the tuner, though it shows how the tuner wires into the system. You may be surprised at the number of components used in the C-2 tuning system. The technology is not antique; it's just a bit dated. You ought to know Zenith used a similar scheme for quite a while before moving on to the simpler design found in later chassis.

Before changing a tuner in such a design, you ought to take a few minutes to check the necessary voltages not only at the tuner but also at the tuner control chip, which in this instance is IC6002. When it receives instructions from the microprocessor, the controller sets the operating parameters of the tuner by supplying it with the correct voltages to tune the channel the viewer wants to see. It also performs standard PLL tuning and band switching functions and contains part of what Zenith calls "the valid signal detection circuit." Auto channel search and "no signal mute" – muted audio when there is no video – are two of those "valid signal detection circuit" functions.

If you suspect a system control problem, go directly to IC6003 and check its "must haves." In other words, check for vcc (+5 volts) at pin 32, vss (ground) at pin 31, reset at pin 7, and the correct oscillator waveform (peak-to-peak value and frequency) at pins 33 and 34. Incidentally, you may check vss and vcc at the same time by using pin 31 as the ground point. If you get a good reading, you know ground and B+ are okay. If these are present, check for AC detect at pin 22. If AC detect is good, scope the data input and output lines to see if something is "hanging them up." A stuck front panel switch or a shorted or leaky capacitor on a data input line (pins 8, 9, 14, 19, 20, 23, 28 and 29) often creates some really weird symptoms.

I have encountered two problems with the C-2 system control that are usually not associated with microprocessors. I probably wouldn't mention them except that I have seen both several times.

First, I just recently serviced two sets that came on and shut down about five seconds after turn-on. Since the regulated B+ was "right on the money" and the retrace pulse checked okay, I thought the problem had to be a component in the x-ray protection circuit breaking down under load. After spending a bit of time trying to locate the component, I came to the conclusion that the shut down problem was not XRP related and set the chassis aside because I didn't know what to check next. I removed the microprocessor a bit later from one of the "defective" modules to try it in a third set that had a different problem. Lo and behold, it too came on and shut down in about five seconds, and that led me to the microprocessor as the cause of the problem. I don't why or how it caused the chassis to shutdown, and the technician to whom I spoke at Zenith didn't know either. But I do know a new microprocessor fixed both televisions.

Second, I serviced three C-2 chassis that had vertical deflection problems caused by IC6003. I found the cause by unsoldering pin 15 of the microprocessor. This pin serves as the input port for the vertical signal used to develop on-screen information. When I lifted it, the raster filled out perfectly. But after it filled out, I noticed that the TV didn't have video or audio and that I couldn't change channels or control the volume. A new microprocessor also fixed these sets.

Sweep Circuits

Even though they are relatively straightforward, the sweep circuits do need a comment or two. I chose to use a Sams as the basis for discussion and borrowed the schematic (figure 1-9) from the PHOTOFACT.

Vertical Sweep

Zenith says the vertical output circuit uses a new IC (new in the early1990s) that improves the interlace characteristics of vertical scan. It uses a negative-going pulse rather than a ramp to develop vertical yoke drive because the negative-going pulse has more noise immunity, and it requires both a 9-volt and a 23-volt B+ supply.

If you do the research, you discover that IC2100 (Zenith part number 221-637) actually belongs to the LA78xx series of vertical output integrated circuits. The exact LA78xx depends, of course, on the size of the picture tube. For example, a 25-inch set uses an LA7838, the generic equivalent being an ECG7039 or NTE7039. There are instances that call for an exact manufacturer's part (an OEM), but the generic part seems to work fine in these circuits. However, I usually use the correct LA78xx for the chassis on which I am working. I must assume you know that you can get them from any number of parts jobbers like MCM.

Pinouts for the LA78xx series of ICs:

pin 1	vcc 1
pin 2	vertical drive input
pin 3	time constant
pin 4	vertical amp control
pin 5	vertical size control
pin 6	ramp generator
pin 7	vertical out ac/dc feedback
pin 8	vcc
pin 9	pump-up out
pin 10	oscillator stop
pin 11	ground or vdd
pin 12	vertical output to yoke
pin 13	power supply for vertical output

Note the diode and capacitor configuration at pins 8, 9, and 13. The arrangement permits about 46 volts to be developed to ensure a quick, smooth vertical retrace. A shorted or open diode (CR2101) or a capacitor (CX2110) that has lost capacitance or developed high ESR effectively inhibits vertical deflection, creating either a line across the screen or decreased vertical height. A quick voltage check should tell you if either is causing your problem because the voltage levels at the corresponding pins will be considerably below specs. Incidentally, an ECG125 makes an excellent replacement for CR2101.

If you have read this chapter from the first, you remember that a defective microprocessor sometimes defeats vertical deflection by pulling the vertical drive pulse low. If you suspect IC6003 as the cause of no vertical deflection, simply unsolder pin 15 and see if the raster fills out. If it does, replace the microprocessor. I suppose there are other televisions whose microprocessors kill vertical deflection, but I know of just one other, RCA's CTC167 chassis.

I have serviced dozens of C-2 chassis and have found them to be remarkably free of vertical deflection problems. I do recall having to replace IC2100 a few times, and except for the microprocessor-caused problems, that's about it. One other thing. I suppose you know the LA78xx series of ICs are notorious for developing "ringing cracks" around their pins. When the cracks develop, the TV naturally loses vertical deflection. Simply resolder the pins to fix the problem.

The C-2 Chassis

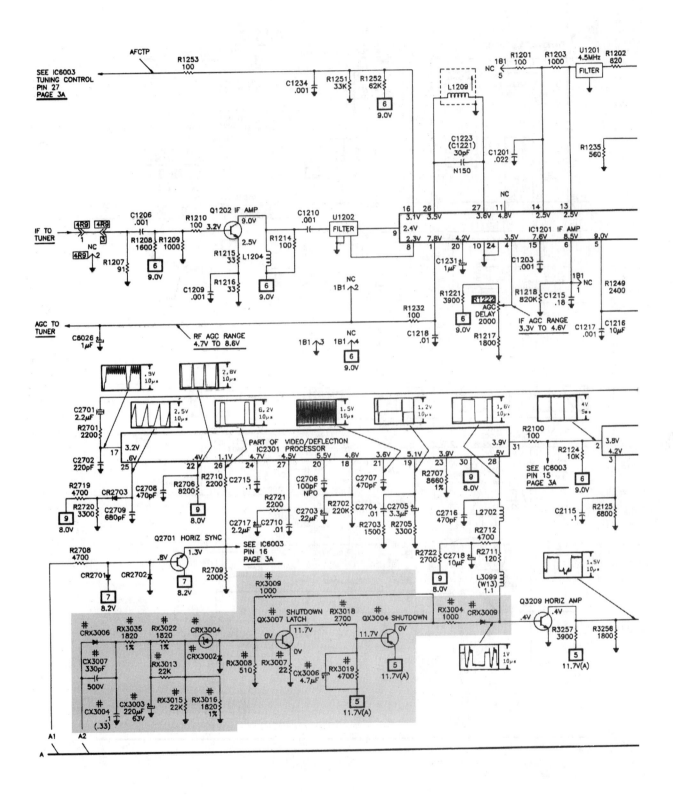

Figure 1-9 C-2 Chassis Schematic

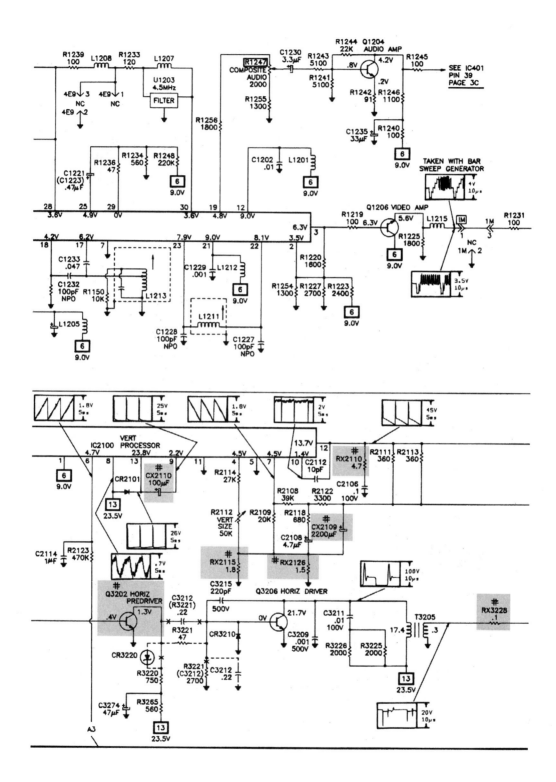

Figure 1-9 C-2 Chassis Schematic (Continued)

The C-2 Chassis

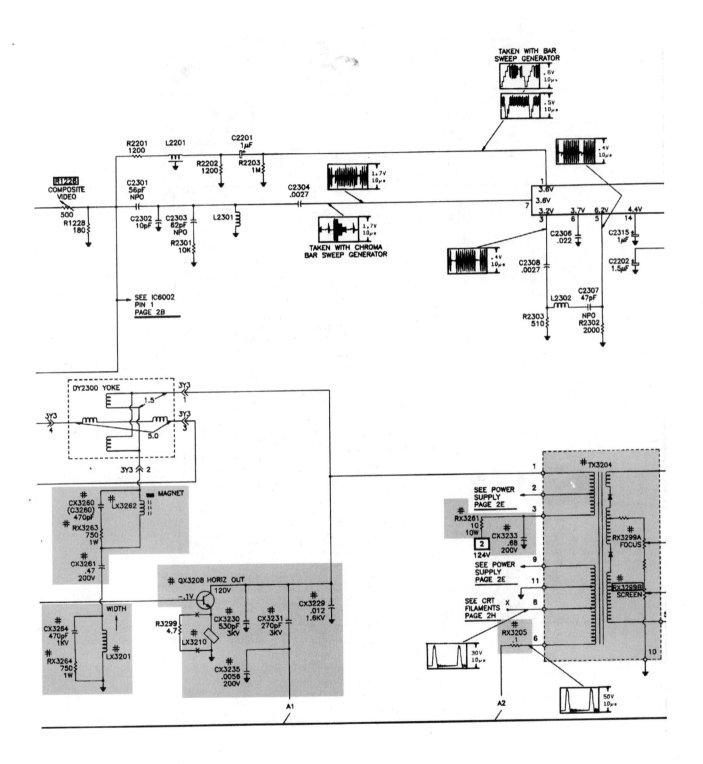

Figure 1-9 C-2 Chassis Schematic (Continued)

Servicing Zenith Televisions

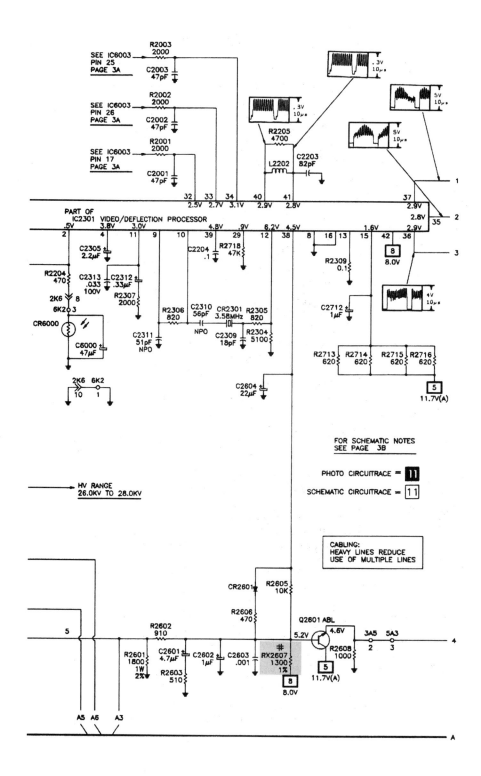

Figure 1-9 C-2 Chassis Schematic (Continued)

Horizontal Sweep

I guess I ought to point out that figure 1-9 depicts the deflection circuits for a 27-inch television. The circuit is virtually the same for the 25-inch sets except for the addition of pincushion correction circuitry. Q3201, Q3202, and LX3201 make up a diode modulator circuit that provides east/west pin correction at the upper and lower portions of the picture and provides an adjustment for the overall horizontal sweep width. The yoke itself accomplishes this correction in the 20-inch and 25-inch sets.

Horizontal drive exits IC2301 at pin 28. If you check it with a scope, you find a nice square wave with a peak-to-peak value of about 1.5 volts. Q3209, Q3202, and Q3206 develop horizontal drive for the horizontal output transistor. Zenith uses such a configuration to develop an out-of- of phase signal for horizontal deflection. An "out-of-phase" signal means the horizontal output transistor (QX3208) is off when the drive circuit is on. Engineers say such a configuration minimizes damage to the circuit if the high voltage shuts down. Moreover, the 50 percent duty cycle produced by the arrangement is sufficient to insure that the horizontal output transistor begins to conduct before the end of the damper diode conduction.

The arrangement works quite well. I seldom have to replace a horizontal output transistor in one of these chassis, and when I do, I often have to replace the flyback as well . The Zenith part number for the output transistor is 121-1141-01. However, an ECG2302, NTE2302, or SK9422 works just as well.

Scan-Derived Voltages

Horizontal deflection produces the usual EHT (extra high tension), focus, and G2 voltages. It also produces the 200-plus volts for the video output transistors, about 35 volts for the diode modulator in the pincushion circuit, and about 23 volts for the vertical output circuit. Other windings generate the voltage necessary to power the filaments of the CRT and supply a pulse to the high-voltage shutdown circuit.

High-Voltage Shutdown

Figure 1-9 depicts a shutdown circuit. It takes a pulse from pin 6 of the flyback and routes it through RX3205 to the anode of CRX3006. CRX3006 and its circuitry develop a DC voltage that is applied to the cathode of zener diode CRX3004. The resulting DC voltage depends on the value of the flyback pulse. If the peak-to-peak value of the pulse increases, the DC voltage applied to the zener diode also increases. If the applied voltage becomes great enough to cause CRX3004 to conduct, QX3007 turns on causing QX3004 to conduct and shunt horizontal drive to ground, shutting the TV down. By the way, QX3007 and QX3004 form a latch circuit. When the circuit latches, you have to remove AC power to unlatch it.

The shutdown circuit is relatively trouble free. On occasion I have had to replace CRX3006, and that's about it. If you suspect a problem with this circuit, defeat it by lifting one end of CRX3009. However, **BE CAREFUL WHEN YOU DEFEAT ANY SHUTDOWN CIRCUIT!** Be absolutely certain the high voltage isn't ramping high by monitoring the peak-to-peak value of the retrace pulse. If your scope can't handle the voltage at the collector of the horizontal output transistor, monitor the pulse at the winding that produces the +200 volts for the video output transistors. By making allowance for its peak-to-peak level, you should get a reasonably good idea whether its peak-to-peak level is too high.

Video Processing

Luminance, chrominance, sync and scan processing, the IM bus interface, and customer controls are very similar to the system used in the C3 chassis. However, the C2 chassis does not contain the video switching, comb filtering, and automatic CRT tracking that the C3 chassis features. Since the C3 chassis is the subject of the next chapter, I intend to defer most of my comments until then.

Let me conclude this section by observing that figure 1-11, which is taken from Zenith's service literature, should give you an idea about how the C-2 chassis processes video.

Video Output Circuit

The video output circuit (figure 1-12) is a separate circuit board tied to the main chassis by means of a series of wire cables. With the exception of an occasional defective video output transistor or bad solder connections, I seldom have to pay attention to it. These C-2 televisions are getting age on them now, and I am seeing more and more cold solder joints crop up on the video output circuit board. I almost condemned one the other day because I thought the picture tube had shorted, but I had the good sense to check the video output transistors before I did. And a good thing too because the bad picture tube turned out to be poor solder around the red drive transistor!

Audio Processing

The C-2 chassis comes equipped with two, different audio systems. Certain ones have a rather full-featured stereo system (figure 1-13) while others have the usual monophonic audio seen in lots of TVs (figure 1-14). Some C-2s also have a jack pack. It certainly isn't sophisticated by today's standards, but it does offer the benefit of audio inputs and outputs (figure 1-15).

Now for a few details. IC451 (figure 1-13) is the audio processor for the stereo versions. It receives input from the stereo decoder (IC401) and the auxiliary input jacks and outputs the audio that the customer has selected. It not only switches between the two audio sources but also controls volume, balance, and tone. After it has been "processed," the audio goes to the audio power amplifier (IC801) and from there to the speakers.

Those chassis that support mono audio naturally have a far simpler audio circuit (figure 1-13) consisting of Q801 (the volume control), Q802 (the mute transistor), and IC801. The microprocessor controls the volume by controlling the conduction of Q801, the output of which goes to pin 7 of the power amplifier through CR801. Q802 mutes the audio when it turns on by making a low impedance path to ground for the signals at pin 5 and 6 of IC801.

The final component in the circuit is IC801. The audio signal exits pin 19 of IC1201 (the video processor) and goes through Q1204 (an audio preamplifier), enroute to the power amplifier. IC801 receives the signal, boosts it, and sends it on to the speaker via connector 4P4.

Please don't confuse this IC801 with IC801 in the stereo circuit because they are entirely different. You might be saying, "Oh, I know that." And, you probably do, but sometimes the obvious needs to be said aloud.

The C-2 Chassis

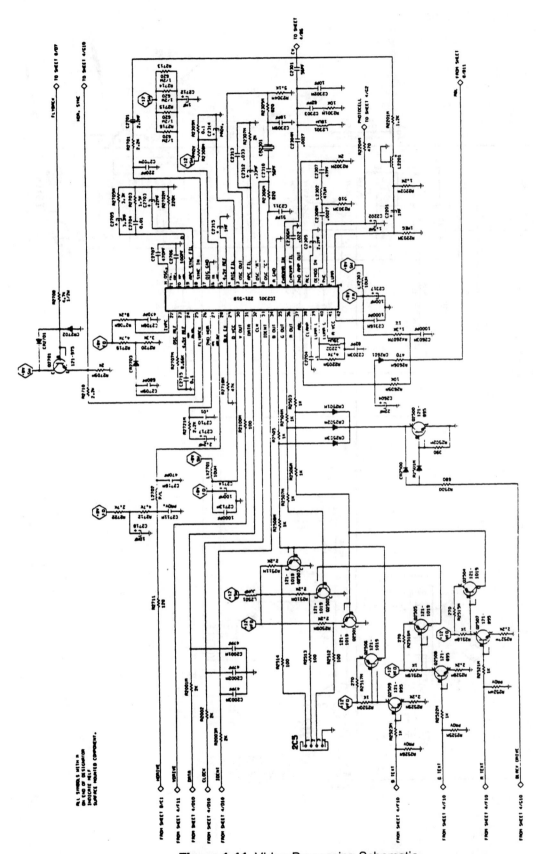

Figure 1-11 Video Processing Schematic

Servicing Zenith Televisions

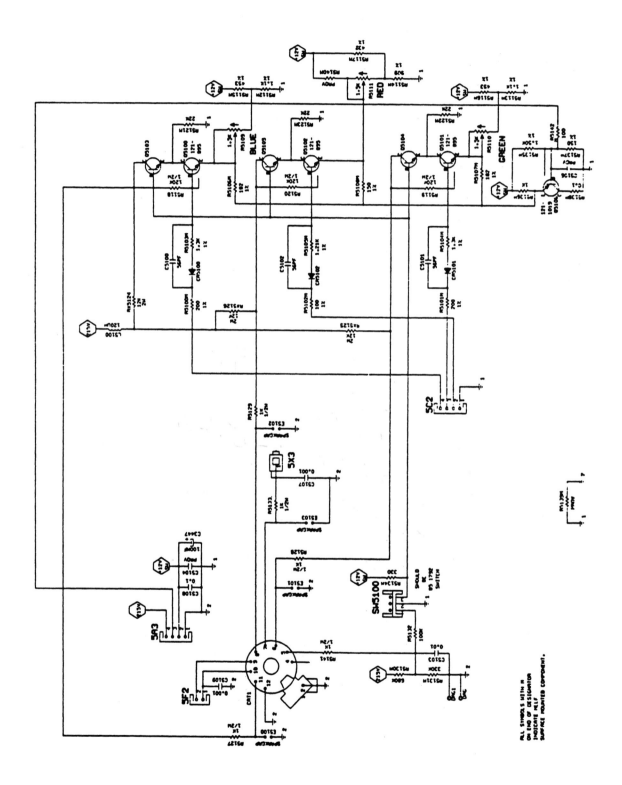

Figure 1-12 The Video Output Circuit

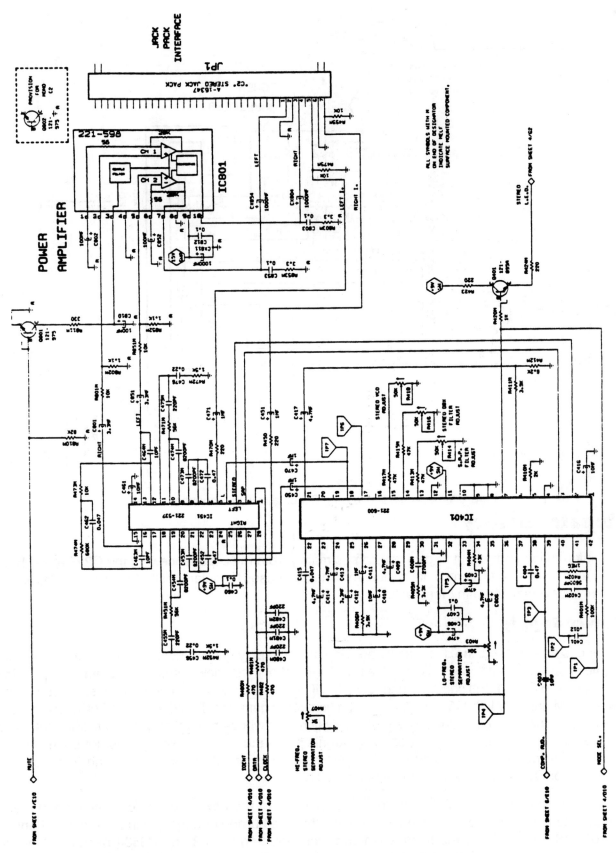

Figure 1-13 Audio Processing Schematic with Full-Featured Stereo

Servicing Zenith Televisions

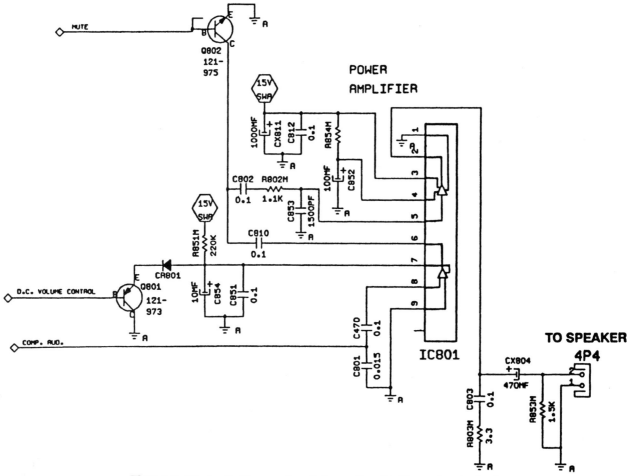

Figure 1-14 Audio Processing Schematic with Monophonic Audio

Repair History

The information I am about to give you comes mainly from my experience with the C2 chassis. But I have also gleaned and stored information from the databases I regularly visit, Zenith service bulletins, and my conversations with other techs.

Power Supply

I intend to defer most of the discussion about the power supply until we get to Chapter Ten. Until then, let me simply say that you should be able to solve most of these power supply problems by replacing ICX3431 and Q3431 and checking and replacing if necessary a few other components. I recommend using OEM parts for ICX3431 and Q3431. If you used a generic STR53041 for ICX3431, make sure it is a good quality one by ordering it from a supply house whose reputation you trust. I bought some from a supply house that offered them at a bargain price and thought I was glad to get them. My joy turned to sorrow when I realized they wouldn't hold up in circuit. I know that some techs use an ECG194 for Q3431, but I really prefer the original part.

Of course, you should check the diodes in the primary of the power supply for opens, shorts, and leaks. Then check the resistors and replace any that are burned or open, paying particular attention to RX3416 and RX3433. You might even consider replacing C3432 and C3434 because they are known troublemakers. If you don't replace them, you really should at least check them.

Be aware that the power supply may have failed because of a shorted picture tube. After you get the power supply up and running, pay close attention to the picture. If the face of the CRT gets white with retrace lines on it, turn the set off immediately to avoid destroying the parts you just replaced.

Dead Set

If the power supply is up and running, begin your troubleshooting at the microprocessor. Check pins 19 and 38 for a voltage change as you issue a series of on and off commands. If the voltage doesn't change, troubleshoot the microprocessor. You should remember that a simple thing like the loss of the 60-hertz signal results in a dead set. For example, if RX6335 (12k at 1/2 watt) opens, the 60-hertz pulse train won't reach the microprocessor, and the TV will be dead. Take a few extra minutes to see if this chip has its "must haves" before you condemn it. I give you these words of wisdom because I have been there and done that and been embarrassed by it more times than I can count!

Let me also point out that a shorted CR3432 also causes a dead set condition because it defeats the standby voltages. It is located at pin 10 of the switchmode transformer. I have even known bad solder around the transformer to cause a dead set.

Vertical Deflection Problems

Most vertical deflection problems can be traced to IC2100 and/or R3242, the resistor fuse in the B+ line to IC2100. I have already discussed the vertical problem caused by the microprocessor and how you can diagnose it.

There have been reports of vertical retrace lines at the top of the picture that are accompanied by a shrinking raster. If you encounter these two symptoms, check CX2110 and CX2109 for loss of capacitance and/or high ESR. I have known these capacitors to cause vertical problems even when they checked good. You might consider replacing them when they check good just in case.

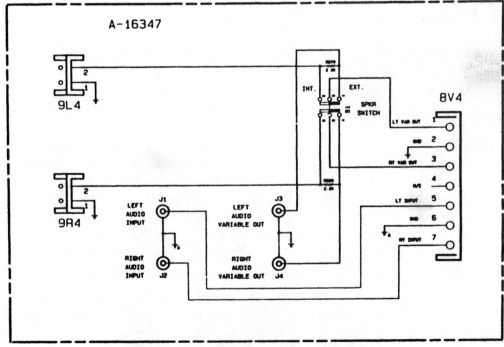

Figure 1-15 Jack Pack Inputs and Outputs

Video Problems

R5133 (1,000 ohms, 1/2 watt) on the CRT board has a history of causing one of two problems. First, if it opens or increases in value, it causes the set to shut down because it permits the CRT to draw too much current after it has been on for a few seconds. It's an easy problem to spot because the screen turns a wicked white color with retrace lines. You might even think the CRT is shorting out, but before you condemn it, check this resistor. Second, a defective R5133 may cause the TV to come on with a bright screen and retrace lines and then go dark after about a half minute of play.

Let me complicate things just a bit by pointing out that a defective RX3277 (2.7 ohms, 1/4 watt) located behind the IHVT also causes the same set of symptoms.

I have seen a set or two that had what looked like a brightness problem because the picture was extremely dim. I found the G-2 voltage low because C5107 had developed a high resistance leakage to ground.

A defective Q2601 may cause a "no video" problem.

Set Won't Turn Off

If you are called upon to service a C-2 that won't turn off, check Q3402. If it is defective, you can replace it with an ECG123AP.

Will Not Autoprogram Or Change Channels

A defective EAROM or microprocessor usually causes these problems. However, I think you will find the microprocessor to be at fault most of the time.

CHAPTER 2
THE C-3 CHASSIS

The C-3 chassis has been in use for about ten years. I first became aware of it in 1989 when I attended a Zenith service meeting in Memphis. Since I have never been particularly good at remembering which Zenith line is which, I can't give you those particulars. I am sure it was in the F, G, and J lines and perhaps more than these three. If C-3 doesn't mean anything, you probably know it as

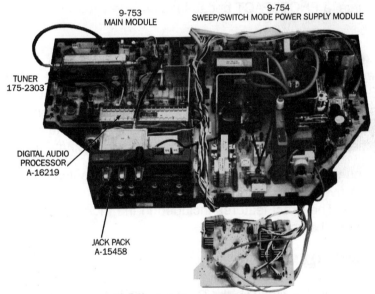

Figure 2-1 The C-3 Chassis

"the dual module monster" that often gives you more trouble than you really want to deal with! Figure 2-1 illustrates a typical configuration and should be familiar to you.

The main modules (small signal panel) used in the C-3 are:

9-753-01	2 x 5W stereo and full-feature jack pack
9-825-01	2 x 1W stereo and full-feature jack pack
9-826-01	2 x 1W stereo and audio/video input and variable audio output only jack pack
9-840-01	mono and RF input only (no jack pack)
9-909-01	same as 9-753-01 except for three resistor value changes in video output for a 31-inch CRT

The sweep-power supply (large signal panel) modules are:

9-754-01	27v, 2 x 5W stereo
9-832-01	31v, 2 x 5W stereo
9-822-01	27v, 2 x 1W stereo or mono
9-823-01	25v, 2 x 1W stereo or mono
9-824-01	20v, 2 x 1W stereo or mono

I find a good interconnection diagram handy when I service one of these sets. So, I have included two: figure 2-3 for the stereo models and figure 2-4 for the mono models.

A component location guide is another handy piece of literature. Figures 2-5 and 2-6 should be adequate for the task. Please understand that there may be variations as you move from one "line" to another, but you should find these illustrations to be quite adequate. Have you noticed that figures 2-5 and 2-6 also have test points, voltage checks, and waveforms on them? But, the best component location guide is, of course, a PHOTOFACT because of its extensive index and grid trace locator.

Features

The C-3 had some neat, innovative features when it came on the market. Those features included auto tracking (like AKB in today's sets), menu tuning, two-input RF switch, source selection, an auxiliary jack pack that had stereo inputs and outputs (figure 2-7), a light sensor, comb filter, noise filter, color sentry, MTS stereo system, parental control, a 178-channel tuning system, and a picture-in-picture option. Zenith offered these features in screen sizes up to and including a 35-inch model.

Even the on-screen menu was innovative. We are accustomed to seeing rather sophistical customer menus these days, but a five-page customer menu in those days was uncommon and gave users more than a few problems. The exact features included in the menu depend on the model year (or "line") to which the television belongs.

Available Literature

I am unable to give you a full bibliography for the C-3 because it doesn't exist. I am working with three booklets that I suppose Zenith still offers if you are interested. They are *Technical Training Program 43*, *Technical Training Program 46* (for the J-Line), and *Technical Training Program: C-3 System 3 and Advanced System 3*. Another booklet, *Technical Training Program: Service Menus, Sentry 2, System and Advanced System 3, System 3 Digital* is also available, but you ought to know that it deals only with service menus. If there is other information about the C-3 chassis than is contained in these publications, I don't know about it.

Zenith's technical literature from this period is a bit skimpy. I guess I ought to clarify what I mean by this. Their explanations of how circuits work are not particularly detailed and their service manuals (schematics) have just "bare-bones" information. Their rationale seemed to have been, "Find the module that's causing the problem, and replace it. Don't try to repair it to the component level. Just replace the module." Most of us don't like to do that and scoffed at the literature Zenith sold. I developed the habit of using a PHOTOFACT and making service notes on it. Oh, there were occasions when I preferred to refer to Zenith's manuals, but I consider them second to a good PHOTOFACT. I use PHOTOFACT number 2877 for most of the work on these chassis, and it is the one to which I refer in this chapter. It won't do for all C-3 chassis, but it is adequate for most of them.

The C-3 Chassis

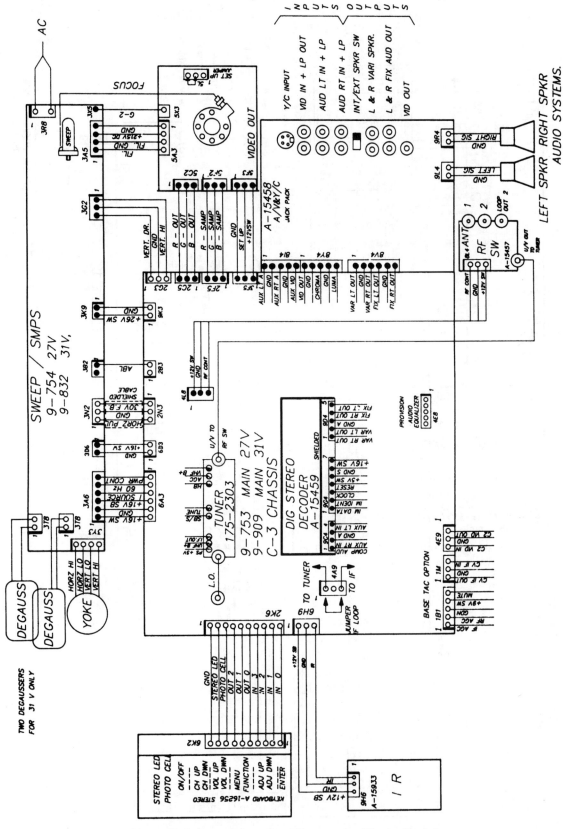

Figure 2-3 Interconnection Diagram for Stereo Model

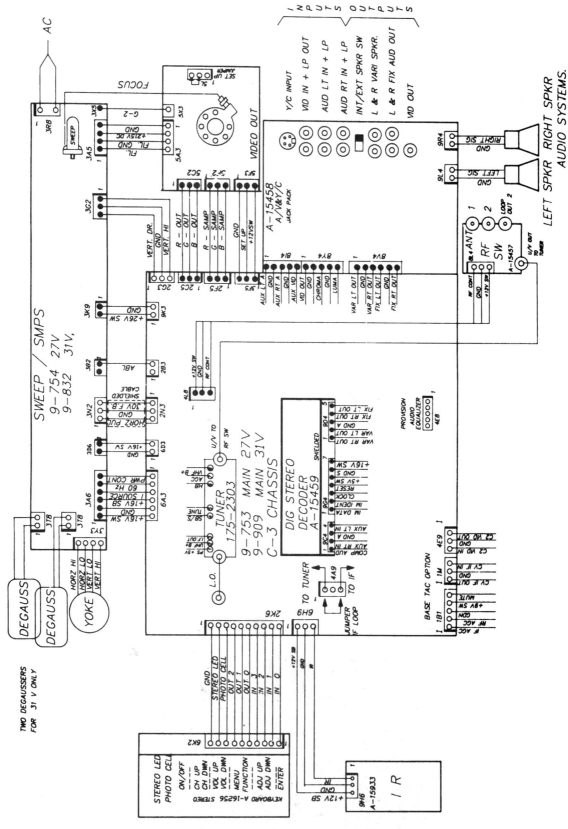

Figure 2-4 Interconnection Diagram for Mono Model

The C-3 Chassis

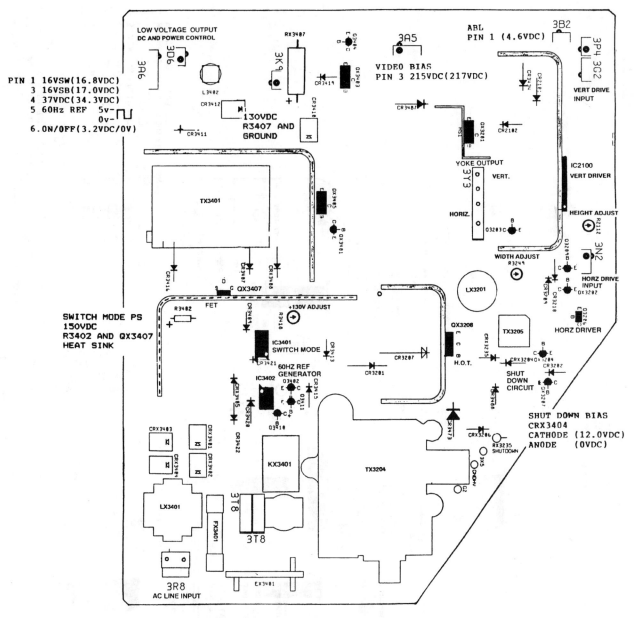

Figure 2-5 Component Location Guide

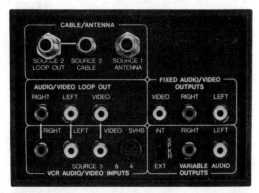

Figure 2-7 Jack Pack Inputs and Outputs

Servicing Zenith Televisions

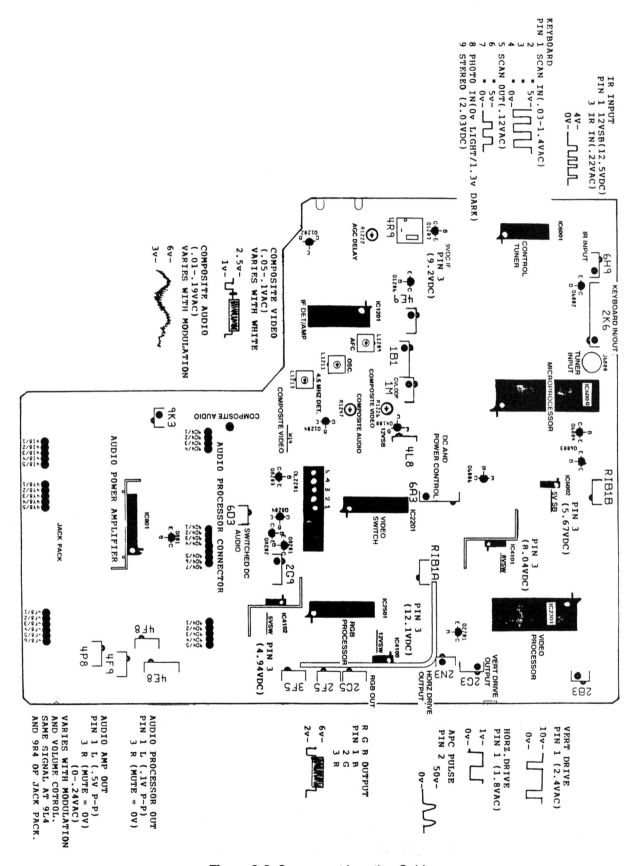

Figure 2-6 Component Location Guide

The C-3 Chassis

Factory Service Menu

The information I am about to give you comes from the J line chassis that came out in 1991, which means some of it won't apply to earlier C-3 chassis. But you should find the information useful regardless of the model year with which you are working.

The Zenith engineers call the factory menu "a ghost key." Press the menu, volume down, and channel down keys on the front of the set at the same time to access it. You can't get into it using the remote control. The service menu works the same as the customer menu except that it doesn't time out. The service menu is two pages long. Use the menu key on the set or the remote control to toggle between the two. A white arrow to the left of an item tells you which adjustment you have chosen to work with. The select keys move the white arrow from one item to the next while the adjust keys let you make the necessary adjustments.

Page One

The top line on page one has a configuration of letters and numbers that look something like this: V-684-138. It may appear as if it is the only version number. But a control called "Factory Mode" is present and pops up when you press the adjust key. "Factory Mode" will be on or off. It is used on the production line only and should always be left off "in the field."

"Store Controls" is the next item and is used to store customer menu controls in the non-volatile EAROM memory.

"Auto Mute" is a new feature of the J Line C-3 chassis and is used to turn on or off the auto audio mute feature of the slice level detect circuit. What does this mean? If auto mute is set to on and if there is a sync glitch, the audio automatically mutes. I have a C-3 set at home. Our cable channel 16 sometimes transmits a signal that includes a few distorted sync pulses. When it picks the signal up, my TV temporarily mutes the audio. And that, ladies and gentlemen, is terribly annoying! So, I turned the auto mute feature off, and now I don't hear the glitch in the audio when I watch the programming on channel 16.

"Brightness" is the next feature. It is used to set the maximum brightness range of the customer menu brightness control. Brightness has a range of 000-015.

"Environment" is the last item on page one. It can be set in one of four positions: N2, N4, P2, and P4. The "N" means "non-PIP," and the "P" means "PIP." The "2" refers to two input sources while "4" refers to four input sources that were available in the Advance System 3 models. For example, "N2" means "no PIP and two signal input sources." "N4" means "no PIP and four signal input sources."

Advanced System 3 projection televisions add two items to page one. The first addition is called "Slice Level" and is used to adjust the sensitivity of the closed caption decoder. Its normal range is 00-31 and is usually set to 19. The second addition is "Hpos" and is used to change the horizontal position of the closed-caption display.

Page Two

Page two has three menu items on it. The first is called "Vert. Sync" and is used for the vertical sync forced mode operation. The second item of page two is "Max. Volume." This function has a range of 00-31. When "Max. Volume" is turned on, it controls the maximum possible volume setting for the customer volume control. And finally, "AC Power On" which allows the set to be turned on when AC power is applied. "AC Power On" should be turned off for normal in home use, but it has some useful applications. If, for example, the customer uses a cable converter box and wants the TV to come on when the cable box turns on, simply go into the service menu and set "AC Power On" to the "on" position.

Comments On C-3 Service Menus

Keep in mind the fact that I have given you the service menu for the J Line only. Earlier C-3 chassis have different items in their service menus, but those items won't be difficult to decipher and set. For example, the F Line menu has "Horiz Pos" that is used to set the horizontal position of the customer menu. It also has "CS Color" and "CS Tint" that are used for setting the color and tint of the color sentry. The F Line sets also have a control called "Environment" that sets the customer menu to match the structure of the jack pack. The choices are M1 to M4 and S1 to S4. "M" refers to mono and "S" to stereo.

G Line televisions are also different. "CS" and "CS Tint" are not present, but an item labeled "SP" is. It monitors the level of the stereo pilot signal. A valid signal registers 10+ while a mono signal registers in the 02 to 03 range. By the way, these are hexadecimal numbers.

Power Supply

The C-3's power supply is both sophisticated and complicated, having quite a number of parts in it. It is so involved that the schematic takes up three pages in the PHOTOFACT I use. I have chosen to reproduce the Sams because it's easier to follow than Zenith's service literature. However, I intend to reproduce several abbreviated block-like diagrams from Zenith's literature because they make working on this complex monster a little simpler.

A failure in either the main module or the power supply results in a dead set. Given the enormous parts count, where do you even begin to find the problem? Zenith has come to the rescue by creating a nice little flow chart (figure 2-8). You should be able to use it and a DMM – a scope would be better – and find the defective module in just a few minutes. Once you have determined which module is causing the problem, simply narrow the field of possibilities until you put your finger on the culprit. My experience leads me to say that you will probably find the power supply to be at fault far more often than the main module. To that end, let's take a very close look at the power supply module. Well, I ought to say sweep and power supply module because it contains not only the power supply but also the horizontal and vertical sweep circuits.

How It Works

Keep figure 2-9 handy because I intend to reference most of my remarks to it. I suggest that you also consult the block diagram in figure 2-10 because it presents the power supply as a series of small blocks that might make the "how it works" discussion a little clearer.

The power supply uses a flyback-style switching regulator to provide most of the necessary DC voltages to operate the television. The voltages include, but are not limited to, +130 volts, +16 volts standby, +16 volts switched, and the audio output amplifier voltage. It is designed around a MOSFET high-speed switching device (Q3407) and an eight-pin (dual in-line) controller IC (ICX3401). The IC is a current-mode pulse-width modulator controller that requires 15 volts B+ to operate. Its pinouts look like this:

The C-3 Chassis

pin 1	comp	pin 2	vfb
pin 3	current sense	pin 4	rt/ct
pin 5	ground	pin 6	output
pin 7	vcc	pin 8	ref

If you want to do additional research on ICX3401, use the manufacturer type number (UC3842AN) and surf a few Web sites for pertinent information.

Zenith's part number for the MOSFET is 121-1190, and the part number for the controller chip is 221-466. I usually shy away from generic parts in circuits like these, but I have in a pinch successfully used an ECG2397 for the MOSFET and an ECG7096 for the IC. Both seem to have worked well because I have never had to repair those sets again for the same problem.

When AC is applied for the first time, the current flowing through R3441, R3403, and CR3405 "kick starts" ICX3401 causing it to output a few square wave pulses of about 12 volts peak-to-peak from pin 6 through R3409 to the gate of QX3407. The switching FET turns on and energizes the primary winding of TX3401, and the secondary windings begin to produce their respective voltages. As the secondary voltages become available, the winding defined by pins 8 and 9 assumes the job of providing run B+ for ICX3401. The run B+ is developed by CRX3408 and C3401 and applied to pin 7 of the IC. The startup path through R3403, R3441, and CRX3405 has now become superfluous.

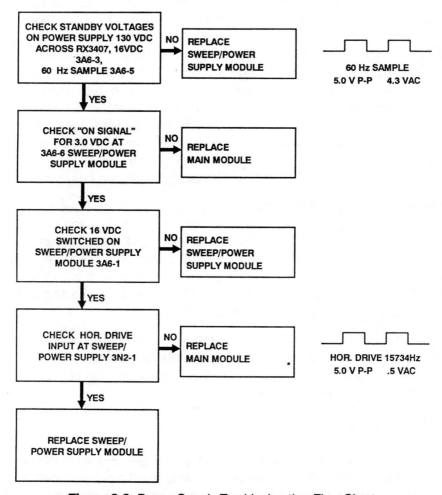

Figure 2-8 Power Supply Troubleshooting Flow Chart

If the power supply fails to start, let the startup voltage be the second voltage you check, the first being the +155 volts DC developed by the bridge rectifier circuit and main filter capacitor. If the power supply fails to start and if the startup B+ and +155 volts are present, I automatically replace ICX3401. On the theory that I might be wrong, I often put the replacement controller chip in an eight-pin DIP socket. However, I am far more right than I am wrong because most of the components in the circuit just don't fail.

There is one exception. So, let me call your attention right now to C3407, a 100-mfd 35-volt capacitor located in the startup circuit. After these sets have been in use for a few years, C3407 gradually begins to lose capacitance. As its capacitance lowers, its ability to provide current to ICX3401 lessens. At a certain point, the loss of capacitance inhibits the action of the IC effectively shuting the set down. The power supply tries to produce run B+, but the voltage fluctuation simply can't produce the necessary current to run the power supply. C3407 has been know to cause another problem, and this one will fool you if you aren't careful. The set starts up and shuts down. The shutdown might be immediately after turn on or occur a few minutes after turn-on. Let me suggest that you replace this little fellow when you encounter a shutdown problem before you do anything else. It just might save you lots of time.

Well, I've gotten a little ahead of myself, and I apologize for that. Let's start the troubleshooting procedure over and see if we can't proceed in a logical manner! The service literature says that you should begin troubleshooting a dead set condition by checking the voltage across RX3407, a large 12k, 3-watt resistor located near the back of the module. Put your DMM or scope leads across it and read the voltage. No voltage indicates a dead power supply, but a voltage that fluctuates, let us say between 125 and 136 volts, indicates the power supply is starting but doesn't generate sufficient current to run. I suggest you use a scope rather than a DMM because the constantly fluctuating voltage causes the trace to move up and down in a way that's hard to miss. Replacing C3407 fixes the problem almost every time. Now, you probably own an ESR meter, and I confess that's one of the handiest test aids I have on my bench, but don't use it here because C3407 will not have developed high ESR. It will have lost capacitance. Simply replace the capacitor. I am certainly not joking when I observe that C3407 causes trouble. I write these words at the end of November (1999), a month during which I repaired no fewer than seven C-3 chassis by replacing one, small capacitor!

If you don't measure any voltage across RX3407, you more than likely have a dead power supply on your hands. A few additional voltage checks should confirm your suspicion and point you in the right direction. For example, do you have 155 volts on the drain of QX3407? What signal, if any, do you have at the gate? Do you have start-up voltage on pin 7 of ICX3401? If there is no drive at the gate of the FET and if the other voltages are correct, replace the controller chip. If the set comes in with a blown fuse and the bridge diodes check out, check the FET next. If it is shorted or leaky replace it with the understanding you may also have to replace ICX3401.

Well, enough of this for now. Let's finish the discussion of the operation of the power supply. The winding defined by pin 8 and 5 of TX3401 develops a feedback voltage that is routed to pin 2 (the pin labeled vfb in the previous table) of the controller. R3418, a variable resistor, provides an adjustment to set the output of the 130-volt line.

Standby Voltages

The +16-volt standby line is used to develop the standby voltages. Figure 2-10 should orient you to it, but you also need to look at figure 2-9 for the details. Note that Sams designates it the +18-volt line. Q4100 develops the +12-volt source while IC6002 develops the +5-volt source for system control.

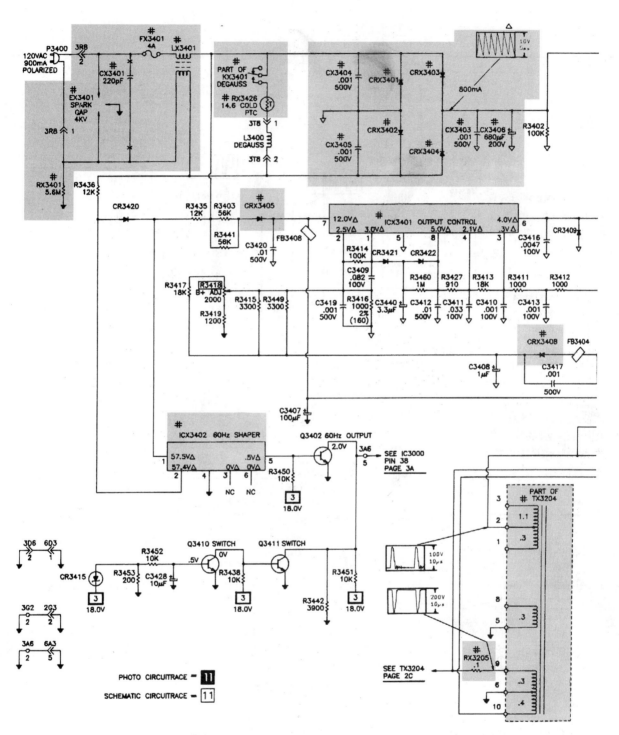

Figure 2-9 C-3 Power Supply Schematic

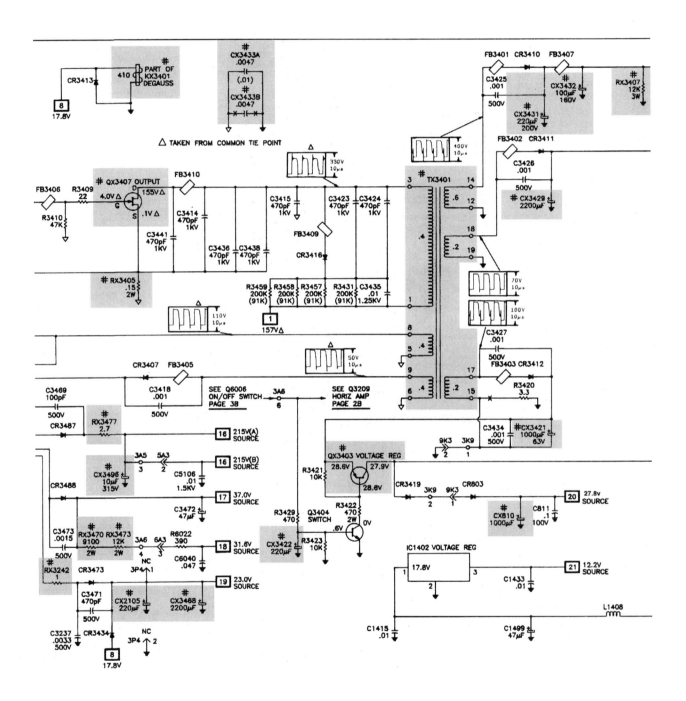

Figure 2-9 C-3 Power Supply Schematic (Continued)

The C-3 Chassis

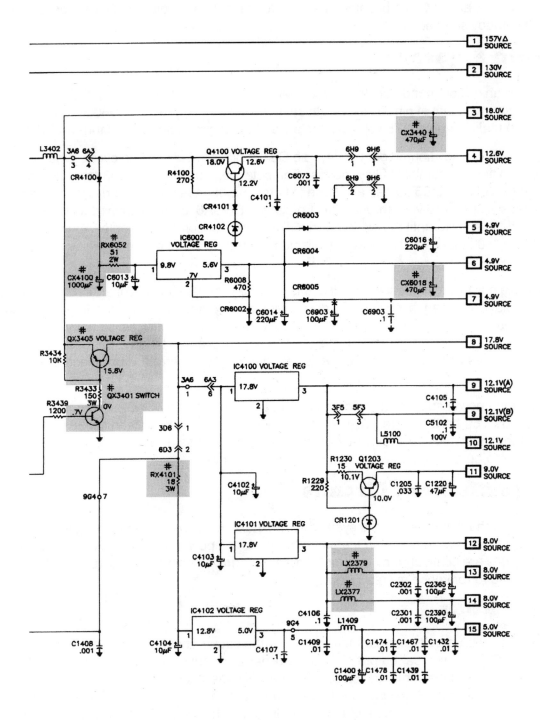

Figure 2-9 C-3 Power Supply Schematic (Continued)

Look for Q4100 and IC6002 on the main module. You can easily find them by following the cable 3A6-6A3 from the power supply module to the main module. While you are looking at the schematic, pay attention to QX3405 and QX3401 because they control the switched voltages that are developed from the +16-volt standby source.

Power-On Sequence

After it receives an on command from the IR receiver or the keyboard (figure 2-11), the microprocessor "generates a ground signal at pin 41" as Zenith says. When pin 41 goes low, Q6006 turns on (figures 2-11 and 2-12) making about 4.75 volts available via pin 1 of connector 6A3 to turn on Q3404.

You have probably already figured out the key role Q3404 plays in the start-up cycle. When it turns on, this transistor initiates a number of key moves. For instance, it turns on Q3403 that turns on Q3401 that turns on Q3405. QX3405 comes alive and initiates the degauss action to demagnetize the CRT and sends the switched +16 volts through pin 1 of cable 3D6-3A6 to the main module where it is used to bring the chassis to life until the sweep derived +23 volts becomes available. You may also have noticed that QX3403 makes a B+ voltage available for the audio circuits (voltage source 20 in Sams literature).

In summary, the power-on sequence goes like this: AC line voltage, startup voltage, run B+ voltage for the ICX3401, standby voltages, power-on signal, switched voltages, and sweep voltages. The power-on sequence illustrates the way a technician ought always to check voltages: standby, switched, and derived. That is, begin your troubleshooting by confirming the proper standby voltages. Then proceed to check for the presence of the switched voltages, and finally confirm the presence of the derived voltages.

Before I leave the power supply, let me comment on the function of ICX3402. Sams calls it a "60Hz shaper." It sends a 60-hertz square wave developed from the AC line voltage to pin 38 of IC6000. The microprocessor must receive those pulses to function. A missing 60-hertz signal causes the microprocessor to be as inoperative as a missing +5 volts does!

Sweep (Deflection) Circuits

You know by now that I usually discuss system control next. However, I am deferring it till later simply because the deflection circuits are on the same board as the power supply. I offer you two views of how these circuits are configured by including figure 2-13 (from Zenith) and figure 2-14 (from Sams). However, I intend to reference my discussion to the latter.

Vertical Deflection

The vertical deflection circuit is remarkably similar to the circuit Zenith uses in the C-2 chassis. For example, IC2100 – the vertical output IC – uses a negative drive pulse from pin 31 of IC2301 to develop vertical drive for the same reason the C-2 chassis uses a negative drive pulse. It also requires +11 volts derived from the 16-volt switched supply and a scan-derived +23 volts to develop drive for the yoke. R2112 regulates vertical height by controlling the charge on C2114 while R2112 regulates vertical linearity by controlling the strength of the vertical feedback pulse. Use a "circle and square" pattern from your signal generator to adjust both controls for the best possible picture.

We aren't often told what function each component in a circuit plays and are sometimes at a loss to understand why a circuit acts the way it does and why the component we installed fixes the problem.

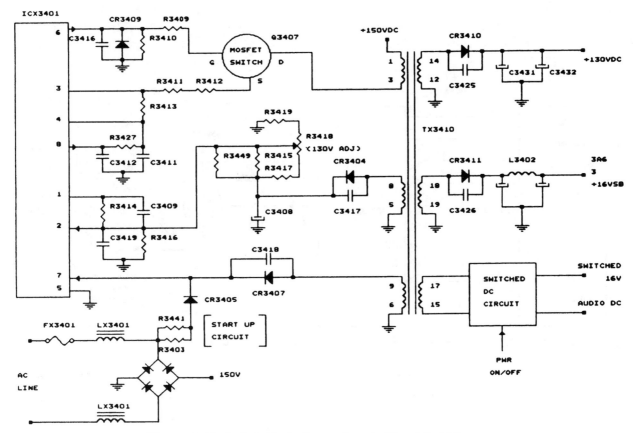

Figure 2-10 Switch Mode Power Supply Stand-By Voltages

I had the same questions about RX2110 (just off pin 11 of IC2100). I do know that I have had an occasion or two to replace it, but I didn't know why the engineers put it in the circuit. I found from my reading that it has been included along with C2106 to dampen any unwanted oscillations that might be present. I do know the circuit doesn't work very well without it.

Horizontal Deflection

The horizontal deflection circuit is vintage Zenith and, like the vertical deflection circuit, is remarkably similar to the one used in the C-2 chassis. The drive pulse exiting IC2301 has about a 50% duty cycle. It is routed first to Q3209, the horizontal amp, and then to Q3202, the horizontal pre-driver. The next stage along the way is the horizontal driver itself. Pay attention to C3212 and R3221 in the base of the horizontal driver transistor because they alter the duty cycle of the horizontal drive waveform. Zenith's engineers tell us the altered duty cycle reduces noise pickup and makes for a cleaner drive waveform.

Horizontal drive goes from the collector of the horizontal driver into the primary of the horizontal drive transformer that couples it to the base of the horizontal output transistor to develop drive for the IHVT. As I said, there is nothing new here. You will note though, the sweep transformer has a reduced load on it because the switching power supply carries most of the DC load for the television. Besides the EHT, G2, and focus voltages, it develops +23 volts for vertical drive, +37 volts for the tuner, and +215 volts for the video output circuits.

The circuit, though relatively trouble-free, does cause a problem or two from time to time. For example, the C-3s do occasionally pop a horizontal output transistor, but something usually causes it to fail. I

Servicing Zenith Televisions

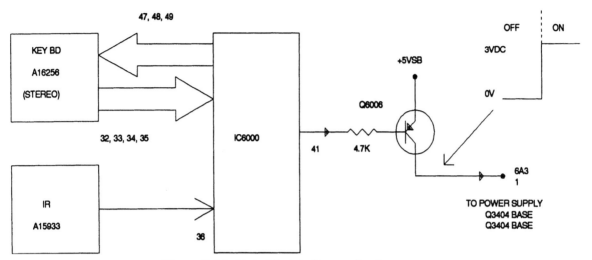

Figure 2-11 Main Module Power-On Sequence

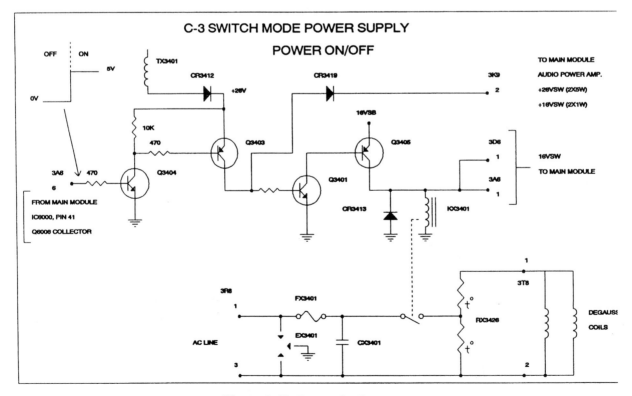

Figure 2-12 Power-On Sequence

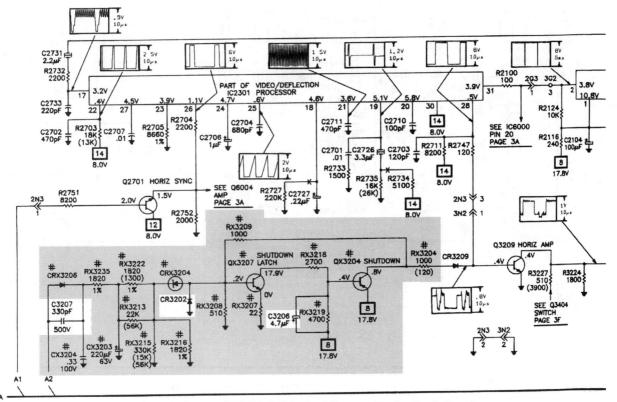

Figure 2-14 PHOTOFACT View of Sweep (Deflection) Circuits

recall maybe one or two of these sets that output transistors alone fixed, but a shorted output transistor usually means that you have to replace the flyback. Zenith's part number for the horizontal output transistor is 121-1148. However, an ECG2302 or NTE2302 or SK9422 work fine. Check the literature or the number on the flyback for its correct part number.

Pincushion Circuit

Q3201, Q3202, coil LX3201 (the balance coil), coil LX3262 (the linear coil), and the yoke comprise the pincushion circuit. Drive for the circuit comes into the base of Q3203 from the "low" side of the vertical winding of the yoke and biases it on. Q3203 then controls the conduction of Q3201 to ensure a pin-corrected current flow through the horizontal yoke. LX3201, the balance coil, maintains a flyback period independent of the pin modulation and prevents modulation of the +130-volt supply. Zenith engineers have provided a way to adjust horizontal width without having to change any components simply by adjusting R3249 in the base circuit of Q3203.

Shutdown Circuit

The shutdown circuit monitors a 40-volt peak voltage taken from pin 9 of the flyback. The circuit rectifies the pulse then filters and applies it to a precision voltage divider that outputs a voltage to the cathode of CR3204, a 12-volt zener. If the peak-to-peak pulse at pin 9 increases beyond the level determined by CR3204, the resulting DC voltage causes the zener to conduct. When it conducts, the voltage at the base of Q3207 rises and turns it on. Its falling collector voltage causes Q3204 to turn on and ground horizontal drive to the base Q3209 effectively inhibiting horizontal drive. Since Q3207 and Q3204 make up a latch circuit, horizontal drive won't reach the base of Q3209 until AC power has been removed and reapplied.

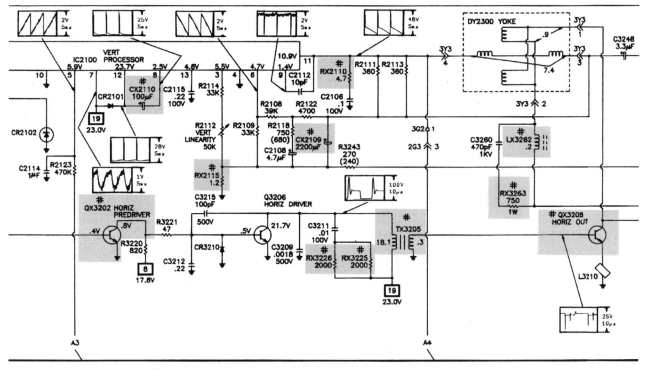

Figure 2-14 PHOTOFACT View of Sweep (Deflection) Circuits (Continued)

System Control

The system control circuit is similar to the one found in the C-2 chassis (figure 2-14a). I would like to offer you more information than I have here, but as far as I can tell, Zenith doesn't discuss it anywhere. However, you don't need much information to troubleshoot it. Follow the procedure laid out for troubleshooting system control problems by first checking the microcontroller's "must haves." If you don't find any abnormalities among them, check the data-in and data-out lines. Don't forget that a single malfunction in system control keeps the TV from working properly. For example, I just recently changed a microprocessor in a set that wouldn't turn on. The chip had everything it needed to work – +5 volts, ground, oscillator, reset, etc – but, it wouldn't execute the on command at pin 41. Just one malfunction shut the TV down.

Video Processing

The "main module" or the "small signal panel" does the video processing. I am including figure 2-15 to give you an idea of its layout. I suppose I should have given it to you earlier because it contains a few key test points, but I decided to insert it here. Notice the waveform and voltages listed on each side of the module. You should find these to be a valuable aid in your troubleshooting ventures.

Nuts and Bolts of Video Processing

Video processing begins with a video signal from either the tuner or the jack pack entering the IF circuit on the main module. The composite video goes first to IC2201 (figure 2-16), a four-section switch responsible for selecting video from one of its four inputs. Figure 2-16a is a block diagram of the switch and gives you a visual idea of how the signal moves through the chip. The customer makes

The C-3 Chassis

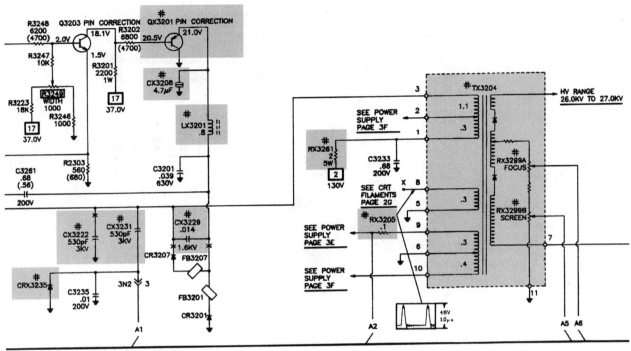

Figure 2-14 PHOTOFACT View of Sweep (Deflection) Circuits (Continued)

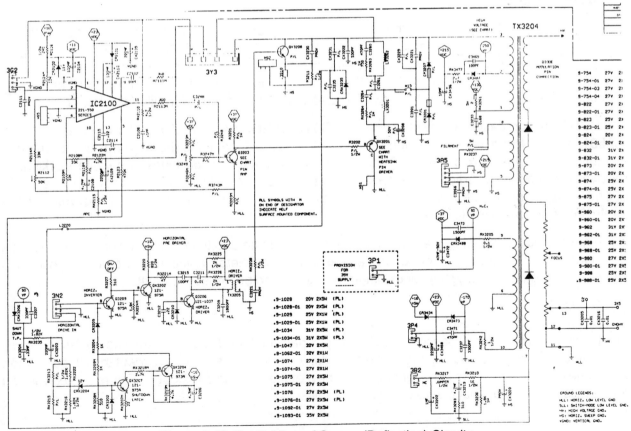

Figure 2-13 Zenith View of Sweep (Deflection) Circuits

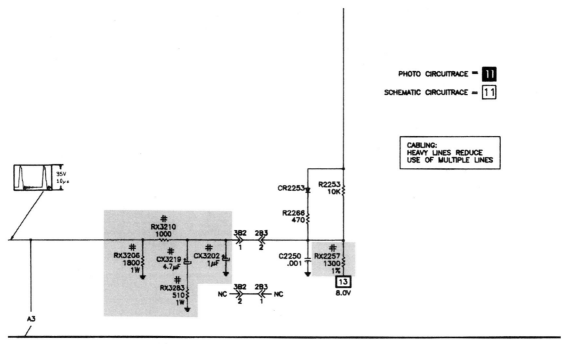

Figure 2-14 PHOTOFACT View of Sweep (Deflection) Circuits (Continued)

the selection from the on-screen menu and system control executes the command by causing IC2201 to respond appropriately.

The output of IC2201 goes either to the PIP module if the set has a PIP and on to the video processor (IC2301) or directly to the video processor if the set doesn't have a PIP. IC2301 uses the composite video signal to develop vertical and horizontal drive, and the RGB video signals. It also has a light sensor input for luma correction if the light level in the room changes. Did you notice the IM bus connections at pins 32, 33 and 34. Look at figures 2-17 and 2-18 for a detailed illustration.

The RGB signal proceeds by way of connector 2C5 to the RGB switch and output driver that is labeled IC2501. IC2501 has several functions in the C-3 chassis. It contains the auto tracking circuit that maintains the proper mix of the R, G, and B signals to ensure good black and white tracking. It receives as feedback a portion of the RGB signals for correction of its amplifiers. It also receives R, G, B, and fastblanking signals from system control to generate OSD information. And, it develops drive for the video output module.

If you have difficulty following the written text – as I do from time to time – take a close look at figures 2-17 and 2-19. The former is, of course, a reasonably complete depiction of the video processing circuit while the latter is a simple block diagram of the video processor IC itself.

Jack Pack

Most of my customers don't pay attention to the jack pack on their TV. Some see it as just another way manufacturers have of confusing them! So, they don't use any of its features except the RF connector! But some do. Therefore, we servicers need to know at least a little bit about it. Figure 2-20 gives you the details for the lines that contain the C-3 chassis.

The C-3 Chassis

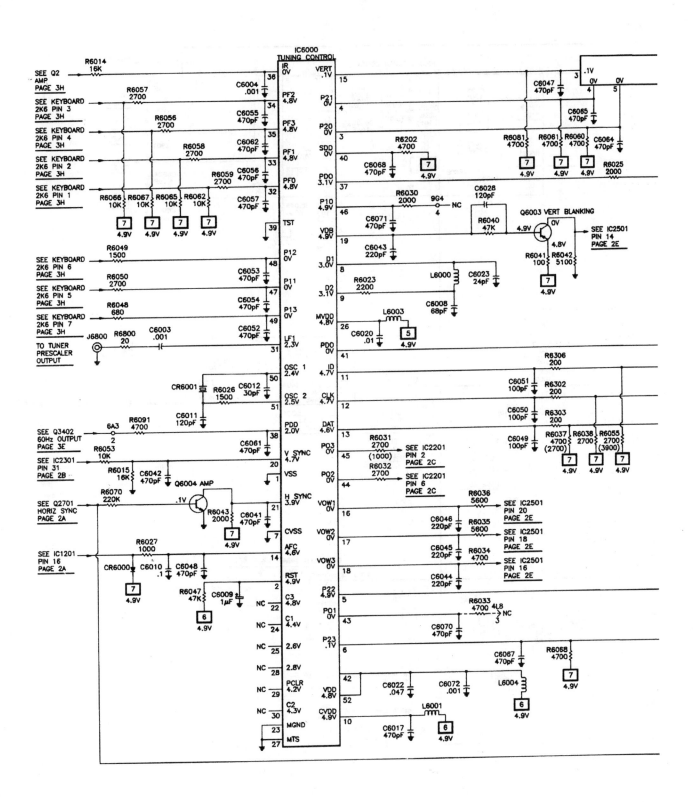

Figure 2-13a System Control Circuit

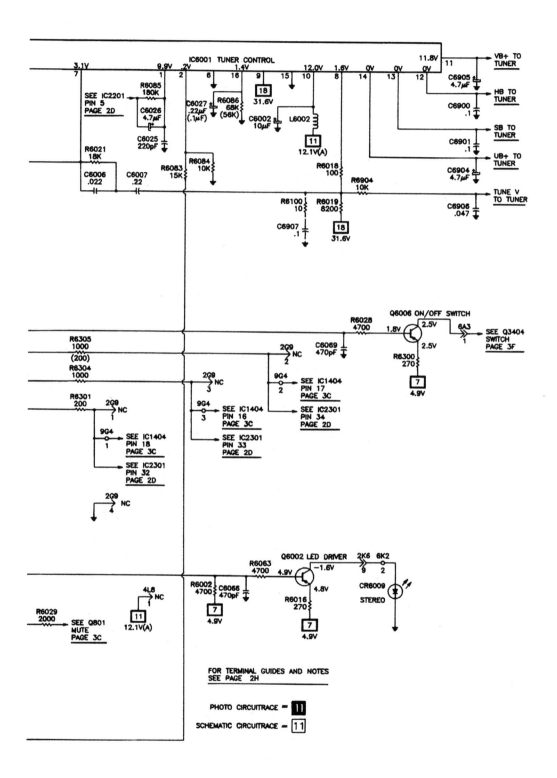

Figure 2-13a System Control Circuit

The C-3 Chassis

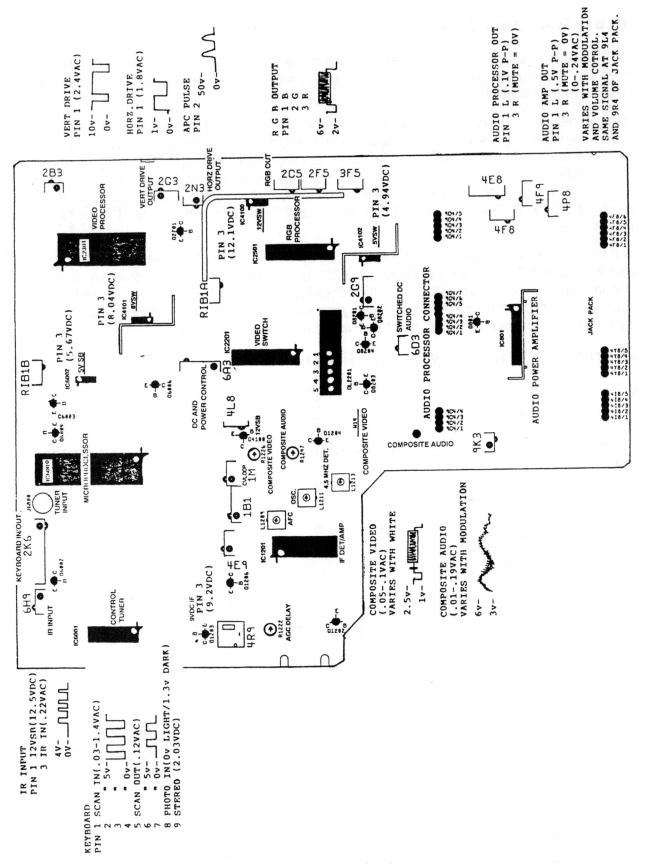

Figure 2-15 Main Module Layout

Picture-In-Picture Module

The PIP feature has always been a major feature of the C-3 chassis. Figure 2-21 illustrates how it "wires into" the chassis while figures 2-21 and 2-22 show you what's inside the little silver box and where its various controls are located.

Let me underscore the fact that the PIP causes a variety of problems because it is in parallel with the IM bus, the video signals, and the +5- and +12-volt regulators. The last one I fixed caused loss of luminance and horizontal tear in the video. At first I thought I had major problems with the main module, but I tried bypassing the PIP. When I did, the problem cleared up. If you suspect it is causing trouble, disconnect all of its cables and use the correctly labeled one to connect the 4D9 and 4C9 connector on the main module. The television now functions as a non-PIP model. Do this first when you have a video problem.

I will forego a technical description of the PIP module on the assumption that it's cheaper to buy a replacement from Zenith or a repair depot like PTS than it is to try to fix a major problem. If you want a technical description, read all about it in the books I have already cited. But if you are interested in repairing the PIP, begin by checking the electrolytic capacitors with an ESR meter and replacing those that are defective. Anyway, it won't hurt to try, and the money you save will be yours.

35-Inch CRT and Dynamic Focus

As you know, the C-3 chassis changed from model year to model year. My discussion of it has, therefore, been somewhat generic with tidbits tossed in to indicate the various changes. When it introduced the J line, Zenith also introduced a set with a 35-inch CRT that featured a dynamic focus assembly to achieve an overall flat focus and a new picture tube that incorporated some construction changes.

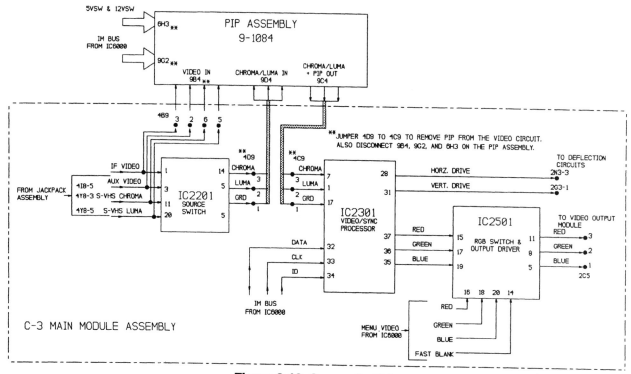

Figure 2-16 Composite Video

Before getting into the dynamic focus, let me briefly discuss the construction changes. The faceplate panel, according to Zenith's engineers, "is a 1R aspheric contour/rectilinear screen design" (Technical Training Program: J-Line Color Television Update, page B-12). It also has straight sides and square corners. An invar shadow mask improved performance over previous designs. The engineers put a lithium silicate faceplate coating on the tube to reduce glare, static charge build-up, and reduce dirt build-up. If you have ever seen one of these monsters, you know the changes have made a difference in its performance.

The final engineering change involved a Coty MDF multielement focus precision in-line gun with dynamic beam firing. It utilizes two adjustable-focus grids to improve spot size for enhanced picture definition and dynamic focus. The engineers decided to apply a dynamic parabola waveform to the F2 focus anode to provide a center-to-edge-spot size uniformity. You know that uniform focus is difficult to obtain because of the curvature of the faceplate. You also know that you can do a pretty good job of focusing the center or the edges, but you can't focus both equally well. It's a case of the one or the other but not both unless you use a higher focus voltage at the edges than you do for the center.

Different focus voltages are possible by superimposing a parabolic waveform onto the DC focus voltage.

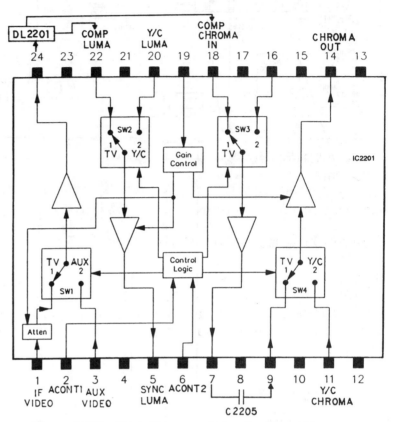

Figure 2-16a Block Diagram of How the Signal Moves Through the Chip

The resulting voltage (DC voltage plus peak-to-peak parabolic signal) is high enough to improve focus at the edges of the screen without making much of a difference in the focus potential at the center of the screen. A dynamic focus assembly makes the new focus scheme possible. The new assembly means additional circuitry and extra connectors added to the sweep circuits, and it somewhat complicates service procedures (figures 2-24a and 2-24b). But the results certainly justify it.

It also means a change in the way we techs adjust focus. First, apply a crosshatchpattern to the set on which you are working. Then adjust the F1 focus control for the smallest horizontal size of the vertical line at the midpoint between the center and the edge. If your signal generator has a "circle and cross pattern," I suggest you begin by using it instead of the crosshatch. Third, adjust the F2 focus control for the smallest vertical size of the horizontal line between the center and the edge. Be prepared to repeat the process several times in order to get the best results.

Servicing Zenith Televisions

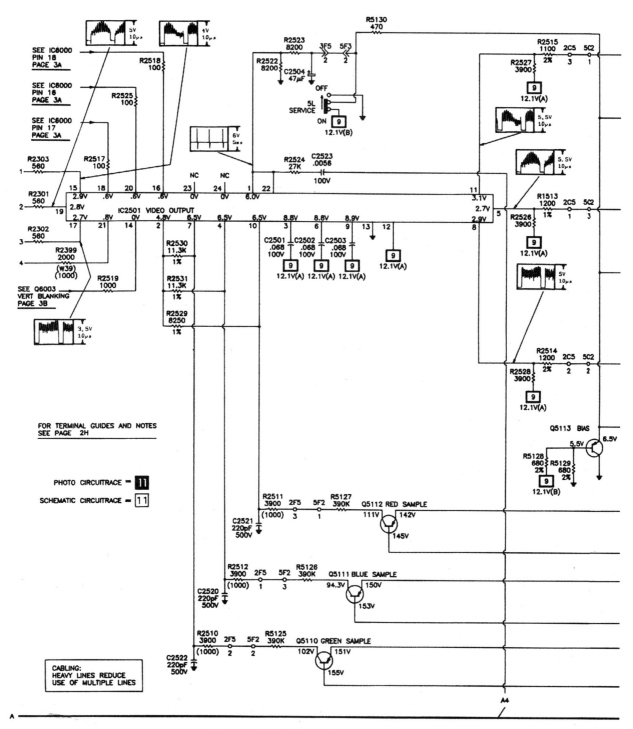

Figure 2-17 Video Output Schematic

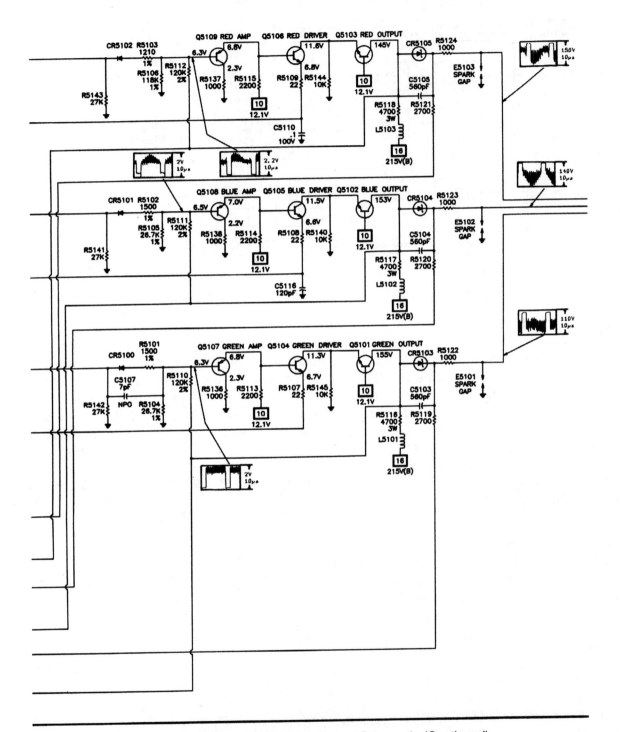

Figure 2-17 Video Output Schematic (Continued)

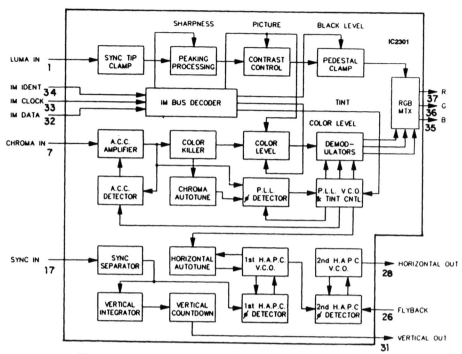

Figure 2-18 IC2301 Video Processor IC Block Diagram

Audio System

I don't have much information about the C-3's audio system, but I will share what I have. It seems that the digital audio processor – the same circuit used on the digital 9-700 module – is the "brain" of the audio system. Figure 2-25a illustrates how it connects to the main module, and figure 2-25b shows you what's inside the silver box. The circuit converts the analog audio signal into a digital one and sends it to IC1405 (figure 2-25b). IC1405 processes the signal for stereo detect, volume, bass, treble, SAP, and converts the digital information back into an analog signal. The analog signal then goes to the power amplifier (IC801) located on a heat sink on the main module and from there to the speakers.

If the television has mono sound only, the main module won't have a digital processor on it, but will have an assembly containing an audio output amplifier that is equipped with a DC volume control.

Surround Sound

The standard surround feature sometimes wears the name "matrix surround sound." It is based on a speaker connected to the positive terminal of the right and left stereo channels. The resulting audio will be the phase difference signal between the right and left stereo signals combining in the

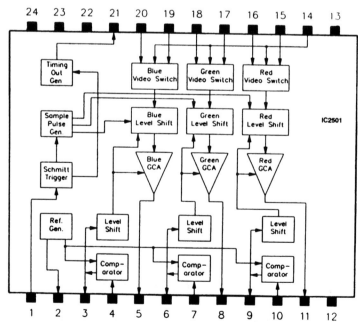

Figure 2-19 IC2501 Video Output IC Block Diagram

The C-3 Chassis

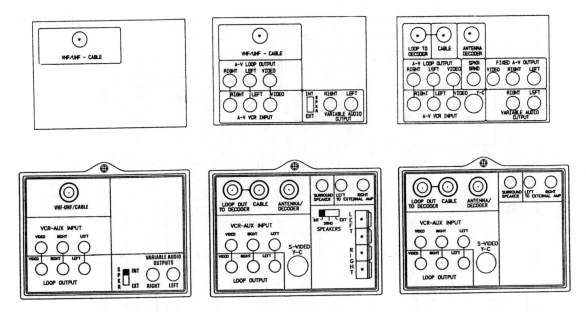

Figure 2-20 F-, G-, and J-Line Jack Pack Assemblies

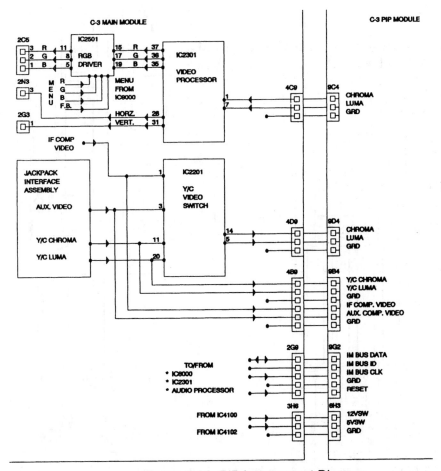

Figure 2-21 PIP Interconnect Diagram

59

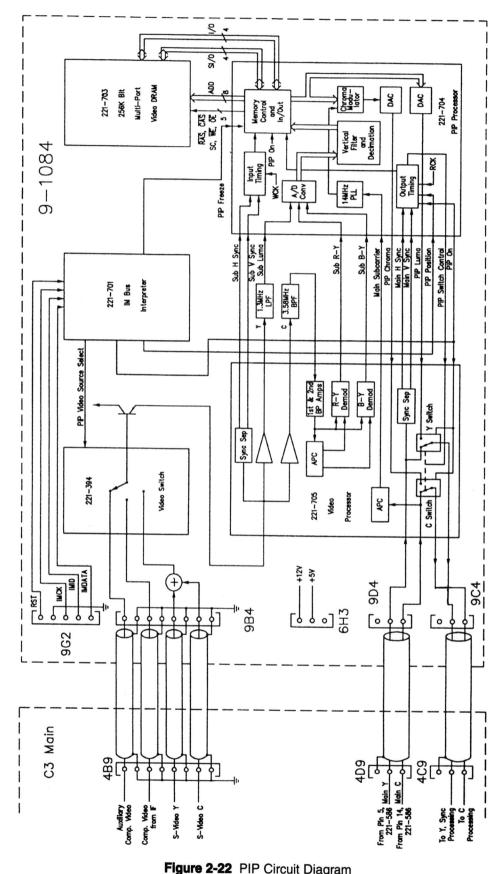

Figure 2-22 PIP Circuit Diagram

The C-3 Chassis

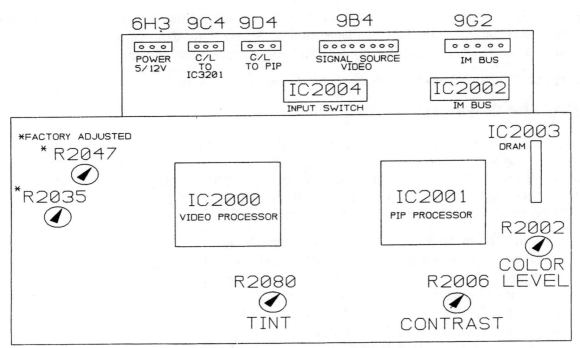

Figure 2-23 PIP Parts and Controls Layout

speaker coil. You should know that this arrangement has no amplifier or volume control and is present only when a stereo signal is present. This often confuses customers who think the surround speakers should be "on" all the time. The fact is they are "on" only if the program which they are watching is being broadcast with stereo audio. The surround sound function is turned on or off via a switch on the jack pack (figure 2-20) itself.

SEQ

SEQ is an abbreviation for "spatial equalization," a feature intended to improve the audio fidelity and stereo effect connected with the video. This is how it works. When you press the SEQ button on the remote or access it via the customer menu, the digital audio processor increases the bass and treble frequencies as well as the stereo sound stage. The amount of change depends upon the setting of the volume control and the content of the audio. Generally speaking, you have to set the volume level higher than you normally do. My wife often has to tell me, "Turn that thing down!" If you have ever experienced SEQ, you know its effect is dramatic, especially if you are watching a high-quality video tape, a laser disc, or availing yourself of the newer DVD technology.

There are, however, some signals upon which it will not work, like mono audio or audio with limited content like newscast dialogue or talk shows. There are also some instances where SEQ works but not very well, as when the volume is turned down, or the TV is in a room that is large, or when the viewer sits too far from the set.

Troubleshooting Tips

I designed this book to help you fix Zenith televisions. However, I confess the C-3 chassis is often a demon to troubleshoot. Let's face it, Zenith didn't design it to be repaired to the component level. The cables aren't long enough to give easy access to the underside of the circuit boards. When you do unwind the cables, you often have to unplug the cable that connects the front controls to the main module, meaning you can't turn the set on to make necessary voltage measurements. Moreover, the components aren't marked on the foil side of the boards. So, you have to guess and grope to find the part you want to check. These realities make fixing the C-3 chassis a real challenge, but you can do it.

Refer to figures 2-4 and 2-5 as you work through the following material.

Verifying the Power Supply

Use these checks to verify the presence of voltages needed to start the power-on sequence. Remember there are two standby voltage regulators on the main module. May I suggest you begin your troubleshooting adventure by measuring the voltage across RX3407? If you measure about 134 volts DC, you know the power supply is up and running and you need to begin your checks with the standby voltages. Read on if you want the complete troubleshooting.

(1) Verify the AC line fuse and voltage in the power supply.
(2) Check for the presence of +150 volts at the anode of CRX3403 and CRX3404.
(3) Make your check at the left end of R3402 using the heat sink of Q3407 as ground.
(3) Check for +16 volts standby at pin 3 of collector 3A6.
(4) Check for +12 volts standby at pin 1 of connector 6H9.
(5) Check for + 5 volts standby at pin 3 of IC6002.
(6) Check for the 60-hertz reference voltage at pin 5 of 3A6. It should be at a 5 volt peak-to-peak level.

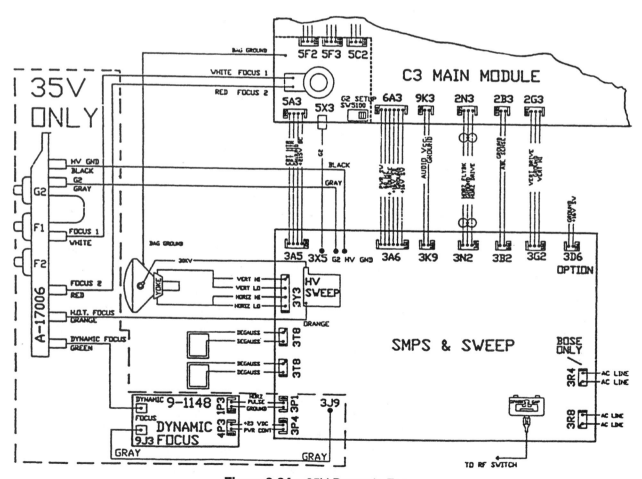

Figure 2-24a 35V Dynamic Focus

The C-3 Chassis

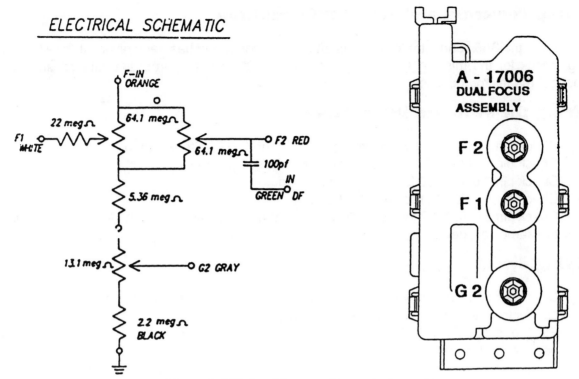

Figure 2-24b Dual Focus w/Dynamic Assembly

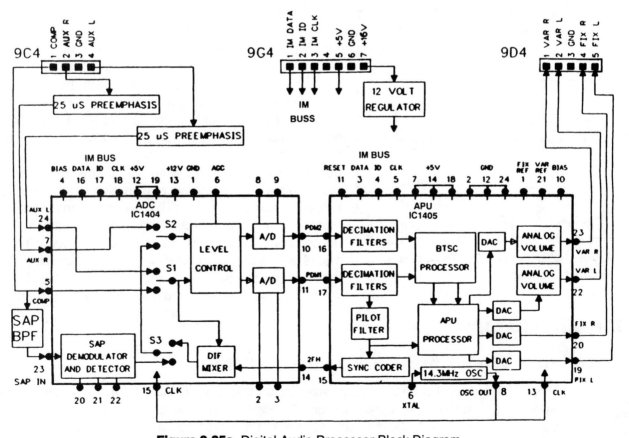

Figure 2-25a Digital Audio Processor Block Diagram

Verifying Power-On and Power-Off Condition

Check pin 6 of the 3A6 connector for +5 volts after an on command has been given. If the +5 volts is missing, check for the missing signal at pin 41 of IC6000. If the appropriate voltage isn't there, troubleshoot the microprocessor.

Verifying the Switched DC Voltages

(1) Check pin 1 of the 3A6 connector for the switched +16 volts switched.
(2) Check pin 1 of the 3D6 connector for +16 volts switched.
(3) Check pin 3 of IC4100 for +12 volts switched.
(4) Check pin 3 of IC4101 for switched +8 volts.
(5) Check pin 3 of IC4102 for switched +5 volts. You should also check the emitter of Q1203 for the same voltage.

Verifying Sweep

(1) Check for horizontal drive (about 1.5 volts peak-to-peak) at pin 1 of connector 3N2. The connector is located on the sweep-power supply module.
(2) Pin 3 of connector 3G2 is a good place to check vertical drive. You should find a 10-volt peak-to-peak negative going pulse.
(3) Check the scan-derived +215 volts at pin 3 of connector 3A5, the scan-derived +37 volts at pin 4 of connector 3A6, and the +23-volt scan-derived voltage at the cathode of CR3434.

These basic checks go a long way toward helping you pinpoint the problem you are trying to fix. I confess there are some problems that are difficult to fix even when you find their cause. When I encounter those problems, I don't hesitate to use a good repair depot like National Electronics or PTS or TNI. For example, I just sent a sweep and power supply module off because it had a blown horizontal output transistor and had a bad flyback. I could easily have changed the parts and would have if I had had them in stock, but I got the whole job done in less than a week for about $50.00. Fortunately, I am able to repair about eight out of ten of the C-3s I put on my bench. And you can too! You should even consider getting your parts from Zenith. Let's give credit where credit is due. I used to have real problems with the parts I ordered from them, and so did you. But I don't anymore. Moreover, their prices rival those of the major repair depots, and I usually get next-day shipment. Even if their prices were a bit higher, next-day delivery service would make up for it. These two features alone make Zenith worth considering when you need a module or a part.

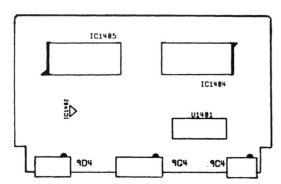

Figure 2-25b Digital Audio Processor Assembly Connectors

Repair History

(1) Picture bowed on right side. Check and replace CX2109.

(2) Picture shaped like a barrel or squeezed in on the sides (reverse brackets). These are symptoms of a failed pincushion circuit. Check and replace, as necessary, the following components: CX2109, CX3407, and QX3201. Pay particular attention to CX3407.

(3) Dead set. As you have gathered from this chapter, lots of things lead to the dead set symptom. Among those things are Q3205, ICX3401, QX3405 (either bad solder around the terminals or just defective), CR3415, and Q3203.

(4) Intermittent no start or won't come on. Replace C3407 (100mfd at 25 volts). See the text for an explanation of its function.

(5) Set won't come on because 16-volt source is low. Replace C3497 (100mfd at 25 volts).

(6) PIP module problems. If you want to try to fix the PIP, begin by checking the parts that commonly fail, namely C2036, C2104, C2080, C2034, and C2102. If you have an ESR meter, check all the capacitors in the module because others may have failed in addition to the ones I mentioned. However – and this is important – inspect the printed circuit board for damage caused by leaky capacitors before you attempt to repair it.

(7) Bright raster with retrace lines. Check RX3477 and CR3487. If they are defective, replace the resistor with a flameproof one and the diode with an ECG552.

CHAPTER 3

THE C-5 CHASSIS

The C-5 series of SENTRY 2 televisions includes 13" mono, 19" mono, 20" stereo, 25" mono and stereo, and 27" stereo sets. It is constructed on a single-sided printed board that is approximately the same size as the C-2 chassis and is found in consumer and certain special market products. New features at the time of manufacture included a single video processing IC, audio-video and RF inputs on mono and stereo models, a crystal-controlled real-time clock, automatic sensing of the environment, and a service menu selected feature that permits the TV to turn on when AC is applied.

We servicers know the C-5 family by its modules: 9-1118-01, 9-1130-01, and 9-1132-01. The family also includes a few modules that are less well known: 9-1129-01, 9-1153, and 9-1154. Perhaps I should say they are not known very well in my service area.

I remember when these televisions first came out more distinctly than the debut of any other Zenith product because they had two memorable problems that caused retailers many headaches but helped us servicers to make a little extra money.

First, the audio momentarily muted on one or more channels under certain conditions and flashed the word "mute" on the screen. These TVs have a feature that causes them to mute the audio when system control fails to detect a valid sync signal. As you certainly know, not all video transmissions are equal. Certain broadcast signals in those days had (and some still have!) distorted horizontal sync pulses. If you looked at them with a scope, you quickly spotted the deformed waveforms. I don't mean that they were terribly deformed, but they were not up to NTSC specs.

Customers who purchased the new C-5 chassis-based televisions began to complain loudly about the muted audio because it was a nuisance. A typical complaint usually took this shape, "The sound goes on and off all the time. I know it has to be the new television because my other ones don't do that." There was no way to explain that the problem lay with the received signal and not the TV. Therefore, Zenith's engineers had to figure out a way to permit the system to accept less-than-uniform signals.

We were instructed to replace R1204M with a 180-ohm 1/4-watt, 5% resistor. This SMD device was located in series with a .47 mfd electrolytic from pin 52 of IC1200 to ground. Next we were then told to change C2210M from a 470pf capacitor to a 1000pf capacitor. C2210M is located off pin 29 of IC1200 to ground. Then we were told to add a 470pf capacitor at location C6003M if one was not there already. Location C6003M is at pin 8 of IC6000 to ground. Finally, if audio mute persisted, we were told to add a 1000pf capacitor rated at 50 volts at location C6003M but to keep the value at 470pf if possible. Field Engineering Memo 91-13 gave the particulars of the modification.

The second problem involved nuisance high-voltage shutdown. We were asked to add a .01 mfd 500-volt ceramic capacitor from TP012 to E1. These points were located right beside jumper W63 at one end of IC1200. Then we were instructed to remove resistors RX3013 and RX3015 in 25" sets if they were installed. Field Engineering Memo 91-12 showed us how to do the modification.

I won't reproduce either Memo 91-13 or 91-12 because they are history now. However, you might find the information useful as you deal with the C-5 chassis.

Servicing Zenith Televisions

Chassis Familiarization and Available Literature

Figure 3-1 depicts the chassis layout, including the cabling that goes to the CRT driver circuit board, the front panel controls, the IR receiver, the yoke, and the CRT. Figure 3-2 is a block diagram of the chassis. Each circuit can be placed into one of the following "blocks": (1) the power supply, (2) the system control block that includes the microprocessor, tuner, infrared receiver, and keyboard, (3) video processor, source switch, horizontal and vertical sweep, and (4) the audio circuits. Keep the block diagram within arm's reach because it is a valuable troubleshooting aid.

I wish I could cite a full bibliography for the C-5 family, but I can't. Zenith just didn't publish much in those days. But if you want to add some literature to your reading list, I recommend *Technical Training Program: C-2 & C-5 Sentry 2*. You might want to check out an article I wrote for the September 1999 issue of *Electronic Servicing And Technology*, "Another Look at a Zenith Favorite." Except for a scrap of information here and there, that's about it.

As you know, I am not using much of Zenith's technical literature for these older sets because it is, at best, skimpy. Fortunately, Zenith changed the way it presents schematics when they began to manufacture "repair to the component level" modules. I intend to use their literature when I get to the new stuff, but for the moment I intend to rely on Sams PHOTOFACTS. I will be using PHOTOFACT number 3181 for the C-5 chassis with the understanding that it won't apply to all variations, but it works for most.

Customer Menu

I don't feel the need to discuss customer menu features in depth, but I do want to comment on one feature that initially caused sales people and some customers a problem. A retailer or a customer might call and complain, "I can't get any channels at all." Or, the complaint might go like this, "I can only get one channel." When we got the TV and turned it on, "aux" invariably flashed on the screen indicating that the auxiliary inputs on the jack pack had been selected as the signal source. Zenith designed these TVs to switch between the tuner signal source and external audio/video source by channel scanning below channel two or above the highest channel saved. The owner's manual clearly pointed the feature out and also said that auto programming needed to be run first, but folks didn't read the manual. Fancy that!

Service Menu

The time has come to discuss the software (the service menu) that drives the C-5 chassis. You gain access to it by pressing and holding the menu button on the remote control until the user menu disappears and then entering the digits 9, 8, 7, 6, and pressing enter. The service menu opens to the first of two pages of menus (figure 3-3).

Page One

Page one opens on a blue field. Early C-5 chassis page one menus contained three items. The first looks as if it is just the software version (version 685-5-161). If you press the adjust key when the arrow is pointed to version 685-5-161, you gain access to the "Factory IR" mode. Make sure it is set to the off position. The second is "Store Controls." It permits the customer menu adjustments (contrast, brightness, color, etc.) to be stored in the nonvolatile EAROM memory. The third item on page one is called "Brightness." It sets the limits of the control used to adjust brightness in the customer menu. "Brightness" has a number indicator whose range falls between 00 and 31.

The C-5 Chassis

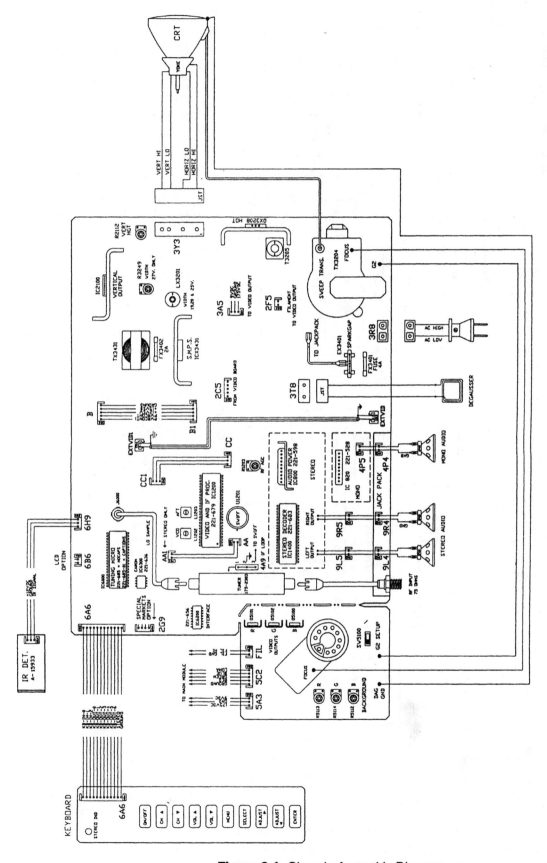

Figure 3-1 Chassis Assembly Diagram

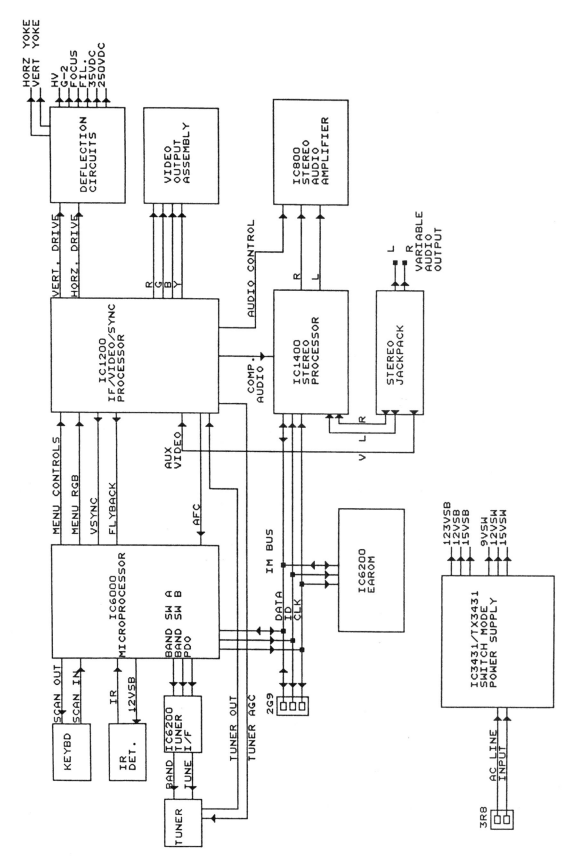

Figure 3-2 System Block Diagram

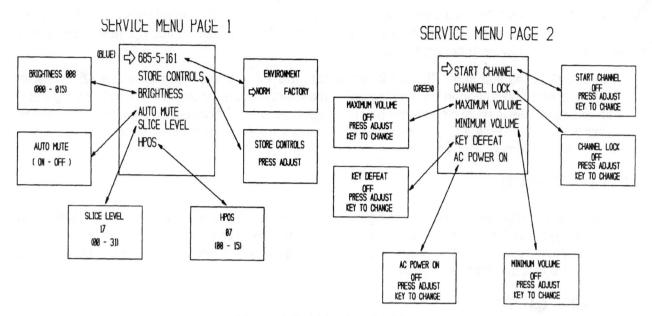

Figure 3-3 J-Line Service Menu

J-Line page one service menus are a bit more complicated, having three additional items. "Auto Mute" allows the service to turn off the auto mute feature I discussed in the previous section. If auto mute is turned off, sync glitches don't affect the audio. It was a helpful feature for the owners of sets who lived in bad signal areas because the modifications I talked about earlier didn't work well there. I had one customer living in a fringe area who decided to trade her Zenith in for a Magnavox and was satisfied with the trade.

The last two items were added to accommodate the closed caption decoder. "Slice Level" permits adjustment of the sensitivity of the closed caption decoder. Its range is 00 to 31, 19 being the usual setting. It is a helpful feature when the closed caption decoder becomes intermittent and needs a bit of tweaking. "Hpos" had to be added to adjust the horizontal position of the closed caption display. Its range of adjustment is from 00 to 15. However, a setting of 07 usually works fine.

Page Two

Page two has six items on it. Each item has two adjustments, on or off. The first item is "Start Channel." If you turn it on, the television uses the current channel as the start-up channel even though the last channel used may have been different.

"Channel Lock" is next on the list. Turning it on locks out all channels other than the one selected at the time the feature is turned on. A power glitch or lightning strike might cause the channel lock feature to change from "off" to "on." If it happens, the customer's complaint might go like this, "I can get only one channel even after I run the auto program." Immediately suspect channel lock as the problem and look at it before you do anything else.

"Maximum Volume" and "Minimum Volume" come next as items three and four. If they are turned on, they limit the range of adjustment accessed by the volume controls on the TV or the remote control. Turning them off gives the user the full volume range the software permits. Don't turn them off unless you have to.

Item five is "Key Defeat." Turn it on, and you disable the menu, select, and adjust keys on the keyboard. "Key Defeat" targets a special market, like hotels and motels. It is also useful in a household full of small children who have wandering hands.

"AC Power On" is the last item. Turning it on permits the TV to come on as soon as AC power is applied. It too has application in certain specialized markets and is useful in the house of a person who, for example, has to have a converter box to access cable TV channels. Just plug the TV into the converter box, set "AC Power On" to the on position, and the TV comes on when the converter box is turned on. A neat feature, don't you think?

Power Supply

You should quickly discover by examining figure 3-4 that the C-5's power supply is almost identical to the one used in the C-2 chassis. As I said then, Zenith continues to use it with a variation or two in their brand-new products. I shall talk in depth about it in Chapter Ten when we get to the Y-Line. In the meantime, let's talk about its two states.

Standby Mode

When it receives AC power, the power supply begins to produces +150 volts, +123 volts, +15 volts standby, and +12 volts standby. IC6400 uses the +12-volt line to produce the +5 volts standby necessary to operate the microprocessor and the IR receiver. There is no 60-hertz detect circuit as there is in the C-2 chassis because a DC-operated oscillator runs the real- time clock.

Full Power Mode

When it receives an on command, the microprocessor pulls pin 49 low sending the low to the base of Q6005. The low turns Q6005 on, and it responds by turning on Q3402. Q3402 turns on the switched voltages (+9 volts, +12 volts, and +15 volts), and they bring the chassis to life.

When the chassis comes to life, the scan-derived voltages come up. They include +23 volts, +35 volts, +215 volts, +250 volts, G-2, focus voltage, and the EHT. The high voltage varies from chassis to chassis because its value depends on the CRT size.

System Control

The brains of system control isIC6000. System control naturally involves more than the microprocessor. Look at figure 3-5 from Sams PHOTOFACT to get an idea about the complexity of the circuit. Figure 3-6 simplifies IC6000 by presenting its functions in a block diagram form.

Zenith describes the microprocessor as a four-bit CMOS device that has been miniaturized to fit into a 52-pin DIP package. It is therefore unlike the microprocessor used in the C-2 family. However, like the C-2's microprocessor, it uses an IM bus to communicate with the EAROM, the tuner controller chip, and the video processing IC. The IM bus configuration calls for clock, data, and "chip select" (or ID) lines for its communication. I thought about expanding my comments on the IM bus but decided against it because Zenith abandoned the IM bus in favor of the industry-wide I2C (I-Square-C or Inter-integrated) bus.

IC6200 is the tuner control chip and is responsible for instructing the tuner via voltage changes to tune whatever channel the microprocessor tells it to. See Chapter 1 for additional details about how the system works.

The C-5 Chassis

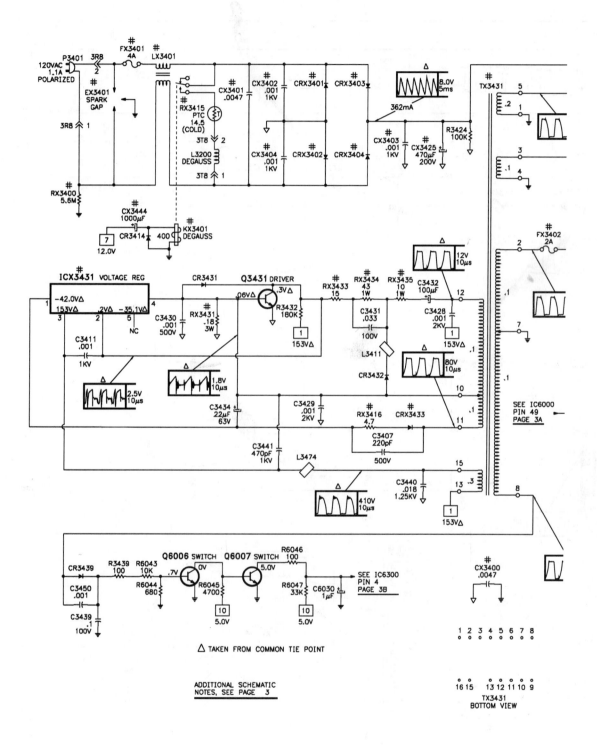

Figure 3-4 Power Supply Schematic

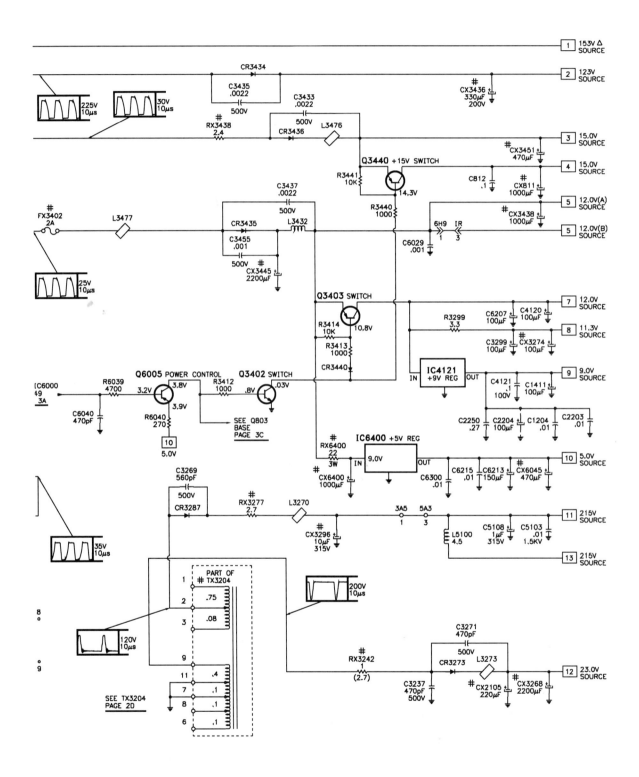

Figure 3-4 Power Supply Schematic (Continued)

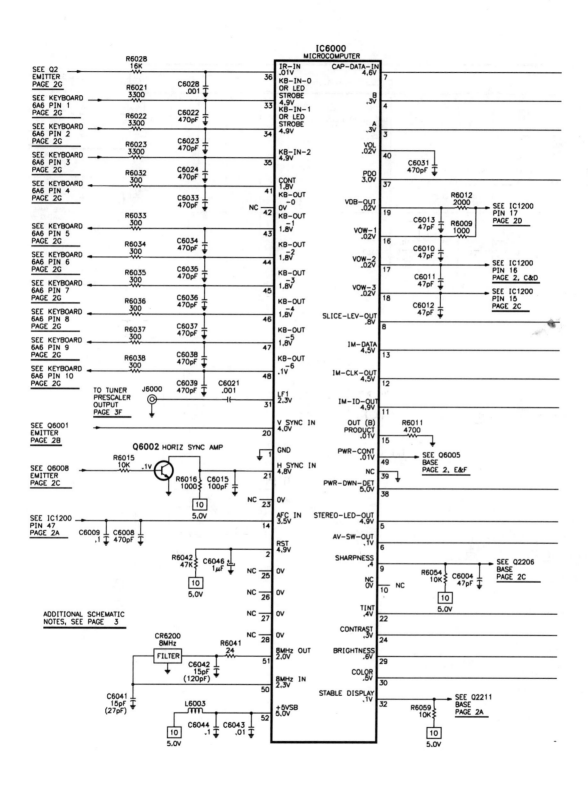

Figure 3-5 System Control Schematic

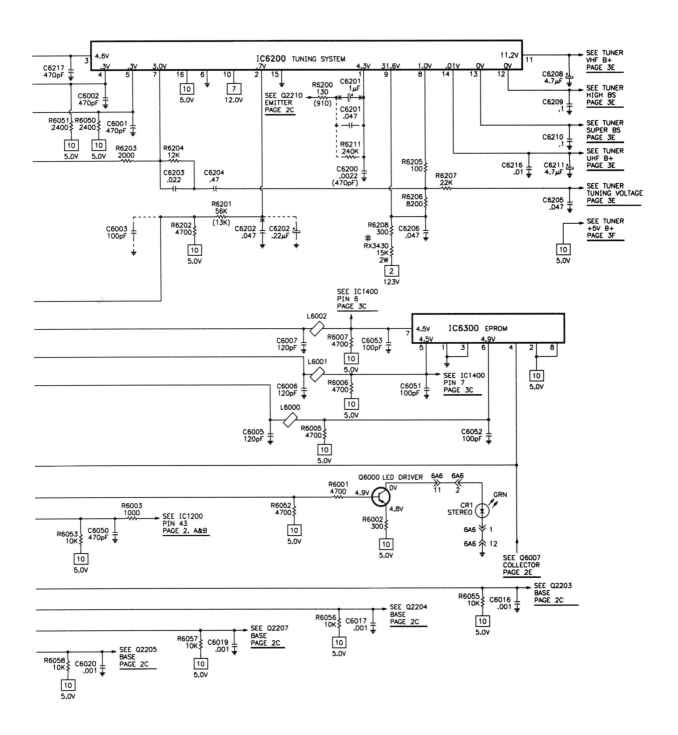

Figure 3-5 System Control Schematic (Continued)

The C-5 Chassis

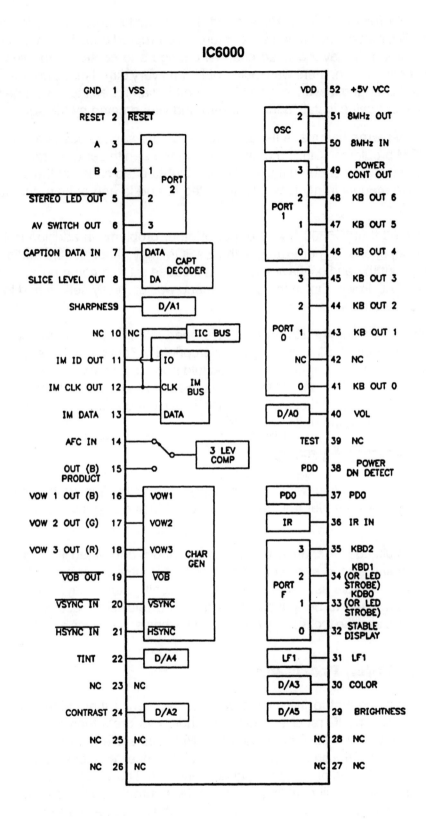

Figure 3-6 System Control Block Diagram

IC6300 is the EAROM (or EEPROM). Believe me, it is capable of causing a lot of really unusual problems. I never will forget the first time I encountered a corrupted one. The TV turned on and off as it should. The on-screen display indicated channel change and control of the volume, but the set produced no video or audio. Oh, there was raster, but it was very dim. I just knew the video processor IC was defective, so I changed it even though I found nothing amiss as I took voltage measurements. I turned the TV on after having installed the new part and was greeted by the same symptoms.

The famous RCA tuner wrap had made its appearance on the service scene along with RCA's now well-known EEPROM problems. Since the symptoms in the Zenith set were remarkably like those I had seen in RCA products, I wondered if the EAROM might be the cause of the difficulty. I snatched one out of a defective module in my junk pile, installed it, and lo and behold, the set worked like a champ!

I have since discovered that IC6300 is at the root of other problems in addition to the no video, no audio, and blank raster scenario. For example, the EAROM may affect just the audio. In other words, everything works except the audio. I have even seen it cause a TV to come on as soon as AC power was applied. No, the "AC Power On" in the service menu was correctly set! It had to be the EAROM because a new one fixed the set.

There is one glitch to replacing the EAROM. I have been told that Zenith will sell you one, but it is blank and must be programmed. Since its program is proprietary, Zenith won't give you the programming information. If my information is correct – and I was told it was by a servicer whom I trust – it puts us techs into a kind of catch-22 situation. My solution has been to save C-5 duds for parts when I can get them. I use the EAROM, horizontal output transistor, flyback and a few other components and make more money than if I returned the duds for credit.

Video Processing System

The video circuit (figure 3-7) takes two pages to display, but I think it is worth the space it takes up. Since block diagrams are a helpful troubleshooting aid, I include figure 3-8 to help you deal with IC1200.

IC1200 performs a multitude of chores, producing the audio and video IF, doing the audio-video source switching, serving as the sync separator, RGB generator-driver and RGB mixer for the on screen menus. It is a veritable assembly line in and of itself. Zenith says the chip incorporates a number of features that used to be found only in high-end television receivers but are now found almost everywhere. We see the technology now on a daily basis, but it was a new feature when the C-5 first came out.

Features of the Video Processor

All customer commands go first to the microprocessor that sends them along to IC1200 as pulse-width modulated signals. The IC receives the commands, interprets them, and executes them in a timely fashion. The horizontal section employs a dual PLL loop system with a horizontal coincidence detector to control the gain of the phase detector. The coincidence detector also makes fast acquisition possible during channel change or when no signal is present. The vertical deflection section utilizes a narrow-range count down circuit enabling the vertical synchronization to operate within +/- 10% of 60 hertz. The system differs from previous circuits in that the absence of a signal causes it to default to 60 hertz. These design changes result in minimal change of vertical size and on-screen menu position when no signal is present.

IC1200 also has a video muting circuit built into the sync separator. If no signal is detected at pin 29, the IC blanks the video. This feature proves its worth by keeping the menu stable and screen blanked when the user selects the auxiliary video mode and no signal is present.

Luminance processing is similar to circuits we have already seen but with the difference that the IC has a built-in luminance delay. It also performs video switching. Auxiliary video goes to pin 40. Pin 43 controls the switching, enabling the user to select between tuner audio and video or external audio and video like the audio-video coming from a VCR.

Now a final word. IC1200 uses three inputs (pins 15, 16, and 17) to implement on screen information. The crucial pin, if there is one, is pin 17 because it is the input port for the video and fast blanking signals. It has three "levels" or "states": 0 to 0.8 volts to disable OSD, 0.8 to 2.0 volts to enable OSD with no blue drive, and 2.0 to 5.0 volts to enable OSD with blue drive.

Video Output

Figure 3-9 is the schematic for the video output module. I have been working on these televisions since they first came on the market and have had few problems with the video output assembly except for solder joints that deteriorate over time. If you suspect solder problems, look carefully at the point where those tiny traces join the foot print where the video output transistors solder onto the board.

Deflection Circuits

Figure 3-10 gives you the particulars of the deflection circuits. If you look at it even carelessly, you see there is nothing really new here. So, I intend to keep my comments to a minimum.

Vertical Deflection

IC2100 receives a drive pulse from IC1200 and outputs vertical drive at pin 12. The IC itself belongs to the popular LA78xx series of vertical output chips. R2112 just off pin 4 is the vertical size control. A scan-derived 23 going to pins 9 and 13 provides the B+ to operate the chip. Note CX2110 and CR2102 at these pins. Remember that they serve as a voltage doubler circuit to provide about +45 volts to move the electron beam from the bottom of the screen to the top as quickly as possible, ensuring a quick vertical retrace. Since they ensure vertical retrace, CX2110 and CR2102 inhibit vertical deflection when either one or both fail. A quick voltage check around the IC tells you if either has failed.

Horizontal Deflection

The 12-volt supply from the main power supply provides B+ for the horizontal section of IC1200. Horizontal pulses are therefore present at the base of Q3209 even when the set is in standby. Horizontal drive becomes available for the horizontal deflection circuit when the switched 12 volts comes on line (source 5 in the Sams). However, the drive is slightly delayed by C3256 to prevent the horizontal output transistor from turning on too early in the scan period.

Now, take a look at figure 3-10 and locate capacitor C2212 just off pin 30 of IC1200. If that 22mfd capacitor fails – and it will – you will definitely have horizontal deflection problems. Descriptions of the symptoms a failed C2212 cause vary from technician to technician. Some talk about "firing lines" in the picture or a "christmas tree" pattern in the raster or a "double image" on the screen. If you encounter a C-5 chassis exhibiting symptoms like these, check C2212. Don't merely check it for leakage or loss of capacitance because these tests might mislead you, but do check it for high ESR. Or better, just take it out and throw it away and put a new one into the circuit.

Servicing Zenith Televisions

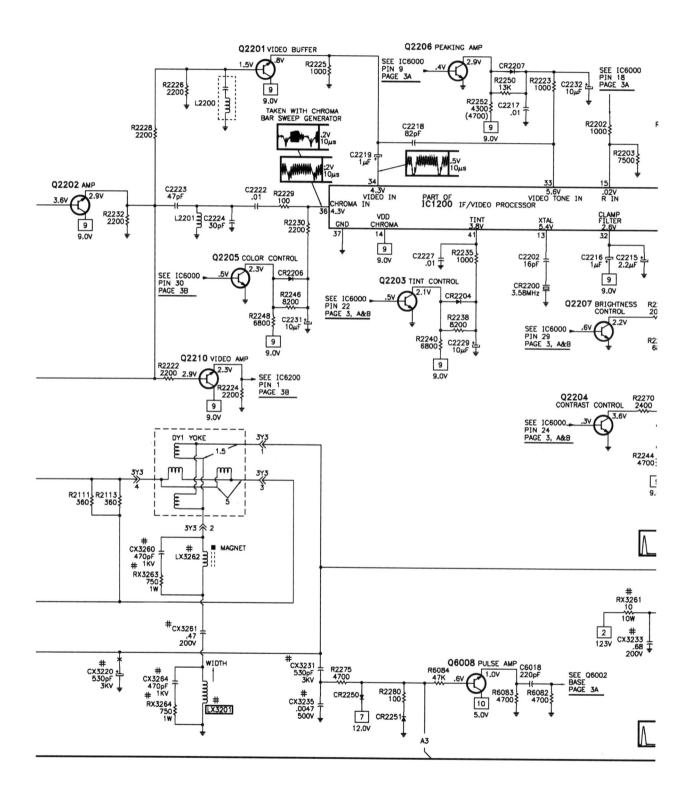

Figure 3-7 Video Processor

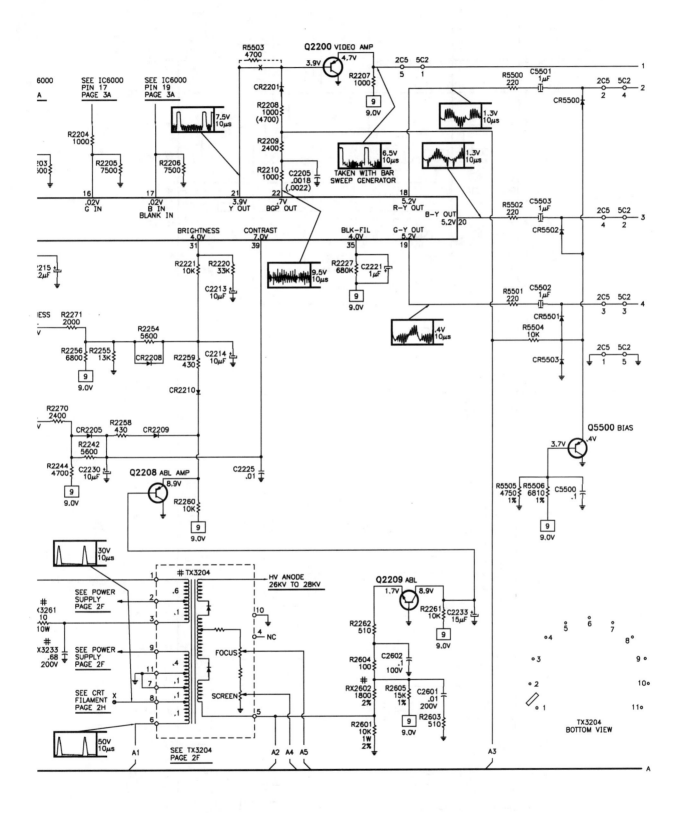

Figure 3-7 Video Processor (Continued)

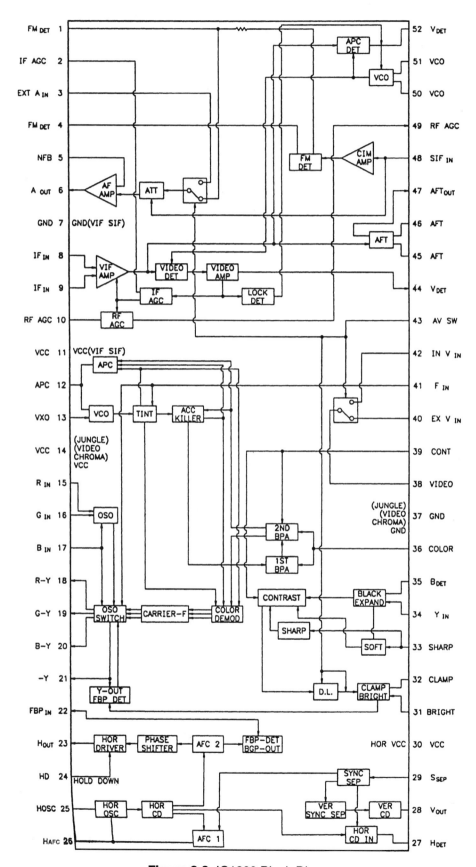

Figure 3-8 IC1200 Block Diagram

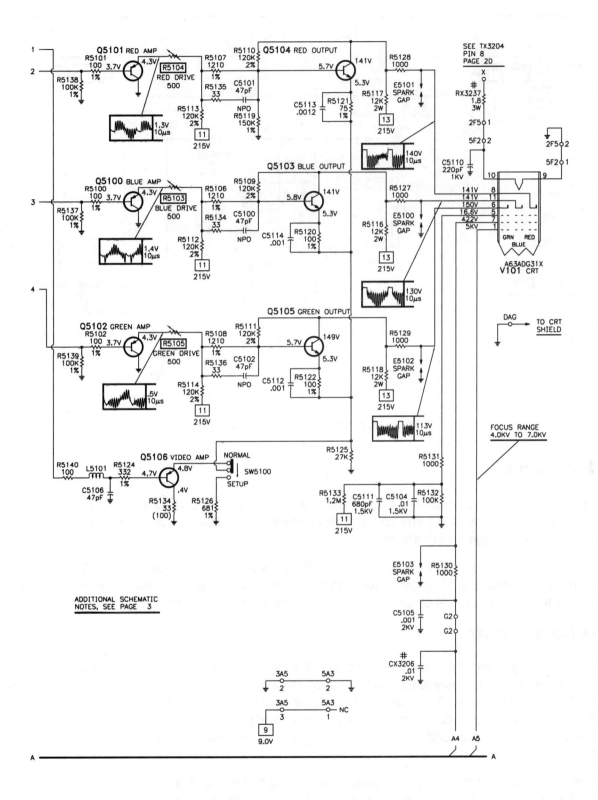

Figure 3-9 Video Output Module Schematic

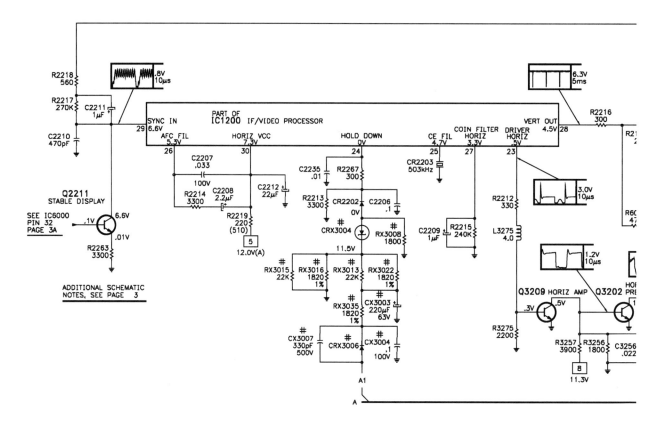

Figure 3-10 Deflection Circuit Schematic

XRP Circuit

Pin 24 of IC1200 is the shutdown input. When the voltage at pin 24 reaches a certain point, a transistor inside the IC conducts and turns off horizontal drive. CRX3004, a 12-volt zener diode, sets the voltage level that triggers XRP shutdown.

The C-5 chassis incorporates a latch circuit into the shutdown circuit, which means AC power has to be removed and reapplied in order to turn the set on after shutdown. If the shutdown really is due to overvoltage, the set will shut off as soon as it receives another on command.

Audio Circuit

Figure 3-11 is the schematic for the audio circuit while figure 3-12 is a block diagram of the audio output integrated circuit. As you see, IC1400 does the stereo audio processing. It receives composite audio from IC1200 at pin 21 and external audio input at pins 35 and 36 and outputs the processed audio at pins 1 and 2. I have looked back over my records to see how many times I have changed it and found just two occasions. Lightning damaged it both times, literally blowing it apart on one occasion.

Left and right channel audio exits it at pins 1 and 2, respectively, and go to the audio amplifier, IC800. This chip delivers up to one watt per channel of audio to the two speakers. The user can switch between external and internal speakers by using a switch on the jack pack (figure 3-13).

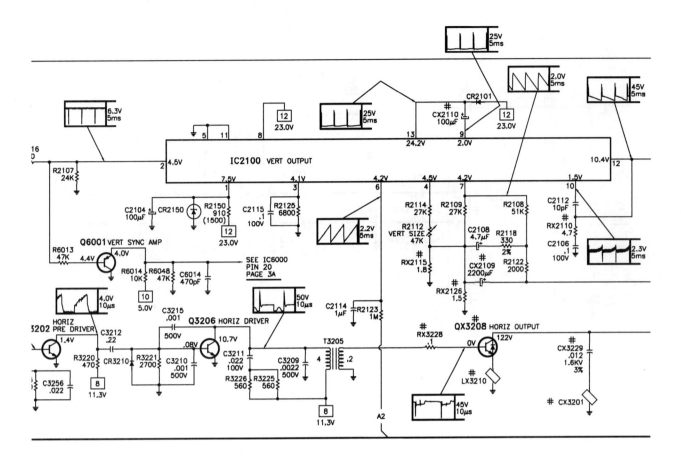

Figure 3-10 Deflection Circuit Schematic (Continued)

If the set has mono audio only, composite audio from IC1200 goes to pin 4 of IC820 and exits at pin 1 and goes directly to the speaker. Unlike other mono sets, these televisions have a provision for auxiliary audio via the jack pack (figure 3-13). The C-5 chassis doesn't require an additional IC to switch between internal and external audio because IC1200 does the switching.

Closed Captions

Closed captions have been a feature of American television receivers since July 1993 because congress mandated it. Zenith stepped ahead of the manufacturing line by releasing closed caption sets in the fall of 1991. Since it is a feature we servicers don't think much about until the circuit fails to work, I thought I would take a little time to explain how the feature works. If you already know how closed captions work, skip this section.

Organizations like CaptionAmerica, the National Captioning Institute, and the Caption Center generate captions and work with broadcasters to transmit them to the TV's decoder. Prerecorded programming has the captions recorded on the video tape. Live programming sends the captions along with the video signal.

The caption display splits the standard TV display area into fifteen rows of up to thirty-two characters per row. Captions normally appear four rows from the top or the bottom of the screen. The so-called "text mode" utilizes the full fifteen rows. You might like to know that the caption mode is defeated in Zenith products when any menu function has been chosen.

Servicing Zenith Televisions

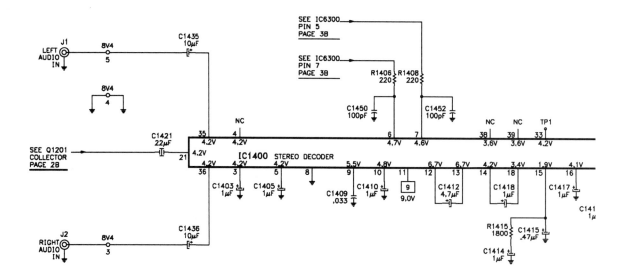

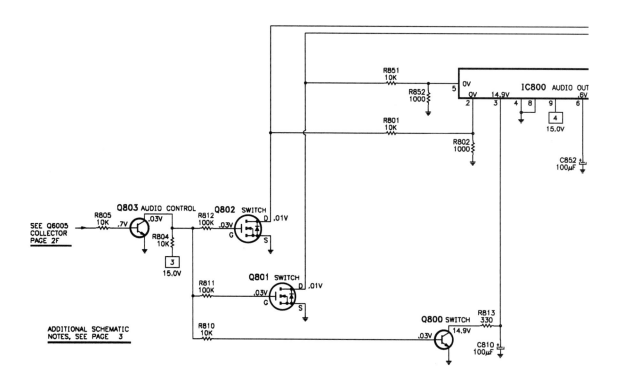

Figure 3-11 Audio Circuit Schematic

The C-5 Chassis

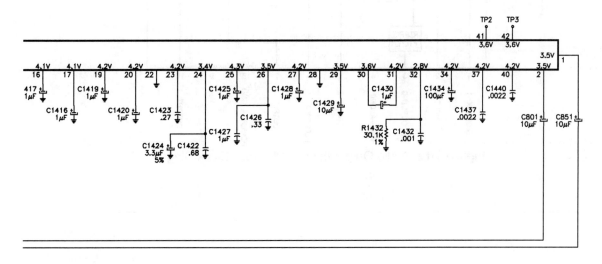

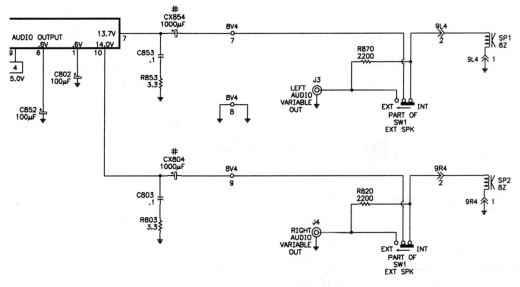

Figure 3-11 Audio Circuit Schematic (Continued)

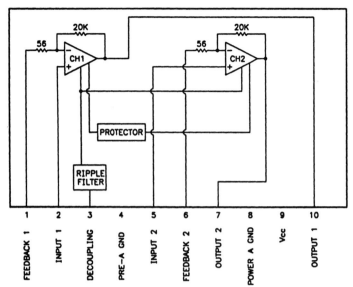

Figure 3-12 Audio Output Integrated Circuit Block Diagram

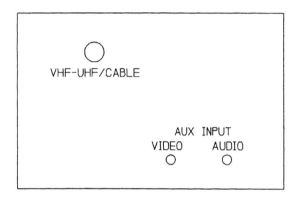
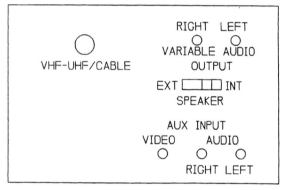

Figure 3-13 Stereo Jack Pack

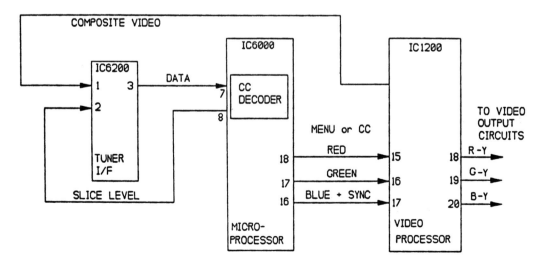

Figure 3-14 Closed Captions Come Directly from the Video Signal

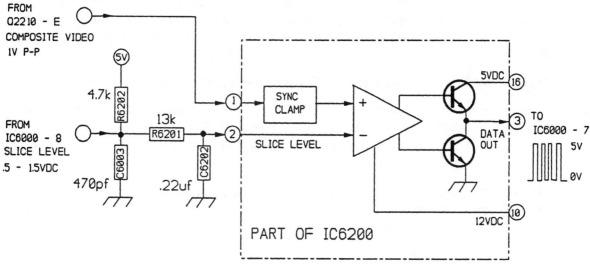

Figure 3-15 Comparator Circuit Built into IC6200

Closed Caption Video Signal

Line twenty-one of the vertical blanking interval has been set aside for closed-caption data. The signal, that conforms to the NTSC standard in every respect, contains a data synchronizing clock signal, a data start bit, and sixteen bits of data. The signal level format is "0" (0 IRE) for the blanking level and "1" for 50% of peak white (50 IRE), and is transmitted at about 480 baud.

The signal for captions comes directly from the video signal (figure 3-14). Composite video from the video detector (IC1200) goes into a comparator circuit built into IC6200 (figure 3-15). The comparator compares the slice level reference from the microprocessor (IC6000) to the video signal. If the caption signal is greater than the slice level, the comparator outputs data to the caption decoder inside IC6000. You probably remember that slice level is set via an adjustment in the service menu. Its range of adjustment is 00 through 31, with settings between 16 and 19 being typical. If the slice level is set too low or too high, the closed-caption decoder will either work intermittently or not at all.

Repair History

The information presented here pertains to the 9-1118, 9-1130, and 9-1132 because I have a very limited experience dealing with the other modules in the C-5 family. The information is listed without any particular order.

(1) Out-of-sync horizontal, firing lines in the picture, or Christmas tree pattern. Replace C2212, 22mfd/16 volts as I discussed a few pages ago. I didn't mention in the previous discussion that a defective C2212 also affects the quality of the audio by introducing a kind of "frying" sound into it. At any rate, you won't be able to overlook the symptoms!

(2) Power on inoperative but standby B+ voltages okay. No voltage at pin 1 of IC6000 because CR3439 has failed. An ECG552 makes an acceptable substitute.

(3) Horizontal jitter. Check and replace if necessary C2211, a 1mfd/50-volt capacitor.

(4) No color. IC1200 was defective. Be sure to check color oscillator crystal before replacing the IC.

(5)	Dead with squeal at turn on. Either the horizontal output transistor or the flyback or both usually cause this problem. If you replace the output transistor, monitor the waveform at its collector for evidence of ringing and other distortions. If these are present, change the flyback.

(6)	Dead. C3434 may cause a dead set. Replace it if you have any question about its reliability.

(7)	Picture shrunk on all four sides. Check pin 1 of ICX3431 for low voltage. If the voltage is low, replace R3431 even if it checks good. ICX3431 (STR53041) has also been known to cause low B+. I have seen one instance when C3432 caused the problem, though that's rare.

(8)	No vertical deflection. Poor solder around the vertical output IC accounts for most of the vertical deflection problems I have seen. However, I have had to replace R3234 and the vertical output IC on several occasions.

(9)	Inoperative channel change. Replace the EAROM. Don't forget the other problems a defective EAROM might cause, namely a TV that comes on at plug in, no audio but good video, and no audio and no video accompanied by a dim raster. I wish I could give you a check that positively verifies a faulty EAROM, but I don't know of one. If you do, let me know.

(10)	Repeat failure of the power supply. If you encounter a C-5 set in which the power supply fails more than once, suspect the picture tube. They may arc internally or short heater-to-cathode, and when they do, the power supply usually goes up in smoke.

(11)	Failure of the power supply. In Chapter One, I indicated the parts that normally fail in the power supply. You might want to reread that section now. If you don't reread it, keep it in mind when you need to repair a C-5 power supply.

(12)	Tuner inoperative. As you know, the C-5 chassis makes use of a rather elaborate tuning system. I suggested in Chapter One that you make a few checks around the tuner and tuner control chip before condemning the tuner. I still believe that's good advice, but I have to admit, I have never had a problem with any component in the tuning system except the tuner. However, in the world of electronics all things are possible. I do know I have a problem keeping good tuners in stock because I use them about as quickly as I accumulate them. If you save your dud modules, you will find them to be an excellent source for parts, for the tuner especially.

(13)	Distorted picture. I am thinking about those few sets that have a kind of noise band in the picture which is, oh, about four inches wide and horizontal in orientation. It may appear as stationary or floating and might change positions if the set is turned off and then back on. As far as I know just one thing causes this problem in Zenith sets, and that is a defective picture tube. Unfortunately, the floating band phenomenon appears in any Zenith chassis, a fact that I will point it out again as our discussion progresses.

CHAPTER 4
THE C-6 CHASSIS

I am not sure why the C-6 chassis has been a favorite of mine, but it has. I used one on my bench to check VCRs and camcorders for several years and gave my son and his bride a 19" version for a wedding present. I even enjoy working on them. It sounds dumb, but it's true.

The modules that make up this family of televisions are: 9-1343-01 (a 9" AC/DC set), 9-1248-01 (13" set), and 9-1244-01 (19" and 20" sets). I have very limited experience dealing with the 9" sets and a somewhat limited experience dealing with the 13" ones. Therefore, I intend to center my discussion on the 19" and 20" products with the understanding that what I say about them is applicable to the other modules in the family.

Figure 4-1 is a photograph of a 9-1244-01 module. If you have ever handled a new module, you know that the video output module is attached to the main printed circuit board as a breakaway item just as it is pictured in the upper right corner of the photograph. Included are figures 4-2a and 4-2b to give an idea about where to find specific test points and components on the chassis.

How does one describe the circuitry found on the 9-1244-01 module? Well, the chassis is built on a single-sided circuit board just like most of the other modules Zenith has manufactured in recent years. However, the C-6 circuit board has just three integrated circuits for all signal, sync, and sweep processing. IC1200 handles the video and audio, sync, and horizontal and vertical processing. The tuner doesn't require a separate prescaler like the previous chassis because the main microprocessor (IC6000) handles tuning and band switching commands and transmits them to the tuner via the control bus. The tuner control chip inside the tuner receives the instructions and executes them. Such an arrangement greatly simplifies the tuning arrangement by reducing the IC count on the main printed circuit board and even permits the use of a physically smaller tuner. The keyboard and IR receiver connect directly to the microprocessor. IC2100 produces vertical drive for the yoke. IC3400, the last of the ICs on the circuit board, is configured as a typical, series-pass regulator supplying about +130 volts to the horizontal output transistor. The chassis is therefore line-connected, meaning you must use an isolation transformer when you service it.

Figure 4-1 9-1244-01 Module

Available Literature

Training literature for the C-6 is limited. *Technical Training Program: C-6 Sentry 2* is the best booklet I have come across, and I recommend it if you feel the need for a bit more information than I am able to provide. I intend to use it and PHOTOFACT 3277 as a source for schematic material.

Service Menu

Discussing the service menu is as much a part of dealing with modern televisions as, for example, the power supply is. So, let's begin talking about the C-6 by starting with the software that runs the chassis (table 4-1). I won't go down the list item by item because by now most entries have become, shall I say, old friends. But, I do need comment on those few that we haven't encountered up to now.

Figure 4-2a Service Adjustments

Access and exit the service menu just as you do for the C-5 chassis. If you need that information, flip back a few pages to the last chapter. After you gain access to the service menu, toggle between its two pages by using the menu button either on the TV itself or the remote control.

(1) Capt Phase. Caption phase is very similar to "Slice Level" discussed in the last chapter. A setting of 95 is typical.

(2) Vert. Position and Horz. Position. These two items determine the position of the closed captions display on the screen. They do not affect anything in the picture or on the screen except the position of the captions.

(3) Caption Search. Don't worry about "Caption Search" because it isn't found in all C-6 service menus. Turn it off and leave it that way if you ever encounter it.

(4) Crystal Frqncy. Found in late-production C-6s, it matches the frequency of the clock crystal to the microprocessor by offering settings of 12.00 MHZ and 12.08 MHZ. If the crystal has the part number 224-143, set this item to 12.00; but if the part number is 224-157, set "Crystal Frqncy" to 12.08. The part number is on the crystal. If you select the wrong frequency, the on-screen clock won't keep correct time and parity errors will appear in the closed captions.

(5) Zen/PL. Zen/PL makes its first appearance in Zenith service menus. It permits switching the codes of the IR receiver between "Zenith" and "Private Label," the latter being chassis Zenith manufactured for wholesalers like Radio Shack, Curtis Mathes, and others. If Zen/PL is set to PL, use remote control codes 21 or 121 to work the set.

(6) AV Jacks Option. Set it in the off position unless the television has a jack pack. When it is set in the on mode, "AUX" appears before the lowest channel and after the highest channel stored in memory.

The C-6 Chassis

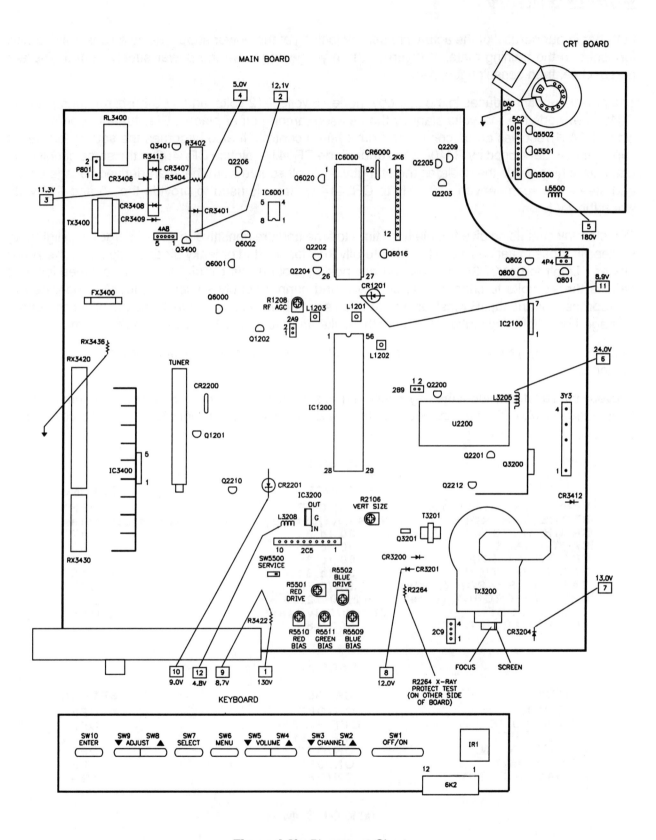

Figure 4-2b Placement Chart

Power Supply

Let's begin our perusal of the actual chassis by looking at the power supply using a schematic Zenith furnished in the training material (figure 4-3). If you want to see the power supply from a different perspective, then consult figure 4-4.

The power supply produces in the standby mode +5 volts, +12 volts, and a 60-hertz reference signal. C3401 develops the +12 volts standby that is used, among other places, to actuate the on/off relay when Q3401 receives an on command from system control. It also becomes the source for the +5 supply that is developed by R3404 and zener diode CR3414. Notice CR3402 in the same circuit. It is a fast switching diode that isolates the standby +12 volt source from the scan-derived voltage circuit and steers the scan-derived +12 volts to CR3414 where it is used to provide the +5 and +12 volts when the TV is on.

The standby circuits do give trouble from time to time and are sensitive to power surges and lightning strikes. In those instances, you should naturally look for shorted or leaky diodes, especially the zener diodes. Don't forget CR3402 while you are poking around this circuit. When it becomes leaky, it causes the +5 volts to check a volt so low it and cannot supply sufficient current to operate the microprocessor and the IR receiver. I have seen several televisions where it was the only component damaged by a lightning strike. Use something like an ECG519 or ECG577 as a replacement.

CR3400, R3401, and C3434 develop the 60-hertz signal the microprocessor uses to develop the on-screen clock.

I haven't said anything about the way Zenith draws schematics up to now. Since I am beginning to use their material, I should familiarize you with the way Zenith engineers designate voltage sources. While

PAGE 1

ITEM	RANGE	SETTING
Version 753-1.38 - 1.61	ON/OFF	OFF
STORE CONTROLS	Press Adjust Key	Activates Store
BRIGHTNESS	00 to 31	17
CAPT PHASE	000 to 254	95
VERT POSITION	00 to 31	5
HORIZ POSITION	00 to 31	15
CAPTION SEARCH (Early)	ON/OFF	OFF
CRYSTAL FRQNCY (Late)	12.00/12.08	12.08
ZEN/PL MODE	Zenith/Priv.label	ZENITH
AUTO SEARCH (LATE)	00-18	2

PAGE 2

ITEM	RANGE	SETTING
CH LOCK	ON/OFF	OFF
MAX VOLUME	OFF-1-30	OFF
MIN VOLUME	OFF-1-31	OFF
AC POWER ON	ON/OFF	OFF
KEY DEFEAT	ON/OFF	OFF
AV JACKS OPTION	ON/OFF	*ON

Table 4-1 Software

The C-6 Chassis

PHOTOFACT use numerical references to tie various components to voltage sources in the power supply, Zenith schematics do it a bit differently. Notice in figure 4-3 the five-sided figure at the anode end of CR3425 with the legend "+12 VSW" in it. The five-sided figure indicates that it depends on the +12 volts switched supply. Now look to the right of that figure and locate the five-sided figure with the pointed or triangular top having the legend "+13 VSB" in it. The pointed or triangular top indicates it is the source for the +13 volts standby. Zenith's notation corresponds to voltage source 2 in figure 4-4.

The full power supply comes up when a high from pin 35 of the microprocessor turns on Q3401. Q3401 responds by causing the relay to actuate and create a path for AC to the bridge rectifiers. The bridge circuit and C3505 then make about 155 volts available to the input of IC3400.

IC3400 is a linear regulator and the workhorse of the power supply. It is an STR30130 (Zenith part number 223-40) and may be replaced by a generic like ECG1777, NTE 1777 or SK9870. Pin 1 is the ground connection, pin 2 is the control pin, pin 3 is the raw B+ input, and pin 4 is the regulated output. Pin 5 is not used. The regulated output follows the voltage at pin 2 to within about a volt. If the voltage at pin 2 is, for instance, 131 volts, the voltage at pin 4 should be approximately 130 volts. The resistors, diode, and capacitor tied to pin 2 set the operating parameters regulated voltage output. C3407 at pin 4 provides additional filtering for the +130 supply.

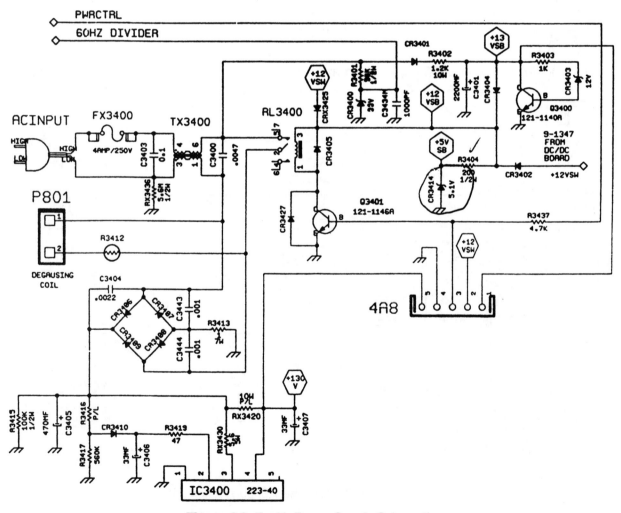

Figure 4-3 Zenith Power Supply Schematic

Servicing Zenith Televisions

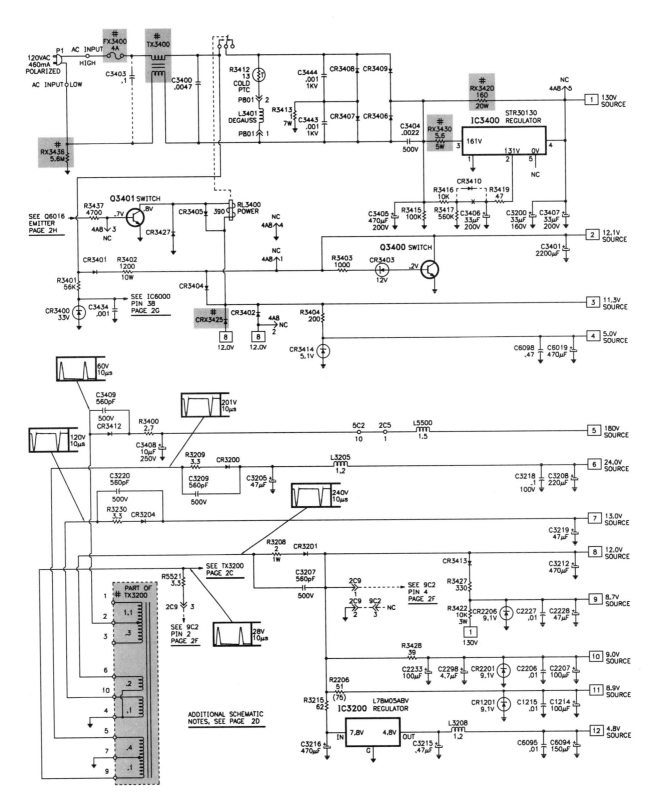

Figure 4-4 PHOTOFACT Power Supply Schematic

The C-6 Chassis

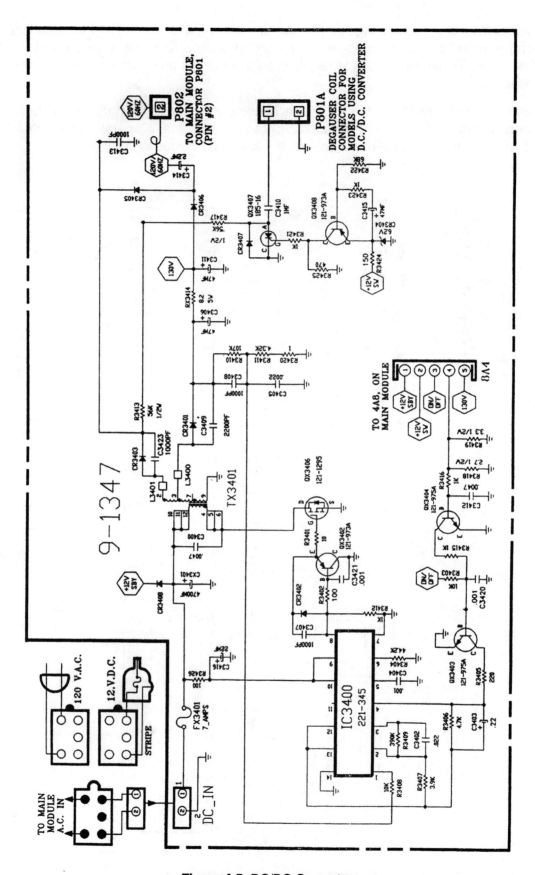

Figure 4-5 DC/DC Converter

Servicing Zenith Televisions

You should remember that the voltage on the collector of the horizontal output transistor is about 150 volts when horizontal drive is not present. After horizontal deflection comes up, the voltage drops to +130 volts. The STR series of voltage regulators don't put out their rated voltage unless they are properly loaded.

When the main supply energizes, the +130 volts become available to the horizontal output transistor and the horizontal start-up circuit. The horizontal oscillator begins to operate, horizontal deflection starts, the scan-derived voltages come up, and the television comes to life.

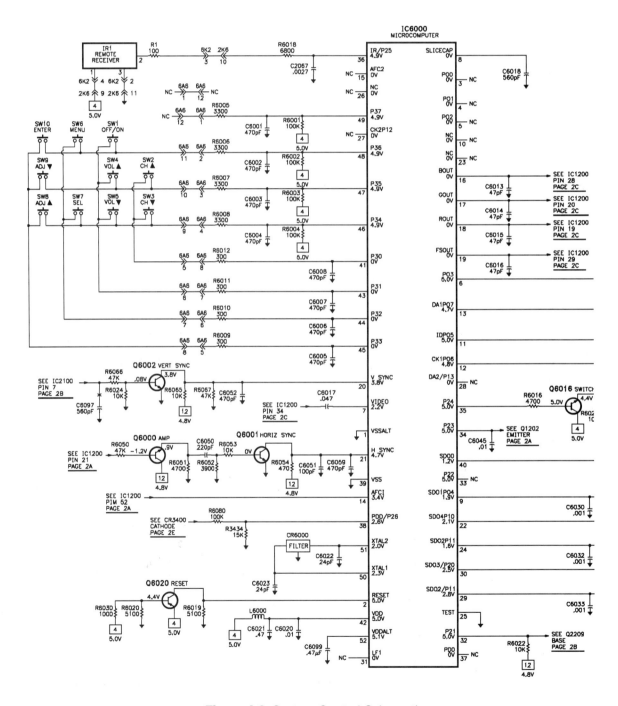

Figure 4-6 System Control Schematic

DC/DC Converter

The DC/DC converter is built onto a separate circuit board and mounted inside the cabinet to provide DC operation for the 9" televisions. It is rather difficult to service because of the way it is constructed and mounted inside the TV, but it is relatively inexpensive to replace as a module. I usually opt to replace it unless I easily spot and quickly repair the problem. Purchase it by ordering module number 9-1347. If you want to service it, use figure 4-5 as a guide.

The converter requires +12 volts to operate. The 12 volts is coupled through CR3408 and becomes the +12 volts standby from which the +5 volts is developed. When the power supply receives an on command, QX3402 and IC3400 turn on and begin the switching activity that energizes TX3401. The diodes and capacitors in the secondary of TX3401 use the energy the transformer generates to develop the run voltage for the set.

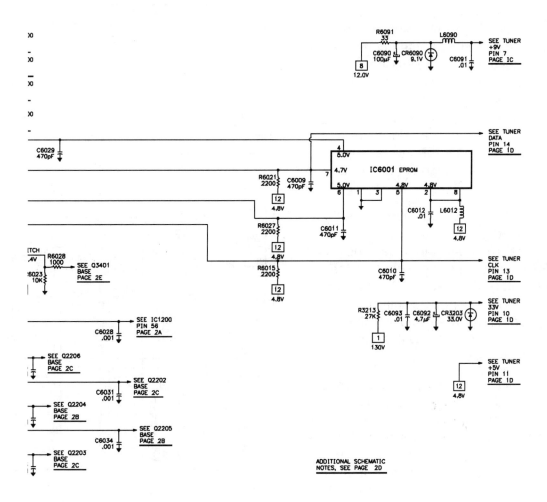

Figure 4-6 System Control Schematic (Continued)

System Control

IC6000 (figure 4-6) is a 52-pin CMOS processor in a DIP package. It receives instructions via the keyboard (pins 41-49) and the IR receiver (pin 36) and outputs those instructions to control brightness, color, contrast, tint, and sharpness at pin 30, 29, 24, 22, and 9 respectively. Those commands go directly to IC1200.

It is not necessary to make additional comments about the system control microprocessor and its circuit because both are similar to those I have already discussed. If you do need to troubleshoot it, use the standard procedure I have often talked about by checking the "must haves" first, +5 volts, ground, reset, and oscillator. I have serviced many of these chassis and if my memory is correct, I have replaced only two defective microprocessors, one because it was completely dead and the other because it wouldn't generate an on-screen display. Lightning had damaged both.

Video Processor and Related Circuitry

Let's talk about a circuit that gives problems, namely the video processor and its related components. Figure 4-7 lays out the video processing circuit the way Zenith does. Figure 4-8 is a bit more specific because it is a block diagram of IC1200, otherwise known as a TA8879N. I keep one or two of these ICs in stock at all times, and I order them by the TA8879N number from one of several suppliers. They usually cost in the neighborhood of $18 to $19.

Of course, IC1200 employs VLSI (Very Large-Scale Integrated) technology, the ability to pack a myriad of components into a very small space. It develops the audio IF, performs sync separator functions, is an RGB generator-driver, an RGB mixer for on screen display, and develops horizontal and vertical drive. The horizontal drive and vertical drive sections work very much the way they work in the video processor of the C-5 chassis.

As a matter of fact, there isn't much difference in the luminance signal processing section either. IF from the tuner enters at pins 9 and 10 and is processed. The resulting video exits at pin 47 where it is sent through an adjacent channel trap and reenters at pin 45. Tint, color, contrast, brightness and sharpness commands from the microprocessor enter at pins 46, 44, 43, 38, and 36, respectively. External video comes in on pin 42. Luma and chroma exit at pins 15 through 18 and go to the CRT.

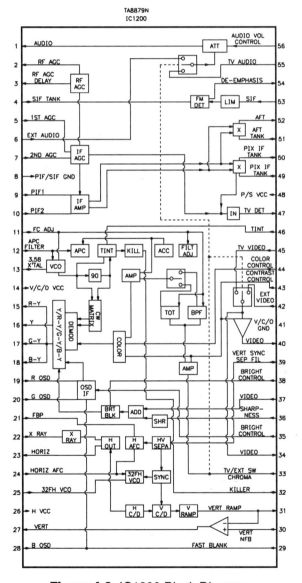

Figure 4-8 IC1200 Block Diagram

The audio IF signal enters at pin 53. It is processed and then exits at pin 1. L1203, the 4.5 MHz quad coil, is tied to pin 4. External audio makes its appearance at pin 6. Make a note that audio and video switching between external and internal sources is internal to the chip, as in figure 4-8.

Now what problems does IC1200 cause?

It's a big, big chip, and has the ability to cause lots of them, but it appears to be prone to some problems as opposed to others. For instance, I have had to replace it several times because it had no luminance output. That seems to be the only damage lightning often causes. I have replaced several because they failed to generate horizontal drive. Be careful when you suspect it is the cause for no horizontal deflection. Make sure that it has +9 volts for horizontal vcc, that the horizontal driver transistor is biased on, and that the horizontal driver transformer is not the culprit. Then, and only then, feel free to replace it. Also, I have replaced it because it has caused a variety of vertical problems. Sometimes it won't develop vertical drive. Sometimes it outputs such a misshapen vertical pulse that it causes the raster to stretch at the top and have very visible retrace lines in it. On one occasion the vertical pulses were insufficient to generate full vertical deflection. I found most of the problems by observing the vertical pulses with a good scope as they exited pin 27 of the chip. A good scope is always the best tool for pinpointing problems like these.

There is one other component in the video circuit prone to give problems, so much so that I always keep a new one on hand. I'm talking about the delay line, U2200 (Zenith part number 105-218). If it fails, it causes a picture to be displayed on the screen that has full color information but no luminance. If you think it is the cause of a no-luminance condition, use a scope to see if the signal enters U2200 but doesn't exit. Confirm your diagnosis by using a jumper to connect its input to its output. The picture will look horrible, but if luminance returns, you know you have found the problem.

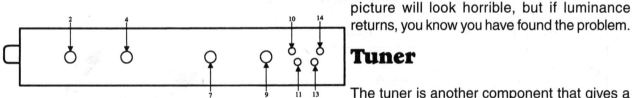

Tuner

The tuner is another component that gives a problem or two now and then. So, I thought you might like to have a bit of information to help you troubleshoot it (figure 4-9). When you suspect a tuner problem, inject an IF signal from a signal generator or a tuner sub into pin 9 of the tuner. If the picture returns, you can be reasonably sure you are fighting a tuner-related problem. If it doesn't return, you need to concentrate your troubleshooting efforts on the components down the line from the tuner. Let me caution you that before you replace a suspected defective tuner, check the "must haves" – +9 volts, +5 volts, +33 volts, AGC (which is developed by IC 1200), and clock and data pulses from the microprocessor.

TUNER VOLTAGE CHART

Pin	Function	VHF Low Band	VHF High Band	UHF Band
2	TUNING	.7V	3.7V	5.8V
4	AGC	6.4V	6.4V	5.4V
7	+9V	8.8V	8.8V	8.9V
9	IF OUT	0V	0V	0V
10	33V	33.0V	33.0V	33.0V
11	+5V	4.8V	4.8V	4.8V
13	CLK	4.8V	4.8V	4.8V
14	DATA	4.7V	4.7V	4.7V

NOTE: Voltages taken with signal.
VHF Low Band voltages taken on channel 2.
VHF High Band voltages taken on channel 7.
UHF Band voltages taken on channel 14.

Figure 4-9 Tuner Information

Video Output Printed Circuit Board

As you have discovered, the video output module is a breakaway portion of the main printed circuit board. Its makeup varies with screen size (figure 4-10). Pins 5 through 9 of connector 5C2 route

Servicing Zenith Televisions

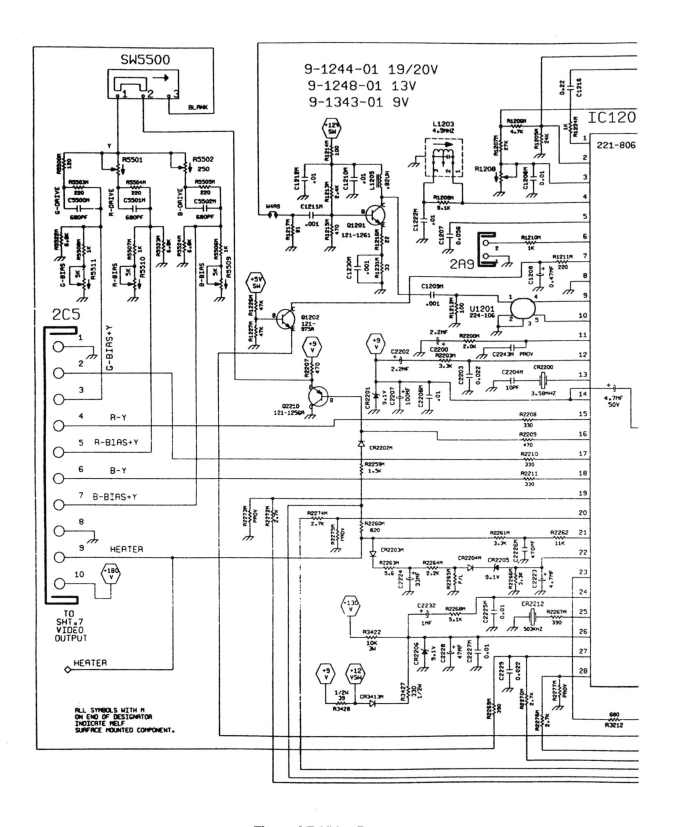

Figure 4-7 Video Processor

The C-6 Chassis

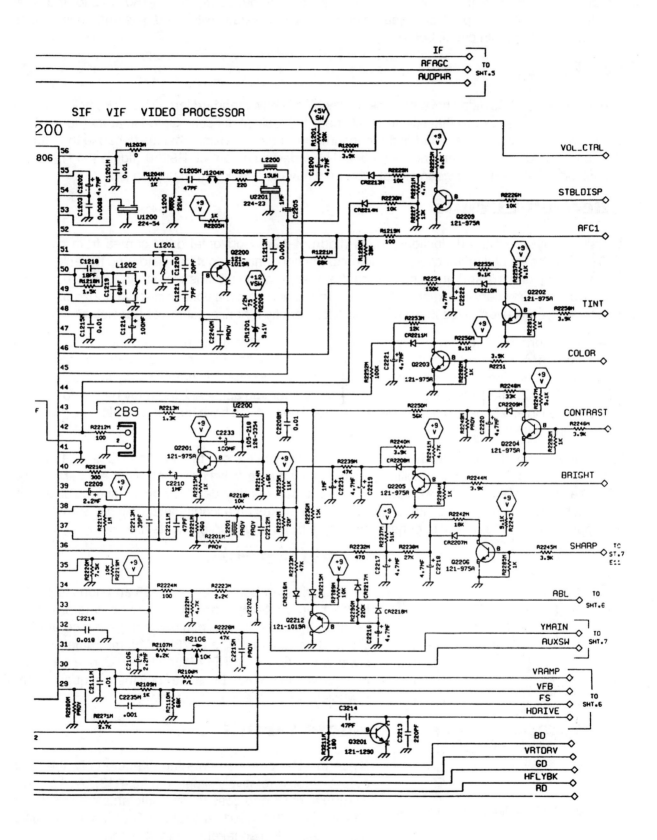

Figure 4-7 Video Processor (Continued)

luminance and chroma signals from IC1200 to the video output board. The +180 volts for the video output transistors arrive at pin 1 of the same cable. The G-2 setup switch and bias and drive controls are mounted on the main circuit board.

Jack Pack

Some early model C-6s didn't have the jack pack which is now standard fare. The jack pack's circuitry is mounted on a small board that sits vertically near the jack pack itself. Opto-isolators (figure 4-11) provide isolation between the world of the consumer and the television chassis because the chassis is "line connected" or at "hot ground" potential. The video gain control, R9210 (figure 4-11), adjusts the amplitude of the external video signal. You may have to tweak it from time to time.

Deflection Circuits

Let's talk briefly about the deflection circuits (figure 4-12). Horizontal drive comes from pin 23 of IC1200, to the horizontal driver circuitry, and on to the horizontal output transistor and flyback. The C-6 chassis really doesn't add anything new to the horizontal circuit. It is just straightforward, vintage Zenith. Based on my service experience, expect to replace an occasional defective horizontal output transistor, a flyback every once in a while, and occasionally a horizontal-driver transistor.

It's easy to determine if you have a horizontal deflection problem when you first apply AC to these chassis. When AC becomes available, the chassis sort of "burps" as horizontal drive comes up, bringing a burst of high voltage that quickly dies down. The chassis is now ready to receive an on command. If horizontal drive doesn't come up shortly after you apply AC, the relay begins to click on-off/on-off/on-off in an almost endless cycle. That's the clue that you have horizontal deflection problems. Get set to find out why.

Look at the horizontal output transistor in figure 4-13. Do you the see the notation, "Q3200, 121-P/L"? The P/L notation is Zenith's way of telling you to consult the parts list (PL) for the correct part number for the chassis on which you are working because the part number depends on the screen size of the TV. Incidentally, a generic horizontal output transistor makes an acceptable substitute. Zenith's TV's don't appear to be as fussy about generic parts as Philips and RCA products are.

Vertical Deflection

Vertical drive leaves IC1200 at pin 28 and goes to pin 4 of IC2100 (figure 4-12) while vertical sync enters at pin 7. Vertical drive to the yoke makes its appearance at pin 6. The vertical output IC is a LA7830. You may safely replace it with an ECG1773, NTE1773, an SK9752, or a LA7830. I usually use an LA7830 because I buy them in quantity from a wholesaler who is a friend of mine.

Audio Circuit

Since the C-6 chassis offers mono audio only, the audio output circuit is relatively simple (figure 4-13). Audio leaves IC1200 at pin 1 and goes to the audio amplifier circuit that consists of Q800, Q801, and Q802. The signal is amplified and routed to the speaker via the 4P4 connector.

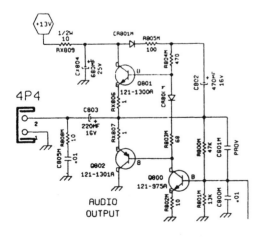

Figure 4-13 Audio Output Circuit

The C-6 Chassis

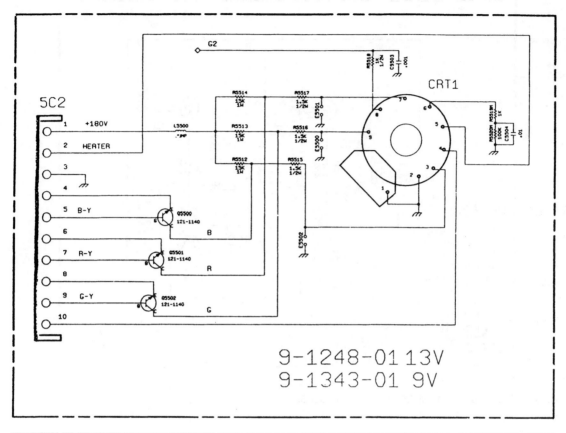

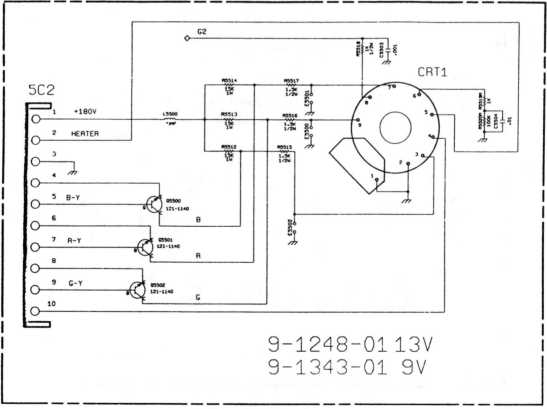

Figure 4-10 Video Output

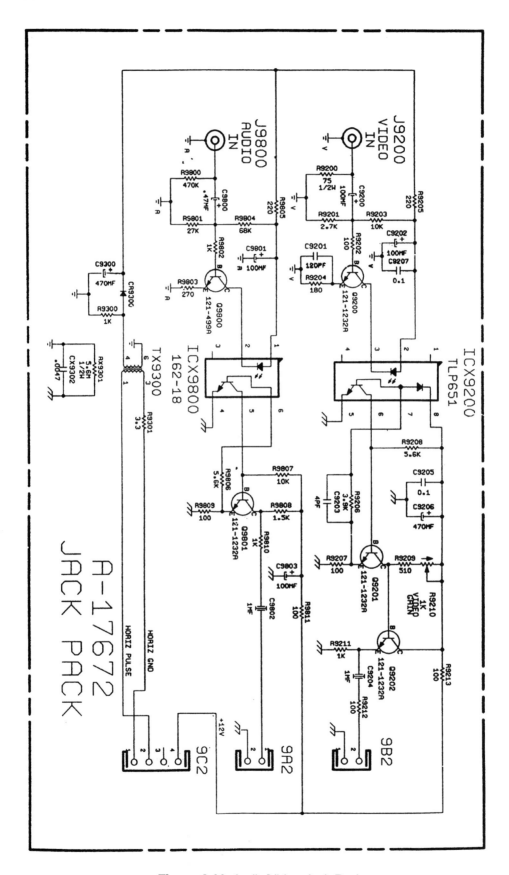

Figure 4-11 Audio/Video Jack Pack

The C-6 Chassis

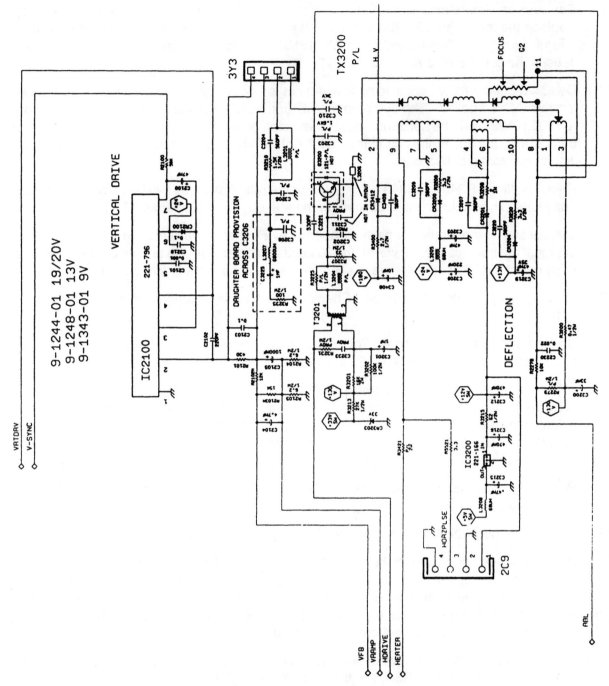

Figure 4-12 Deflection Circuits

Repair History

The information I am about to give you comes mainly from my experience with the 9-1244 module. Of course, I have gathered bits and pieces of information from other techs and my reading, all of which I have added to my database. It appears here in no particular order.

(1) Dead set with standby voltages okay. Defective relay.

(2) Can turn off picture and sound but raster stays on all the time. Again, a defective relay. Replace the relay with a Zenith original, part number 195-138 or use a degauss relay from one of your "junk" Zenith modules. A shorted collector-to-emitter relay driver transistor causes the same problem.

(3) Dead set but the relay begins to click when set is first plugged in and continues to click as long as AC is applied. The symptom points immediately to lack of horizontal deflection. IC1200 (TA8879N), the horizontal driver transistor, poor solder around the horizontal driver transformer, the horizontal output transistor, or the flyback are the most frequent causes for lack of horizontal deflection.

(4) Relay chatter. CR3401, CR3404, CR 3400, or the relay itself may cause the relay to chatter. I have also known defective capacitors in the standby circuit to cause it to chatter.

(5) Vertical sweep inoperative or partially operative. Poor solder around the pins of IC2100 is the most common cause. However, the IC itself can be at fault. Be sure to check for vertical drive out of IC1200 because I have seen it cause a variety of vertical output problems.

(5) TV won't stay locked on a station. Check L1202 for misalignment. It is located close to IC1200.

(6) Bright screen with retrace lines. You may think the picture tube is bad, but check Q2210 before you condemn it. An ECG159 is an excellent replacement for Q2210.

(7) TV will not respond to an on command even though standby voltages are good. Check relay driver transistor, Q3401. For some reason, it is especially susceptible to damage from lightning strikes and power surges. If it is open or leaky, replace it with an ECG, NTE, or SK equivalent. Like other circuits, the diode in parallel with the relay coil can also short out and keep the relay from energizing. Be sure to check it while you are at it.

(8) Poor audio. The audio may be low or distorted or both. Check the adjustment of L1203. If you can't adjust it, replace it.

(9) No audio. Check Q800 (Zenith part 121-975).

(10) Picture changes color especially when the TV is jarred. Check for bad solder around the legs of the video driver transistors on the CRT circuit board. You have been around enough to know that a loose shadow mask also causes the same symptom. I mention this because the cold solder connections on the video output module have almost fooled me a couple of times. So check those connections before you condemn the picture tube.

(11) High voltage goes on and off when the set has been turned off. I owe this one to Zenith's "Z Tips." Check for a leaky CR3401 in the +13-volt standby line.

(12) Poor regulation. A failed voltage regulator is usually at fault. However R3419 (47 ohms at 1/2 watt) in the biasing network for the voltage regulator might cause you to think the regulator has failed. It can open or increase in value.

(13) No high voltage but power supply seems to be okay. R416 in the ABL circuit may have opened.

CHAPTER 5
THE C-7 CHASSIS

First, a little preface.

I argued with myself about whether the C-7 chassis ought to be included in the main body of this book or in an appendix because I checked my records and found that I have fewer repair entries for it than any other Zenith product on the market. The more I thought about it, the more I realized that I just hadn't seen very many in the shop. I checked with some of my friends who reported they had seen quite a few and encouraged me to pay at least a little attention to it. On the hunch that your experience may not reflect mine, I decided to devote a few pages to the C-7.

Zenith put the C-7 chassis into its Sentry 2 Line. It made its appearance in the 1992-1993 model year and was available in 13-inch, 19-inch, and 20-inch sets. You, of course, are more familiar with the module numbers than the other designations. So let me put it like this. The 13-inch sets use the 9-1227-01 while the 19- and 20-inch sets used the 9-1228-01.

As far as I can tell, the C-7 is a derivation of the C-5 chassis but with significant differences in the power supply and the microprocessor as well as the tuner and horizontal driver circuits. The video processing circuits, however, remain relatively unchanged except for the use of a new video-processing chip. The chassis is a hot chassis, meaning it doesn't utilize a hot ground-cold ground configuration. Thus, it requires an antenna isolation block to keep the hot ground of the chassis separated from the world of the consumer. It also means that you must use an isolation transformer when you service it.

If you are unfamiliar with the C-7's layout, take a few moments to study figures 5-1 and 5-2. Figure 5-1 is a photograph of a new chassis before it was installed in a TV. The video output CBA hasn't even been broken away from the main board. Figure 5-2 is a graphic representation showing how the external parts "wire into" the chassis and where the major components are physically located. However, I suspect you'll find figure 5-3 (from PHOTOFACT 3208) a bit more helpful because its information is more detailed and easier to read.

Figure 5-1 Photograph of New Chassis Before Installation

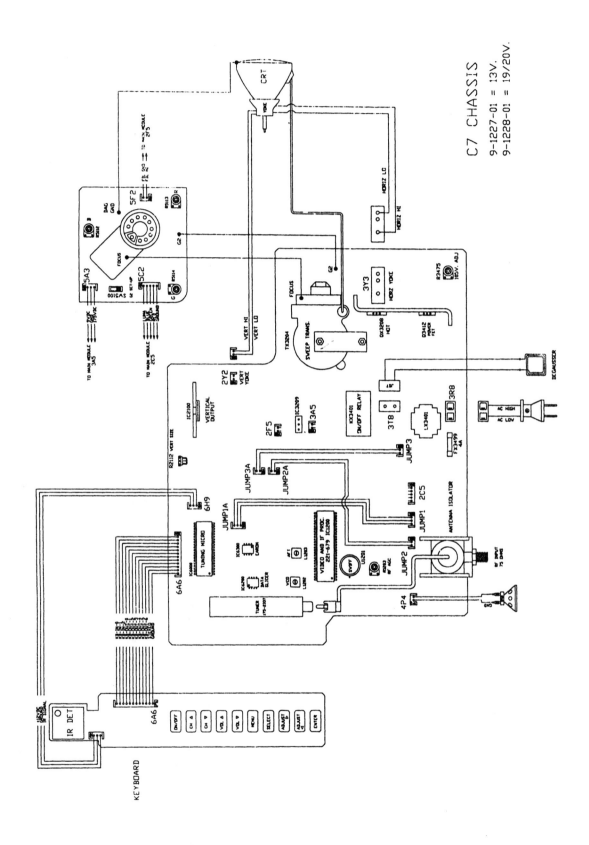

Figure 5-2 C-7 Major Parts / Interconnect Diagram

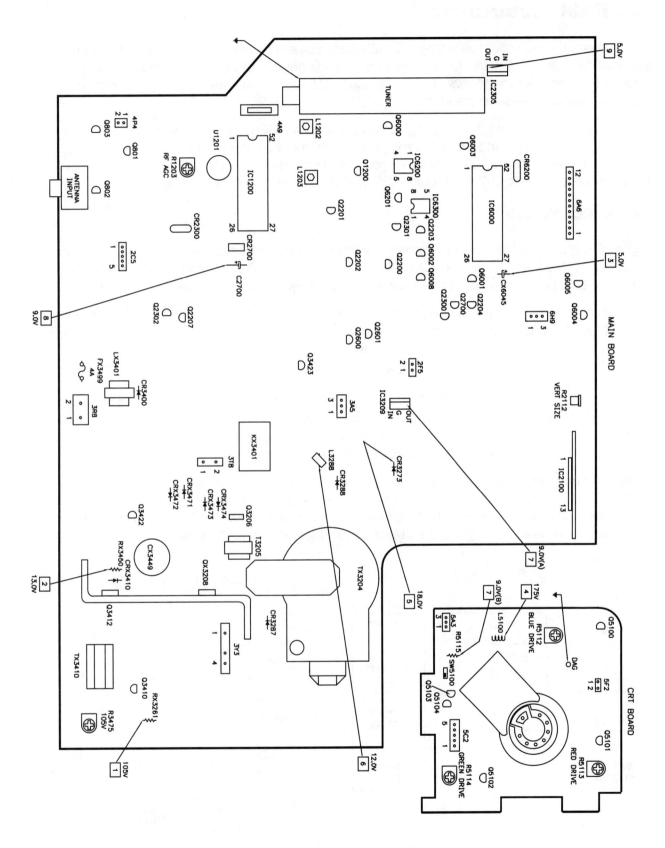

Figure 5-3 Main Board

Servicing Zenith Televisions

Available Literature

If I am correct – and I could be wrong – Zenith really doesn't have a technical discussion to guide the servicer through the circuitry. I found two pages in one service book (*TP 47: 1993 J2-Line*) and one page in another (*Service Menus: D Line Through J2 Line*). If you know of a booklet or other source I have overlooked, please let me know.

As I have done in the past, I'm going to point you to a PHOTFACT as the best service literature available. I am not belittling Zenith's service literature from this period. It's okay if you're just going to sort of "putter around," but it lacks a little if you want to troubleshoot a module to the component level. So when I need to fix a C-7, I fish out PHOTOFACT 3208 for model SS1917BS.

Service Menu

Figure 5-4 gives you the layout of the service menu. You gain access in exactly the same way as I described for the last two chassis. Given the fact that there aren't many new features in this two-page service menu, I shall comment only on those that are new (or relatively new).

"Auto Mute" on the first page is one of the "relatively new" items. Zenith learned an important lesson when it faced the flack the C-5 chassis caused. Do you remember the comment that the auto mute function permits the servicer to turn off the audio mute feature of the slice level detector? When it is turned off, auto mute keeps the TV from muting the audio when the video circuit thinks it isn't receiving a valid snyc signal. All broadcast signals are not equal. If the received signal doesn't have a format that pretty much corresponds to the NTSC standard, the slice level detector thinks no signal is present and asks the microprocessor to mute audio when in fact there is a perfectly good picture on the face of the CRT. Since nobody wants to put up with audio that constantly pops on and off, "Auto Mute" provides the servicer with a means of defeating the function and enhancing viewer pleasure.

"Channel Lock" on page two permits the servicer to lock out all channels except the one selected when the feature is turned on. I know we have discussed the feature, but it's important enough to repeat part of that discussion here. Lightning and power surges do strange things to the newer televisions. For example, I got a TV in last month that was locked on channel 69 and ignored all

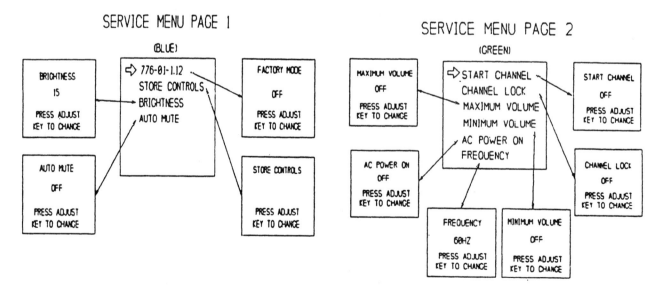

Figure 5-4 Service Menu Layout

commands to change to another channel. Unplug it and plug it back in; turn it off and then back on. Do what you would, the TV always came on set for channel 69. What was the problem? Lightning had caused a spike on the AC line that did no damage to the TV except to turn channel lock on! I accessed the service menu, turned it off, wrote a bill, and sent the TV home.

"Key Defeat" on page two is used to disable the "menu," "select," "enter," and "adjust" keys on the keyboard. These functions, however, still work on the remote control. Most hotel/motel sets utilize the "Key Defeat" feature. It is also handy in a home where there are small children who punch, press, and get little fingers on everything in sight.

"Frequency" on page two adapts the real-time clock to the line frequency that in the USA is 60 hertz. If the frequency is set incorrectly then the on-screen clock cannot keep accurate time by our standards.

Service Adjustments

The chassis has just a few service adjustments (figure 5-3) to which you need to pay attention. R1203 is the AGC delay. You shouldn't have to adjust it unless you replace the video output IC, the tuner, or another component in the circuit. R3475 adjusts the level of the regulated B+ produced by the power supply. R5113 and R5114 adjust the red and green screen voltages to the picture tube. And, R2112 adjusts vertical size.

Power Supply

Zenith introduced a new power supply when it manufactured the C-7 chassis (figure 5-5). Don't think it's a sort of "one chassis wonder" because Zenith uses almost the same circuit in the C-11 chassis, the subject of Chapter 9. When we get to Chapter 9, the power supply is discussed in some depth. In the meantime, let me make a few brief comments that I hope makes the circuit reasonably intelligible.

R3424, CR3491, and related components serve as a 60-hertz detect circuit that alerts the microprocessor to a power failure and provides a 60-hertz signal to operate the on-screen clock. CR3489, RX3425, CR3400, R3426, CR3493, and their associated components develop the +5-volt standby for the microprocessor and IR receiver.

When it receives an on command, the microprocessor outputs a signal that causes the relay (KX3401) to actuate (close contacts). Those closed contacts create a path for AC to the bridge rectifiers. The bridge circuit and CX3449 develop about 150 volts DC that is applied to Q3412, an FET used in a buck regulator switching power supply. When it begins switching action, the full power supply comes on-line with an output of +105 volts for horizontal deflection and a +13-volt source for the chassis. Scan-derived voltages then provide the remainder of voltages necessary to operate the TV.

There is nothing complicated about the power supply. It has even proved to be relatively trouble-free. When it has caused problems, the problems have been easy to solve. There is, however, a kind of anomaly to which I should call your attention. It is possible to have raster but no audio. Voltage source 2 in figure 5-5 supplies B+ to the audio circuit. If Q3412 fails to oscillate, B+ for the deflection circuit will be present at an inflated level, but the +13 volts won't be present at any level. Your tip that something is wrong is the set's propinquity to shut down because B+ to the horizontal deflection circuit is high. When you meet such a set of symptoms – high B+ and no audio – immediately suspect a leaky Q3412. Look in the repair history section for a more complete discussion.

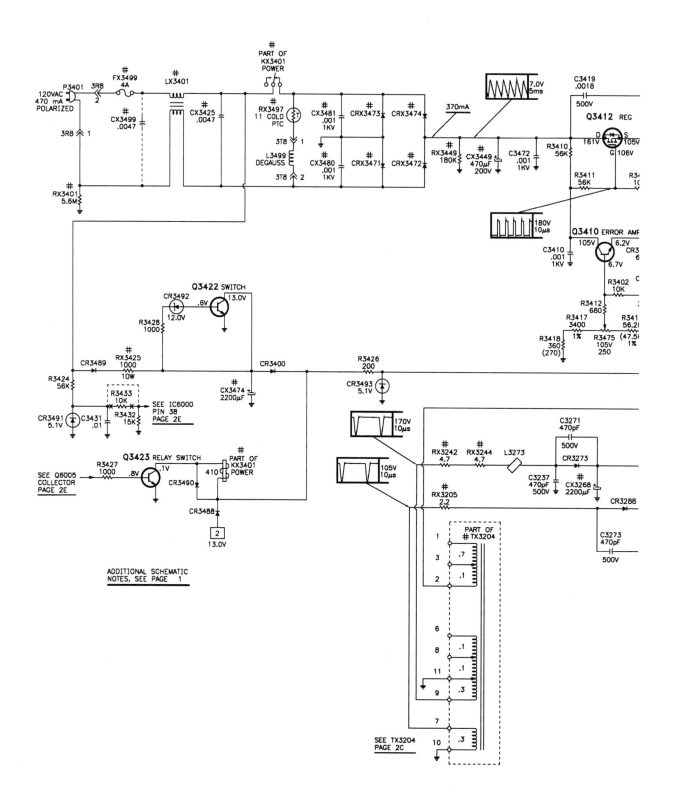

Figure 5-5 Power Supply Schematic

The C-7 Chassis

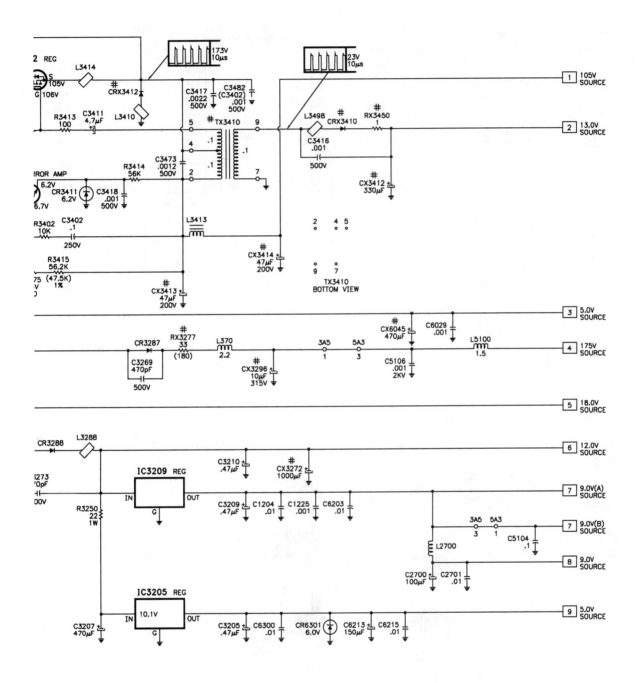

Figure 5-5 Power Supply Schematic (Continued)

Servicing Zenith Televisions

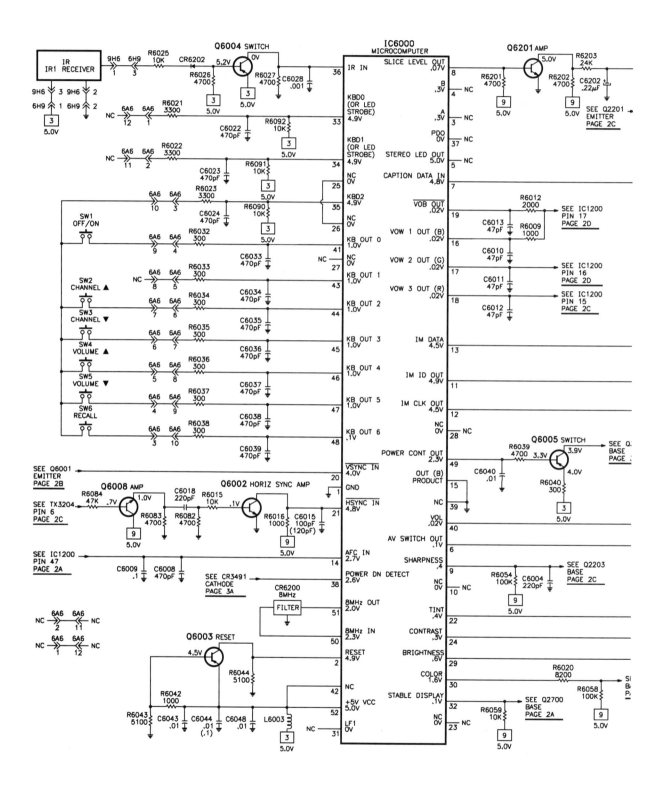

Figure 5-6 Microprocessor Schematic

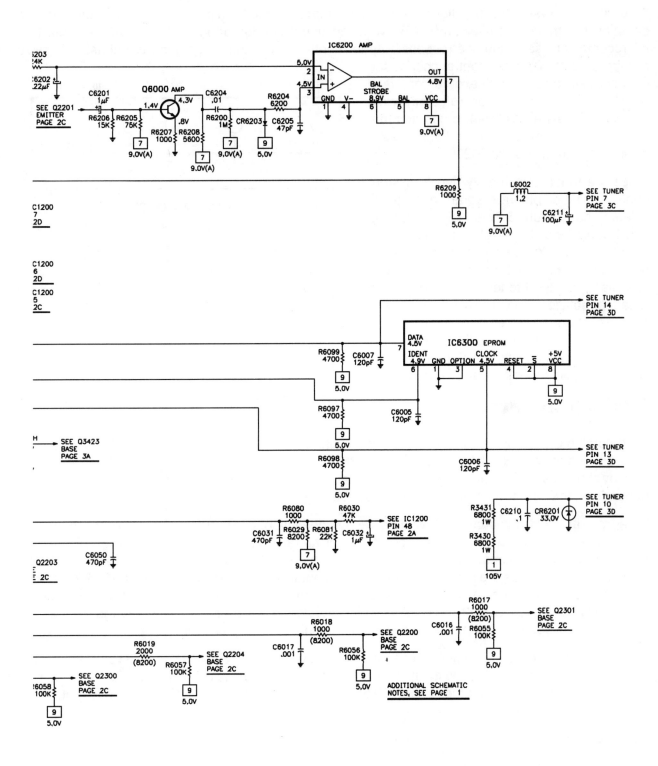

Figure 5-6 Microprocessor Schematic (Continued)

System Control and Tuner Functions

A quick look at system control (figure 5-6) tells you that the circuit is different, but not that different. It contains the microcontroller and EAROM in the usual configuration. Like the C-6 chassis, the microcontroller itself assumes responsibility for control of the tuner, meaning the chassis doesn't require an external tuner control chip as the C-5 chassis does. Unlike other chassis, however, this system control employs a separate IC as the slice level comparator that also serves to detect sync and works as a closed caption decoder (IC6200).

Since the tuner is intimately tied to system control, a tuner pinout diagram and tuner voltage chart (figure 5-7) is included to help you troubleshoot tuner-related problems. I trust you have figured out that this tuner is remarkably similar to the one used in the C-6 chassis. Do I need to tell you to bypass the antenna isolation block to rule it out as the source of a tuning problem before you change the tuner? Oh, I know I'm telling you to do the obvious. Sometimes, though, you and I need someone to point out the obvious, don't we?

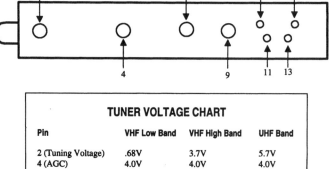

Figure 5-7 Tuner Voltages and Terminal Guide

TUNER VOLTAGE CHART

Pin	VHF Low Band	VHF High Band	UHF Band
2 (Tuning Voltage)	.68V	3.7V	5.7V
4 (AGC)	4.0V	4.0V	4.0V
7 (9V)9.0V	9.0V	9.0V	9.0V
9 (IF Out)	0V	0V	0V
10 (33V)	33V	33V	33V
11 (5V)	5.0V	5.0V	5.0V
13 (CLK)	4.5V	4.5V	4.5V
14 (DATA)	4.5V	4.5V	4.5V

Note: VHF Low Band voltages taken on channel 2.
VHF High Band voltages taken on channel 7.
UHF Band voltages taken on channel 14.

Deflection Circuits

I find nothing especially new in the deflection circuits (figure 5-8). For example, IC2100, the vertical output IC belongs to the LA78xx series of vertical output ICs and works just like those we have already discussed. Like the other chassis we have talked about, the exact part number for IC2100 depends on the screen size of the TV you are repairing.

When you are confronted by a C-7 with no vertical deflection, check first for poor solder around the pins of IC2100. Then check for B+. Poor solder or an open resistor in the scan-derived B+ line to IC2100 accounts for almost all of the vertical deflection problems I have fixed. Q3206, QX3208, TX3204, and T3205 are the major players in the horizontal deflection circuit. According to my records, I have had to replace one defective horizontal output transistor. It is designated QX3208 and crosses to the now familiar ECG2302 or NTE2302. The horizontal deflection circuit is vintage Zenith and should present you with few problems that you can't quickly spot and easily solve.

Horizontal deflection is responsible for generating +12 volts DC, +18 volts DC, +175 volts DC (for the video output transistors), G2 and focus voltages, and the high voltage (EHT) for the picture tube. The +12-volt line is the source for the +5VSW ("+5 volts switched") and the +9 VSW ("+9 volts switched") via IC3205 and IC3209.

Let's move now to the x-ray protection circuit.

The overvoltage shutdown circuit (x-ray protection circuit) takes a pulse off the flyback and applies it to the anode of CRX3006 (figure 5-8). The pulse is rectified, filtered, and applied to a precision voltage divider network the output of which is tied to the cathode of a 12-volt zener (CRX3004). The resulting DC voltage is routed through CR2702 to pin 24 of IC1200. When the value of the applied pulse exceeds the threshold of CRX3004, the voltage input to pin 24 triggers the shutdown circuit inside IC1200, shutting the set down.

Video Circuit

Zenith does use a new video processor chip for IC1200 (an LA7670), but it works like the other video processors we have discussed. The Zenith part number is 221-679, but use an ECG7054 or NTE7054 if you have them because they appear to work just as well as an OEM does.

If you want to peek inside IC1200, take a look at the block diagram in figure 5-9. I believe you'll find it to be a valuable tool if you have to troubleshoot the chip or its circuit for a suspected failure.

Since the video output circuit has been known to cause problems, I include the schematic in figure 5-10. If you ever run across one of these TV's exhibiting symptoms of no luminance, pay attention to Q5104 and Q5103 because they do fail. When they fail, these transistors cause the screen of the picture tube to become excessively bright with heavy retrace lines visible in the picture. I have only seen the problem once, but my friends tell me its a common C-7 problem. Check these transistors before you think about condemning the picture tube.

Audio Circuit

I have decided not to reproduce the schematic for the audio output circuit because it is very much like the monophonic circuits we have already covered.

Repair History

I am afraid this section is going to be like the training material Zenith furnishes in that it won't be very long. When I think back over the last several years, I have to confess the C-7 family hasn't come in very often for repair. I don't believe it's more reliable than other Zenith products. It's just that it hasn't been particularly popular – or perhaps available – in my part of the country. My database has a few entries, just two of which merit mentioning! Of course the repairs could have been so routine that I didn't even reference them, which is often the case when I do elementary repairs for, let us say, lightning damage. Why clutter up a good database with obvious information? I include the C-7 because you may have seen lots of them. After all I can't judge every situation by mine, can I?

First, I have a reference to a set that came on and immediately shut down because the regulated B+ was running too high. I used a variac to turn the AC down to keep the set up and running and quickly noticed there was no audio. I cured the problem by replacing Q3412. Q3412 is a 2SK526. You can replace it using Zenith part number 121-1299, an ECG2900, or a 2SK526. As far as I can tell, the one works as well as the other.

Second, I have a reference to a TV that had a very bright screen with heavy retrace lines in the picture. Three possibilities popped into my mind: a shorted picture tube, low B+ to the video driver transistors, or no luminance. In this instance, Q5103 (reference) and Q5401 (video amp) showed emitter to collector leakage. I fixed the TV by replacing both with ECG159 substitutes.

I mentioned in the text of the chapter servicing some C-7s that had vertical deflection problems. Please make a note of the fact that poor solder around the pins of the vertical deflection chip or an open resistor in its B+ source account for most of the vertical problems I have serviced and heard about. You need to use a magnifying glass to spot some of cracks that often appear in the solder around the pins and use only a voltmeter to find the open resistor.

Servicing Zenith Televisions

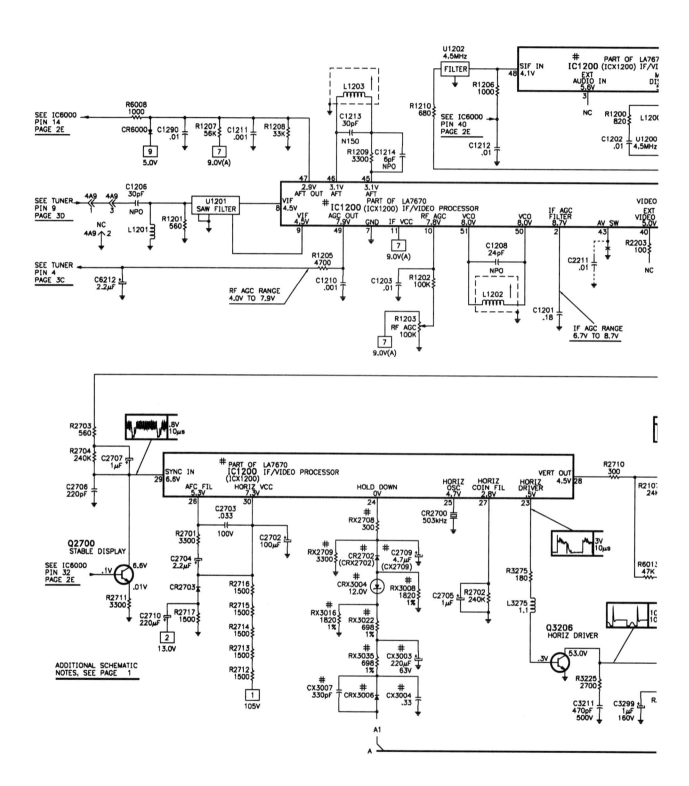

Figure 5-8 Television Schematic

The C-7 Chassis

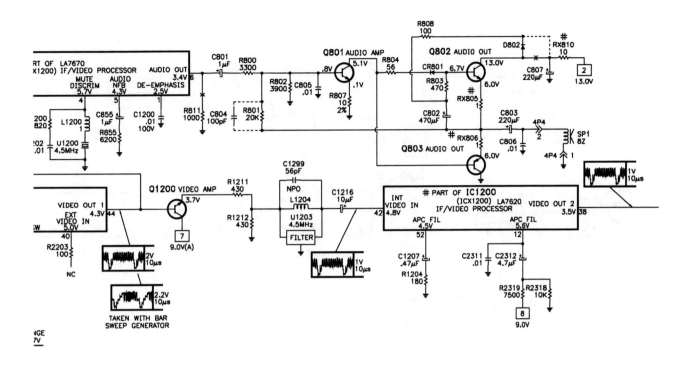

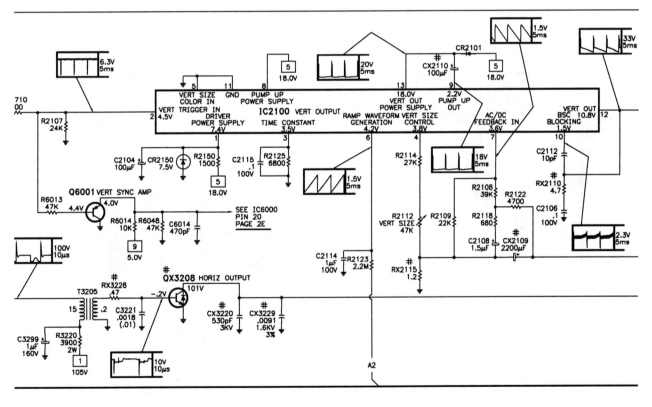

Figure 5-8 Television Schematic (Continiued)

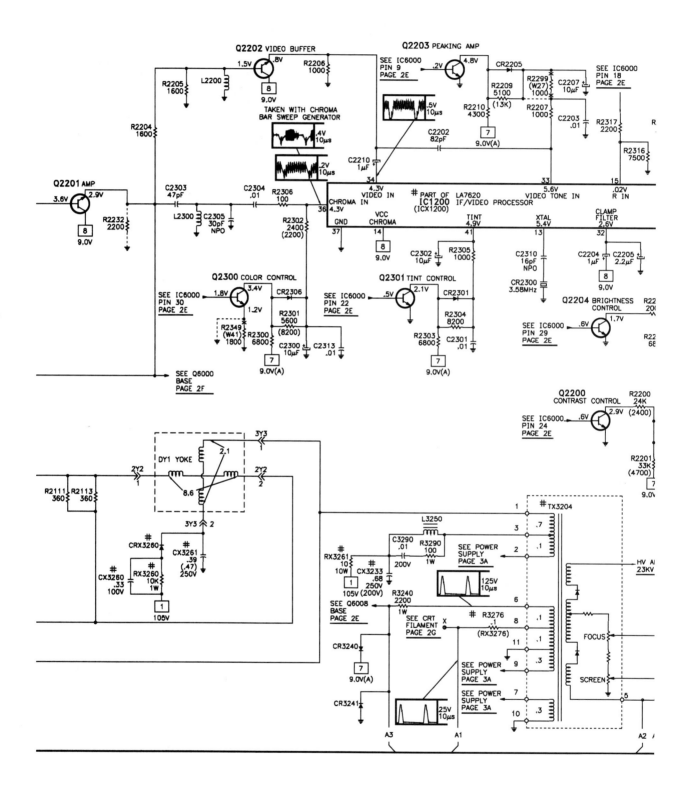

Figure 5-8 Television Schematic (Continued)

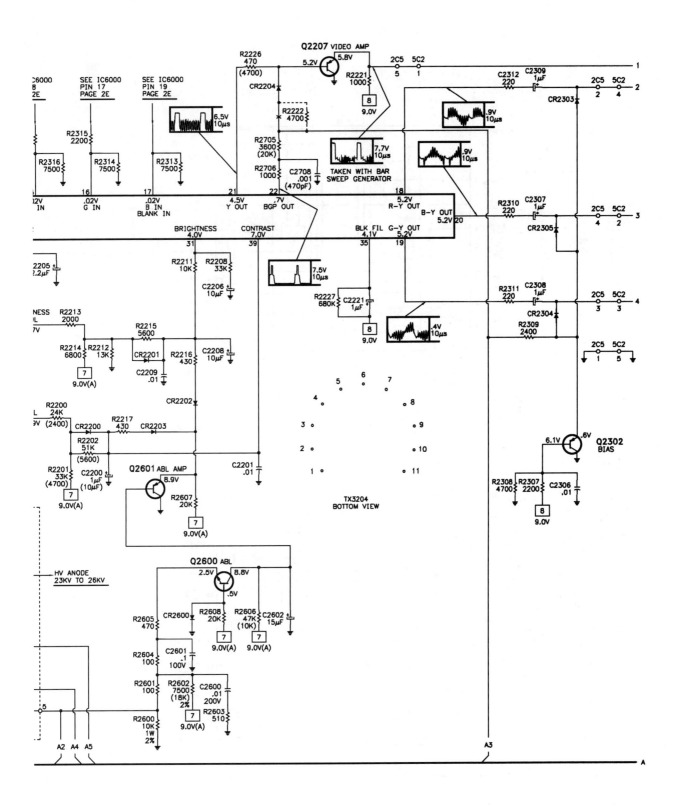

Figure 5-8 Television Schematic (Continued)

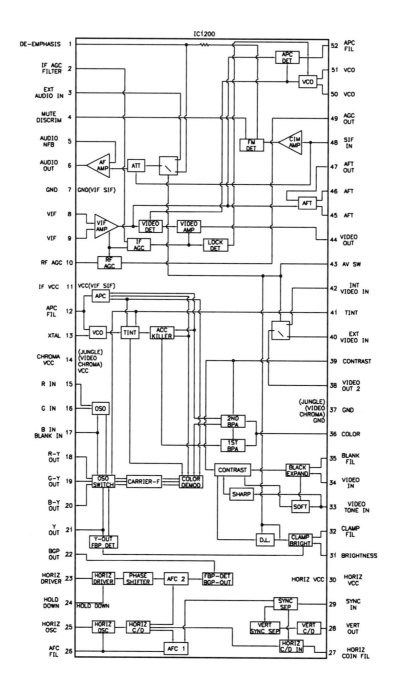

Figure 5-9 IC1200 Video Processor Chip

The C-7 Chassis

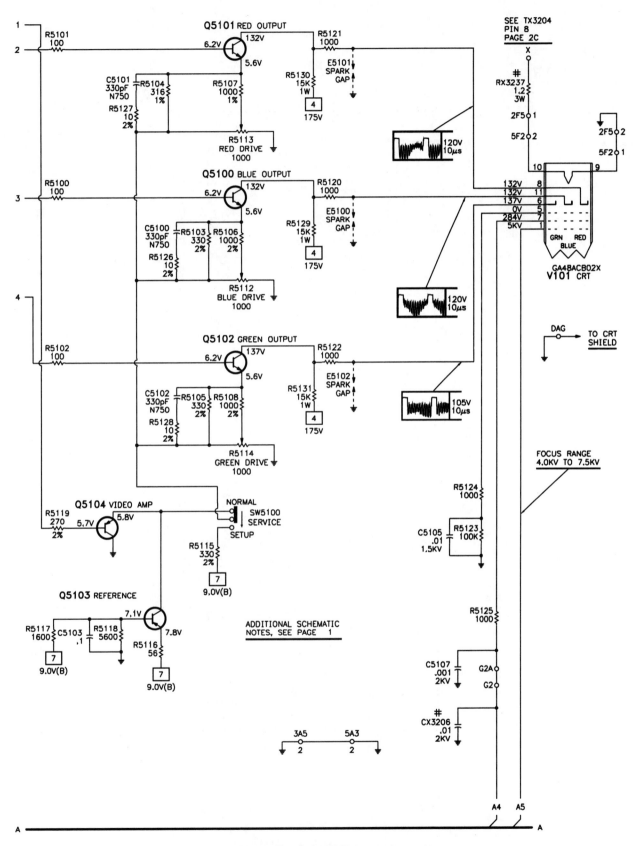

Figure 5-10 CRT Schematic

CHAPTER 6
THE C-8 CHASSIS

Televisions that use the C-8 chassis were marketed under the rubric "Advanced Video Imaging," and used the two-module concept Zenith introduced several years when it put the C-3 chassis on the market (See Chapter 2). The engineers refer to the two boards as the "small signal module" (the main module) and the large signal module (the sweep and power supply). But the similarity between the C-3 and C-8 ends here because most of the electronics are new. For example, the 27" models came equipped with a new picture tube that enhanced both the contrast and the reds in the picture and reduced viewed surface light reflection. The C-8s were also equipped with an improved PIP and video processing system. Certain models even came equipped with Dolby Surround Audio, Starsight, and a sophisticated jack pack. The audio system included a three-channel five-watt output configuration that gave the viewer a choice of mono, stereo, SEQ, SAP, and an amplified matrix surround output, all controlled by the volume control on the audio menu.

You can probably see similarities between the C-8 and the C-3 by looking at a photograph of a major components location guide (figure 6-2). But, there are differences even here. The C-8, according to Zenith, sports a number of service-related improvements that include fewer interconnecting cables, a soldered-in PIP, a soldered-in and shielded IF module assembly, and a data-controlled tuner assembly. The video output module in the C-3 chassis came attached to the main module and could not be replaced as a separate unit. The C-8's video output module is detachable. Simply remove the 5C2 and 5F2 connectors and a couple of other wires, and the module comes away from the main chassis. The new arrangement has been helpful because we servicers can replace it without having to replace the small signal (main) module.

Selected models feature Starsight that, as you know, contains detailed television programming, an "electronic" TV Guide so to speak. The system allows the consumer to view programming information using several methods of sorting and grouping. It even permits the consumer to make a list of programs and record the listed programs with a VCR by using an IR transmitter that turns the VCR on at the proper time!

Available Literature

"How it works" descriptions of the circuitry exist, but not in abundance. *Technical Training Program TP48: L-Line* has about a dozen pages devoted to the C-8, and aside from a few pages in *Technical Training Program: R Line 1995-1996*, that's about all I have been able to find. Zenith's rationale seems to have been, "Trace the problem to one of the three modules (main, power supply, video output) and replace it." Along with the Zenith literature, I use PHOTOFACT 3679.

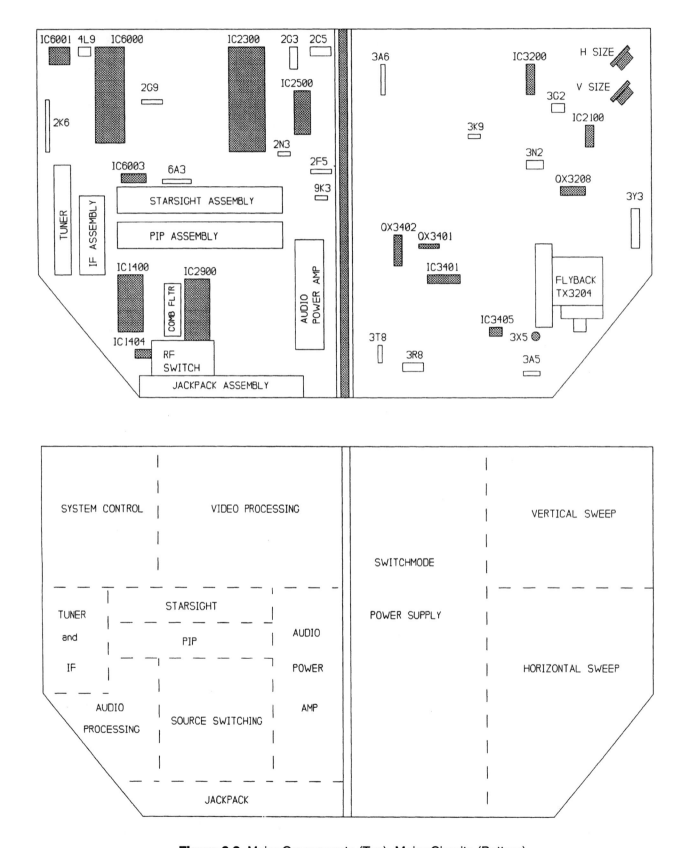

Figure 6-2 Major Components (Top), Major Circuits (Bottom)

Troubleshooting Techniques

The circuits have been radically redesigned, but the troubleshooting techniques really don't differ that much from the ones we learned in Chapter 2. Therefore, refer to the troubleshooting charts I gave you there and use them in conjunction with figure 6-3, figure 6-4, and figure 6-5. To demonstrate how these guides work, figure 6-5 shows you how to verify the operation of the power supply and how to check for the "must have" voltages and waveforms at connectors 6A3/3A6, 2G3/3G2, and 2N3/3N2.

I expect to make frequent reference to Zenith's technical material even though it is scant, but I intend to depend on Sams for schematic information. The Sams information I have chosen is PHOTOFACT 3679 for model SR2789DT. It won't have all the information for all the models in which the C-8 modules appear, but it should do. As far as I know, Howard W. Sams does not provide a PHOTOFACT for every model in the C-8 family. Therefore, I suggest you do as I have done and just choose one to which you refer and on which you write service notes. Even though it won't contain all the data for all the C-8s on the market, it will do far more than just get you by.

C-8 Modules

Most of the techs I know file literature and tech tips by module, not by chassis. I have adopted the chassis approach to servicing Zenith TVs because it permits me to deal with several modules under one heading. But then I have already told you that, haven't I? Here, then, as far as I can tell, are most of the modules that comprise the C-8 family. I can't guarantee that I have listed all of them, but I think I have. At least, I have listed the most popular ones.

Main Modules	**Sweep-Power Supply**
9-1521	9-1268
9-1522	9-1269
9-1523	9-1270
9-1524	
9-1556	
9-1426	
9-1427	

The main module (and its "attachments") used in a particular television depends on the features incorporated in the set. As you recall from the previous few paragraphs, these televisions come packed with options. The sweep and power supply module follows suit with one exception. The particular one used depends not only on the main module, but also the size of the picture tube.

I have said several times that repairing these modules is difficult, not just because of the complex circuitry, but because getting to various components to check voltages and waveforms is difficult and sometimes downright impossible! So, I have an argument with myself almost every time I face a major problem: "Do I try to fix it, or do I ship it to a repair depot?" Let me give you an example. I fixed one the other day that needed a fuse in the +130-volt line, a horizontal output transistor, and a flyback. The flyback alone cost around $32. I could have shipped it to a repair depot I often use and gotten the job done for $49.95 and saved myself a real headache and a few skinned knuckles, but for some perverse reason I ordered the flyback and fixed it myself. I don't have a rule of thumb to determine whether I repair it or ship it. I must confess that I usually make the decision based on my mood!

These comments aside, I hope to give you enough information to help you fix them if you want to. At the very least, the information in this chapter should help you trace the problem to one module or the other, and then you can decide whether you want to pursue the problem to the component level or ship the module for repair.

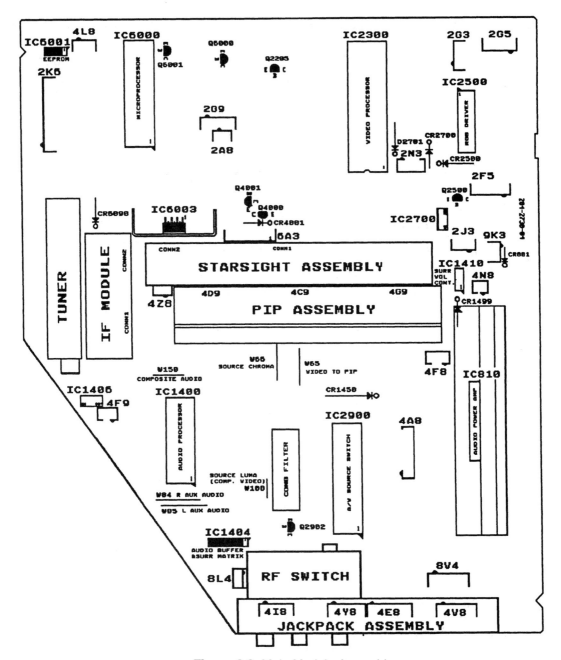

Figure 6-3 Main Module Assembly

Customer Menu

The customer menu is among the features the engineers have changed. Each page in the menu has its own heading – Setup, Video, PIP, and Audio – and each page lists all of the available adjustments pertinent to its heading (figure 6-6). There are, in other words, no submenus.

Service Menu

The service menu (table 6-1) is by far the most extensive and most complicated one we have looked at so far, containing 30 items listed on two pages. To get to it, hold down the menu button on the remote control until the customer menu disappears and then press in quick succession 9-8-7-6-enter.

Or you may access it by holding down the menu button on the front of the set until the customer menu disappears and then quickly pressing adjust right and channel up at the same time.

The service menu isn't as daunting as it appears. As a matter of fact, you need to concern yourself with adjusting just 13 of the 30 items. Set the remaining 17 to their recommended positions and leave them alone unless the engineers recommend that they be changed.

When you access the menu, you should notice black horizontal bars at the top and the bottom of the service menu display. These bars help you to center the menu and caption display by adjusting the "Vert Pos" and "Horz Pos" items. Please remember that these adjustments have nothing to do with centering the video display, just the position of the menu and captions.

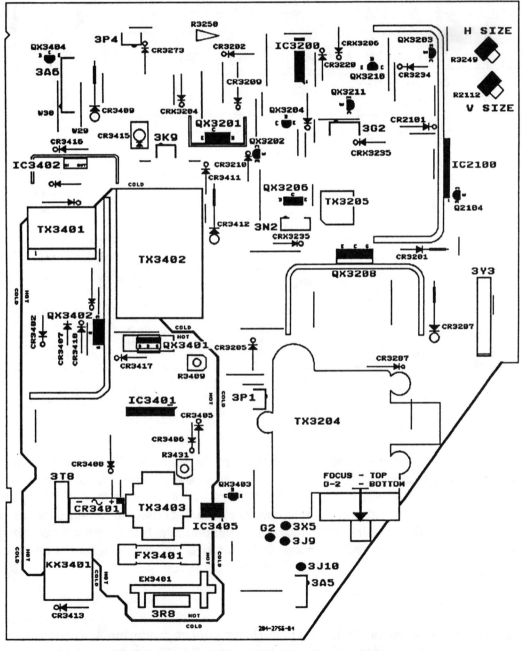

Figure 6-4 Power Supply/Sweep Module Assembly

Servicing Zenith Televisions

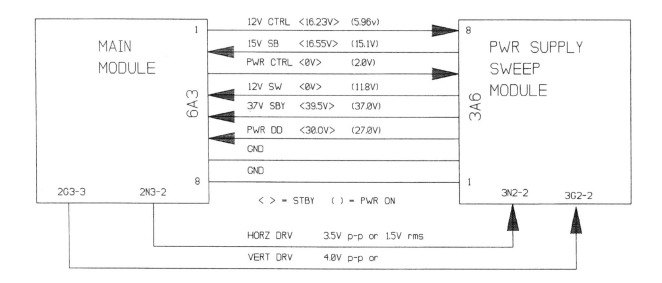

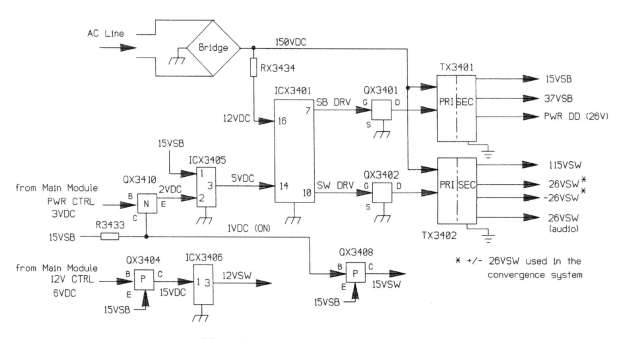

Figure 6-5 Direct View Power Control

If you are familiar with Zenith service menus, you know that "Factory Mode" is used in the manufacturing process and has no application in the field. It should therefore be left off. If you turn it on and leave it on, you create a situation where strange things might randomly happen to the TV. For example, a simple act like turning the set off and on again could cause the customer menu settings to change randomly. The C-8 chassis isn't like other TVs Zenith has manufactured because you don't have to access the service menu to turn Factory Mode off if it has been left on. Just set the on-screen clock, and the Factory Mode automatically switches from "on" to "off."

Service Adjustments

The C-8 has five field adjustments that aren't found in the service menu: the AGC adjustment (R1207) on the IF module, vertical size (R2112) and horizontal size (R3249) on the sweep module, and focus

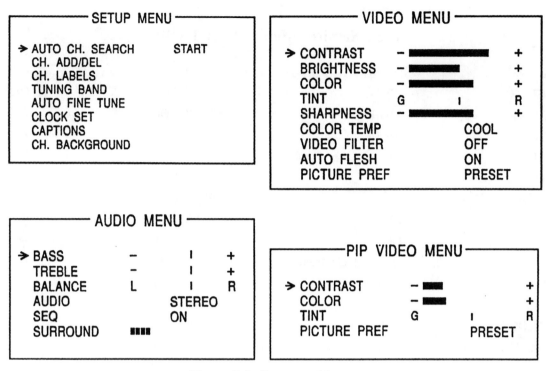

Figure 6-6 Customer Menus

and G-2 on the flyback. You need to know that color tracking cannot be set manually. The RGB Driver and Auto Track (IC2500) circuit not only establishes, but also maintains, color tracking automatically. To put it simply, it sets the red, green, and blue drives to the CRT for proper gray scale operation.

The remaining service adjustments are found in the service menu.

(1) "Preset Px," "Theatr Px," and "PIP px" store changes to "Picture Pref" in the customer video menu. The first two settings allow two preset video setups for different room lighting conditions to be stored in permanent memory. Even though you have stored two settings, the word "Stord" appears just once in the service menu after the last parameter has been stored. Make certain that the correct "Picture Pref" has been selected before you instruct the TV to store the selected information. There are three of them: "Preset," "Theatr," and "Custom." The literature also says you should set the volume to normal listening level before you execute the store command.

(2) "Zen/PL" switches the IR receiver mode of operation as well as the configuration of the customer menu between the standard "Zenith" and "Private Label." If you accidentally switch it to "PL," you will have to change the code used by the remote control to 21 or 121 for it to work the TV. However, the keyboard still functions normally.

(3) "A/V Lock," if set in the on position, turns on a blue screen for the video display and blocks all other video and audio inputs. This adjustment has obviously been included because it has certain special commercial uses.

(4) "Chan Lock," "Max Volum," "Min Volum," "AC On," and "Key Deft" work as they have in other chassis. If these items are turned on, they may cause a situation that mimics a module failure. Be sure to check them if the TV controls fail to work as they normally do.

C-8 Service Menu Page 1 (Blue)

784 - 2 1.43	(No Display Unless Selected)*	AC ON	OFF
PRESET PX	STORE	KEY DEFT	OFF
THEATR PX	STORE	PROJ MODE	OFF
PIP PR PX	STORE	AUTO SRCH	2
BRIGHTNESS	15	PIP TEST	0 *
CAP PHASE	220	PIP HPOS	4*
VERT POS	10	PIP VRAMW	8 *
HORZ POS	37	PIP VRAMR	4 *
DOLBY	OFF	PIP FR CT	45 *
PIP AVAIL	ON (if PIP is present)	PIP GRAY	64 *
ZEN / PL	ZENTH	PIP SWTCH	10 *
A / V LOCK	OFF	RGB BRITE	54
CHAN LOCK	OFF	RGB CONT	54
MAX VOLUM	OFF	COLOR LIM	3 *
MIN VOLUM	OFF	SUB COLOR	16 *

C-8 Service Menu Page 2 (Green)

OSD CONT	0 *	COLOR DET	1 *
VMY GAIN	0 *	SHRP TRAK	2 *
SUB CONT	20 *	WPS	1*
R CUTOFF	127 *	YNR OFF	3 *
G CUTOFF	127 *	YNR ON	1 *
B CUTOFF	127 *	GAMM STRT	0 *
G DRIVE	54 *	GAMM GAIN	0 *
B DRIVE	65 *	BLK EXPND	0 *
COLOR CORR	0 *	SRT	1 *
RY PHASE	1 *	BLE PULLN	0 *
RY AMP	1 *	HORZ PHSE	19
GY AMP	1 *	GATE PHSE	1 *
GY PHASE	0 *	HORZ AFC	1 *
FLSH PULL	1 *	VERT FREQ	0 *
HI BR COL	2 *	VERT PHSE	3 *

02 / 02 / 93 - B

Notes: 1. (*) These controls are not intended to be adjusted in the field. They should be left set to the recommended values . Adjust the others only if required.

2. The FACTORY MODE should be left OFF in the field. If this feature is on it will affect customer menu preset and custom settings. You can turn FACTORY MODE "OFF" from the service menu or by setting the clock time in the customer menu.

Table 6-1 Service Menu

Let's look at two other parameters before leaving the service menu. If "Horz AFC" goes to 0 and/or if "Vert Freq" goes to 3, the video sync will loosen but not affect the sync of the customer or service menu. You may think you have a module problem because strange things happen in the picture (i.e. the video portion) without affecting the menus. Check these settings before you change either module.

As you know, Zenith often keeps a particular chassis configuration through more than one model line. The service menu I have just discussed pertains to the L-Line. If you are working on a C-8 that belongs to the R-Line, you will find a longer service menu (table 6-2). However, most of the adjustments that concern you remain essentially the same. By the way, the R-Line service menu is also displayed differently. It pops up one item at a time instead of a page at a time.

If you need a more detailed explanation of the items in the service menu, I suggest you spend a little time studying table 6-3 which explains each item in the C-8 R-Line service menu. I have taken it from *Technical Training Program: R Line 1995-96*. Pay particular attention to item 17, "Band/AFC", and note that there are eight possible settings, the first three of which help auto search to work in special signal conditions.

Power Supply

Even a cursory glance informs you that the C-8 power supply is different. I include two different schematic representations for this complicated circuit, assuming that two perspectives are better than one. One is from Sams (figure 6-7a), and the other is from Zenith (6-7b). As far as I know, the C-8 is the first direct view chassis Zenith manufactured that uses a "dual mode switching power supply." I call it a dual mode switching supply, a term borrowed from Philips, because one switching regulator (IC3401) controls both the standby and full power operation. If you are familiar with Zenith products, you know that Zenith has used it in their projection units for several years. It should, therefore, be at least a bit familiar to you.

When AC becomes available, the bridge rectifier circuit and main filter capacitor develop about 150 volts DC. This voltage is routed to pin 16 of ICX3401 through a 56k, 1-watt resistor, to the drain of QX3401 through TX3401, and to QX3402 through TX3402. ICX3401 begins to output a series of drive pulses at pin 7 when the voltage at pin 16 rises above +14 volts DC. The drive pulses at pin 7 cause QX3401 to begin switching action. The switching action energizes the primary of TX3401 that couples the energy to the secondary where the various diodes and capacitors develop the voltages necessary to put the television into its standby mode.

The voltages developed by the standby power supply include the following: +37 volts, -15 volts, +15 volts, and +9.1 volts (voltage sources 6, 7, 8, 9, 15, 16, and 17 in PHOTOFACT 3679 of figure 6-7b). You need to be aware that the number of standby voltages may vary from model line to model line. PHOTOFACT 3679 reproduces an R-Line C-8. A C-8 chassis from an L-Line might have fewer. I ask that you be aware of the differences even though the differences don't affect the operation of the power supply. While we're talking about standby voltages, note that IC6003, a TDA8137, develops both a reset voltage and +5 volts for the microcomputer (and the IR receiver) from the +15 volt standby source.

Now shift your attention to QX3403 and ICX3405. When the microcomputer outputs a high at pin 35, QX3403 turns on completing a ground path for the LED inside ICX3405. The LED begins to emit light causing the transistor portion to turn on. When the transistor inside ICX3405 turns on, the portion of ICX3401 that emits drive pulses from pin 10 to QX3402 turns on. TX3402 begins to develop voltage sources 2, 3, 4, and 5. The full power section of the power supply is now fully operational, and the TV comes to life. Incidentally, the on command goes from the microcomputer to the power supply module via cable 6A3/3A6.

The power on command from the microcomputer also brings the switched voltages on line (sources 10, 11, 12, 13, 14, 22, and 23 in figure 6-7). When they come on-line, the power on sequence is complete. The voltage sources labeled 18, 19, 20, and 21 are scan-derived. I wish I could tell you more about the operation of the C-8 power supply, but I can't because I don't have sufficient information. However, a few details sort of jump out of the schematic at you. For example, if the standby section doesn't come on-line, begin troubleshooting by checking for the +14 volts startup voltage on pin 16. If it is missing, check RX3434. If the +14 volts ramps up and down, check the components that develop B+ for the chip (RX3416, CRX3403, and CX34060). Don't overlook the fact that R3418 or QX3401 and/or associated components could be defective. For instance, if R3420 in the source of QX3401 opens, ICX3401 will be deprived of run B+, causing the voltage at pin 14 to rise and fall.

If the full power section fails to come on, check for the on command at the base of QX3403. If it is there, does the voltage at pin 14 of ICX3401 change? If it doesn't, you probably have a defective optoisolator. If it does, think in terms of a defective ICX3401. If the IC turns on, do pulses reach the gate of QX3402? By asking a few questions and making a few checks, you should be able to locate the source of your difficulty in rather short order. Remember, it's just a power supply!

Now I should mention what I consider an important service note. I have seen quite a few C-8 chassis that had audio but not even a hint of raster. When I first encountered these symptoms, I began my search for the problem by checking the filaments of the picture tube and saw that they weren't glowing. Since the audio circuits worked, I knew the main power supply had come on because voltage source 5 (figure 6-7a) provides power to the audio output circuits. When I looked over the circuit board, I found poor solder connections at CR3412, the diode in the 130-volt line. Resoldering the diode fixed the set. If you service a C-8 that has this problem or one that has a habit of coming on playing for a few minutes and losing picture but not audio, check the solder at both ends of CR3412.

#	Menu Control	Direct View Glass - Comb	Direct View 3 Line - Comb
0	FACT MODE	0	0
1	PRESET PX	1	1
2	THEATR PX	2	2
3	BRIGHT RF	30	30
4	BRIGHT AX	20	20
5	VERT POS	15	15
6	HORZ POS	35	35
7	KEYDEFEAT	0	0
8	CHAN LOCK	0	0
9	MIN VOL	0	0
10	MAX VOL	79	79
11	AC PWR ON	0	0
12	AV LOCK	0	0
13	AUD MODE	0	0
14	PIP AVAIL	2	2
15	ZENITH/PL	0	0
16	PROJ MODE	0	0
17	BAND AFC	0	0
18	PIP L DELY	16	16
19	PIP VPOS	9	9
20	PIP HPOS	4	4
21	PIP SWTCH	5	5
22	OSD RGB	19	19
23	SUB COLOR	80	80
24	SUB CONT	20	20
25	G DRIVE	55	55
26	B DRIVE	55	55
27	COLR DEC	185	185
28	COLR EMPH	76	76
29	YNR OFF R	2	2
30	YNR OFF A	2	2
31	YNR ON	0	0
32	Y ENHANCE	2	2
33	HORZ PHSE	19	19
34	H/V FREQ	163	35
35	VEL MODUL	2	2

Table 6-2 R-Line Service Menu

R Line Color TV Update & Notes

C - 8 R Line Service Menu Explanation

The R Line service menu shows only one menu item at a time. On the next two pages you are given the explainations of each of the service menu items. These menu items should not be changed unless directed to do so.

0 - Factory Mode - When this control is set to 1 special factory setup is enabled. This control should always be set to 0 in the field.

1 - Preset Picture - This item is usde to store the current "PRESET CUSTOM" video menu settings into memory as the "PRESET" values.

2 - Theater Picture - This item is usde to store the current "THEATER CUSTOM" video menu settings into memory as the "THEATER" values.

3 - Brightness RF - This is half of the brightness register. It sets the minimum level that brightness can go to when an RF source is selected. The adjustment range is 0 - 63.

4 - Brightness AUX - This is half of the brightness register. It sets the minimum level that brightness can go to when a baseband auxiliary source (audio/video) is selected. The adjustment range is 0 - 63.

5 - Vertical Position - This control sets the vertical position of the on screen displays (OSD). The adjuestment range is 0 - 31.

6 - Horizontal Position - This controls sets the horizontal position of the on screen displays (OSD). The adjustment range is 0 - 75.

7 - Key Defeat - When this control is set to 1 it disables the MENU, ENTER, SELECT, and ADJUST keys on the set's keyboard. It does not affect the remote control hand-transmitter's control of these items.

8 - Channel Lock - When this control is set to 1 it locks the tuner to receive only the currently selected channel. This affects both of the RF sources and disables "AUTO SEARCH" when set to 1.

9 - Minimum Volume - When this control is set to 1 it will limit the minimum volume to the currently set volume level.

10 - Maximum Volume - When this control is set to 1 it will limit the maximum volume to the currently set volume level. The adjustment range is 0 - 79. O is minimum volume and 79 is maximum volume. The setting number changes when the volume control is changed.

11 - AC Power On - When this control is set to 1 the set will turn on when AC power is connected to the set. When this controls is set to 1 it also disables the ON/OFF control functions from the keyboard or from a remote control hand-transmitter.

12 - AV Lock - When this control is set to 1 it locks the display to a blue screen and menu display only. All other video or source use and audio is disabled. This feature can be set and cleared without going into the service menu. Using a remote control you must press and hold the MENU key, until the channel/time display appears, and press digits 9, 1, 1, and then press the ENTER key. From the service menu you may also set this control to 1 to turn AV Lock on and to 0 to turn it off. Either method will allow you to toggle the AV Lock on or off.

13 - Audio Mode - This control sets up the audio system and has three possible settings. O is for standard ZENITH audio, 1 is for Dolby surround sets, and 2 is for Bose home theater audio (which only has an non-amplified audio output for an external audio system).

14 - PIP Available - This control sets the feature level of PIP and has three possible settings. O is to disable the PIP feature, 1 is for standard (single frame) PIP, and 2 is for advanced (multiple frame) PIP mode of operation.

15 - Zenith/Private Label - When this control is set to 0 the IR receive mode is set to decode normal Zenith IR codes. When this control is set to 1 the IR receive mode is set to decode Private Label codes (MBR code "21" or "121").

16 - Projection Mode - This control should be set to 0 for all direct view product and set to 1 for all projection product. When set to 1 the PROJO SETUP feature is enabled on the customer menu's SETUP menu.

17 - Band/AFC - This control allows you to manually set the tuning band and AFC selection for a special signal condition in which AUTO SEARCH does not select the correct BAND or AFC setting.

There are eight possible settings for this control.

0 - Broadcast band, AFC off.

1 - CATV band, AFC on.

2 - HRC CATV band, AFC on.

3 - IRC CATV band, AFC on.

4 - Broadcast band, AFC on.

5 - CATV band, AFC off.

6 - HRC CATV band, AFC off.

7 - IRC CATV band, AFC off.

Table 6-3 R-Line Service Menu Explained

R Line Color TV Update & Notes

C - 8 R Line Service Menu Explanation

18 - PIP Luma Delay - Values are 0 - 31.

19 - PIP Vertical Position - This control sets the vertical position of the PIP frame. The values are 0 - 31.

20 - PIP Horizontal Position - This control sets the horizontal position of the PIP frame. The adjustment range is 0 - 31.

21 - PIP Switch - This control sets the PIP to main video switch rate. The adjustment range is 0 - 31.

22 - OSD RGB Brightness/Contrast - This feature adjusts the brightness and contrast of the on screen displays. The adjustment range is 0 - 92.

23 - Color Limiter and Sub Color Level - This control sets the color limiter and sub color control. The adjustment range is 0 - 127. The bit map settings are listed below.

bits 0 - 4 sub color control - adjustment range 0 - 31

bits 5 - 6 color limiter level - adjustment range 0 - 3

  bit 7   DO NOT SET THIS ITEM

24 - Sub Contrast - This control sets the operating range and offset for contrast. The adjustment range is 0 - 31.

25 - Green Drive - This controls sets the level of green video drive. The adjustment range is 0 - 127.

26 - Blue Drive - This control sets the level of blue video drive. The adjustment range is 0 - 127.

27 - Color Decoder - The adjustment range is 0 - 253. The controls bit map is listed below.

bit 0 - fleash pull in.

bit 1 - DO NOT SET.

bit 2 - GY relative phase.

bit 3 - GY relative amp.

bit 4 - RY relative amp.

bits 5/6 - RY relative phase.

bit 7 - color gamma correction.

28 - Color Emphasis - The adjustment range is 0 - 127. The control bit map is listed below.

bit 0 - white peak supression.

bits 1/2 - sharpness tracking.

bits 3/4 - color detail.

bits 5/6 - high brite color.

bit 7 - DO NOT SET.

29 - Luma Noise Reduction Video Filter Off RF - This control sets the Luma noise reduction value when an RF source is selected and the customer Video menu video filter feature is OFF. The adjustment range is 0 - 3.

30 - Luma Noise Reduction Video Filter Off AUX - This control sets the Luma noise reduction value when an AUX (audio/video) source is selected and the customer Video menu video filter feature is OFF. The adjustment range is 0 - 3.

31 - Luma Noise Reduction Video Filter On - This control sets the Luma noise reduction value when the customer Video menu video filter feature is ON. The adjustment range is 0 - 3.

32 - Luma Enhanced Control - The adjustment range is 0 - 63. The control bit map is listed below.

bit 0 - black level expansion pull in.

bit 1 - SRT (super real time transient).

bit 2 - black level expansion.

bit 3 - luminance gamma correction gain.

bits 4/5 - luminance gamma correction start.

bits 6/7 - DO NOT SET.

33 - Horizontal Phase - This control sets the horizontal phase of the deflection. Changing this control will move the picture horizontally. The adjustment range is 0 - 31.

34 - Horizontal/Vertical Frequency - This control sets the horizontal and vertical PLLs. The adjustment range is 0 - 254.

35 - Velocity Modulation - This control should be set to 0 if "Velocity Modulation" is not a feature of the set. The adjustment range is 0 - 3.

Table 6-3 R-Line Service Menu Explained (Continued)

I have since learned that these same symptoms point to lack of horizontal deflection. If the diode is correctly soldered, I begin by checking FX3403 in the +130 volt line (figure 6-7a). If it is open, I check the horizontal output transistor because it is almost certainly shorted. However, something usually causes it to short, and that "something" is quite often the flyback.

System Control

Of course, the microcomputer (IC6000) is the "brains" of system control (figure 6-8) with the EAROM (IC6001) serving as external memory. It receives input via the front panel controls at pins 36 and 40 through 49 and the infrared remote receiver that is connected to pin 37. Do I need to point out that it controls the rest of the C-8's circuits: the PIP if there is one; the audio processor (IC1400); the source switch (IC2900); the video processor (IC2300); and the tuner, via the serial bus that consists of clock and data lines? It rounds out its activity by generating on-screen data and outputting it at pins 16 through 19. The OSD information arrives at pins 28 through 31 of IC2300.

Troubleshooting the Microprocessor

If you suspect system control problems, go directly to IC6000 and proceed to troubleshoot it in a systematic, logical manner. What do I mean by this? Begin by checking the "must haves," but don't stop there. Using figure 6-8 or a similar schematic as a guide, check for VDD (+5 volts) at pin 42 and VSS (ground) at pin 1. Do both quickly by placing the negative lead of your DMM on pin one and the positive lead on pin 42. You should find almost exactly +5 volts, and it should be stable and ripple-free. A voltage that is considerably higher or lower than +5 volts will cause the microcomputer to do some really crazy things. So, begin by confirming VDD. Proceed to pin 2 and check for proper reset voltage. You should find +5 volts DC. If the reset voltage is missing, check pins 5 and 7 of IC6003 for +5 volts. If it is missing at pin 5, IC6003 is probably defective and needs to be replaced. If it is present, check the path between it and pin 2 of IC6000 for an open circuit condition. Next, check the oscillator at pins 50 and 51. You are looking for a clean sine wave at about a 5-volt peak-to-peak value. Don't forget to check its frequency which should be 12.08 Mhz. The literature says it should be 12.083916 Mhz, but I think 12.08 Mhz is close enough!

Most of the literature I have read calls these four parameters the microprocessor's "must haves": VDD, VSS, reset, and oscillator. I agree, and I disagree. Begin with these four, but don't stop there. Experience has taught me to check the data in and data out pins before I condemn a microprocessor and begin the tedious and difficult processing of removing it and installing a new one. If a tact switch or a disc ceramic capacitor on one of the data input lines becomes leaky, the line goes low and causes the microprocessor to execute a faulty command that in turn hangs up the system. On extremely rare occasions, a pull-up resistor might open causing an incorrect voltage on one of the input pins. For example, if R6002 at pin 48 opened, there would be no +5 volts at pin 48, and the microcomputer could not execute any commands given to it over this circuit. The moral is you may think you have a faulty system control IC when in fact you have a faulty component tied to a data input or output line. The same phenomenon can occur on one or more of the data output lines. Therefore, take a few extra moments and make these checks even though they may seem trivial. The time and money you save are yours.

Tuner

Zenith likes to talk about the tuner (figure 6-9) as a part of system control, and I suppose that's as good a place as any to discuss it. The tuner is controlled by the I2C bus and is quite similar to the tuners first used in the C-5 and C-7 chassis. The microprocessor, instead of a separate tuner control

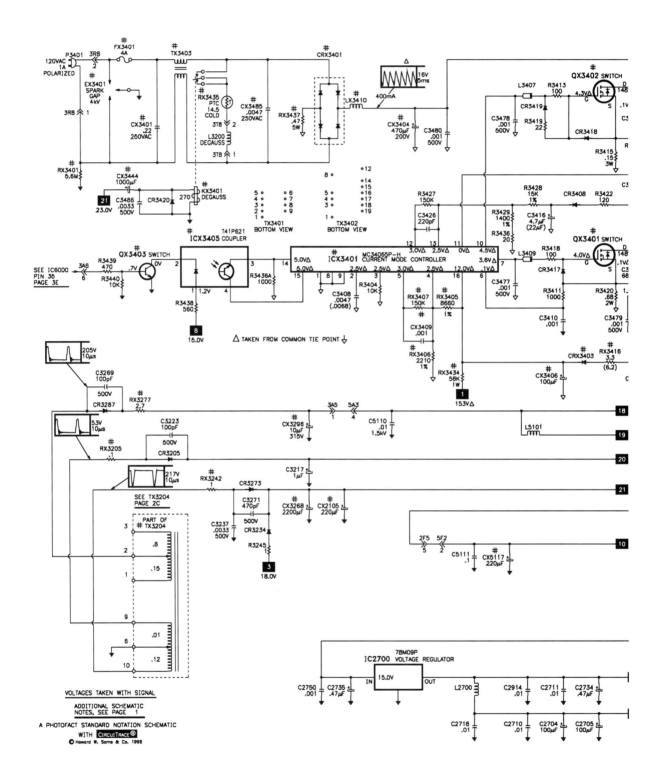

Figure 6-7a PHOTOFACT Power Supply Schematic

The C-8 Chassis

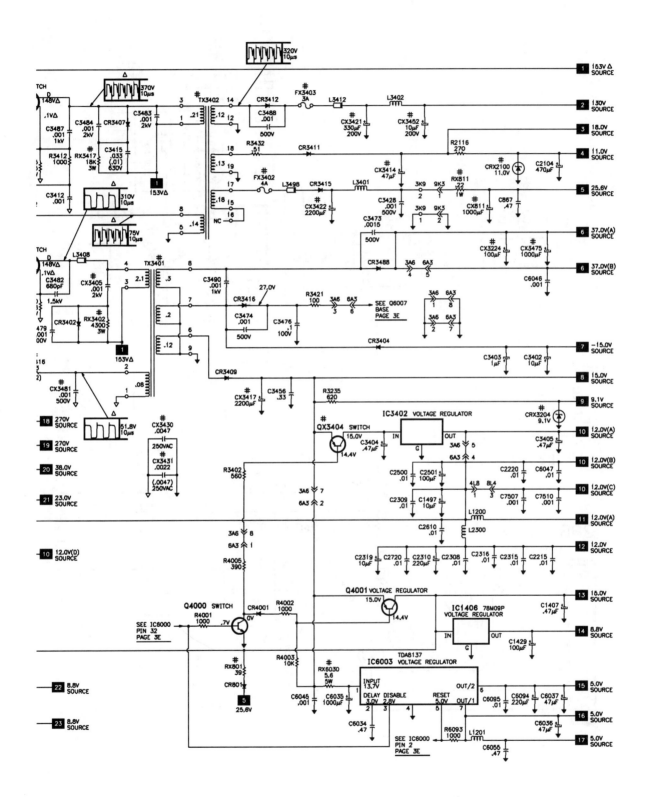

Figure 6-7a PHOTOFACT Power Supply Schematic (Continued)

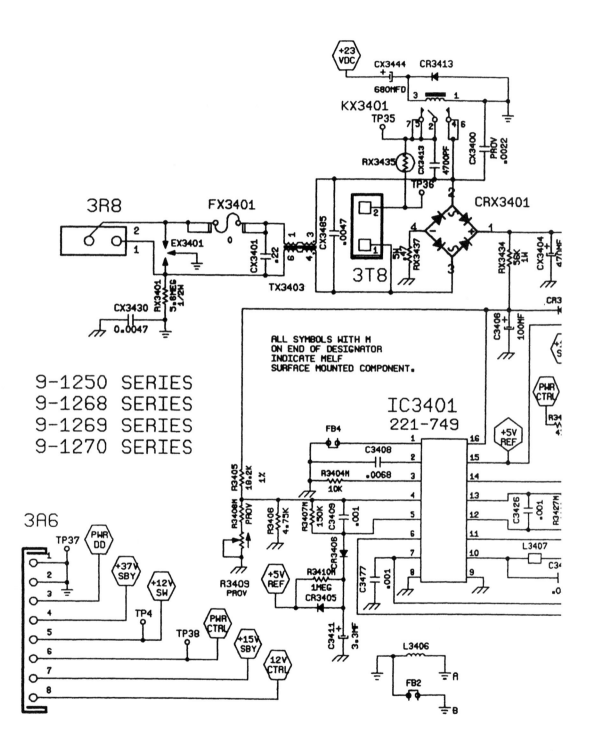

Figure 6-7b Zenith Power Supply Schematic

The C-8 Chassis

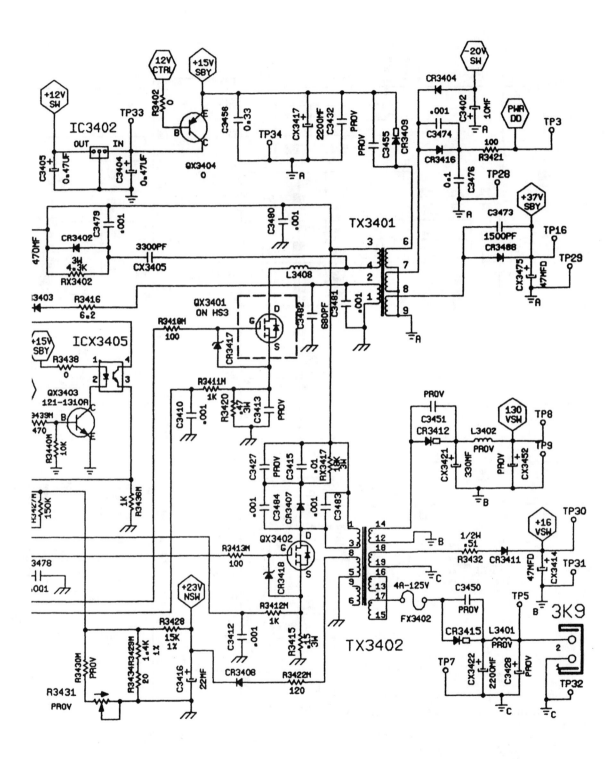

Figure 6-7b Zenith Power Supply Schematic (Continued)

Servicing Zenith Televisions

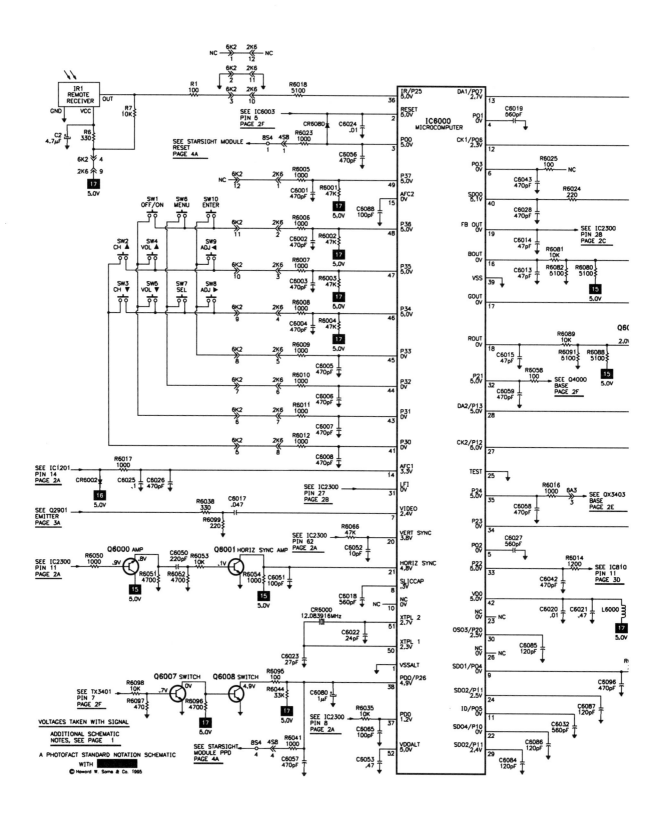

Figure 6-8 System Control Schematic

The C-8 Chassis

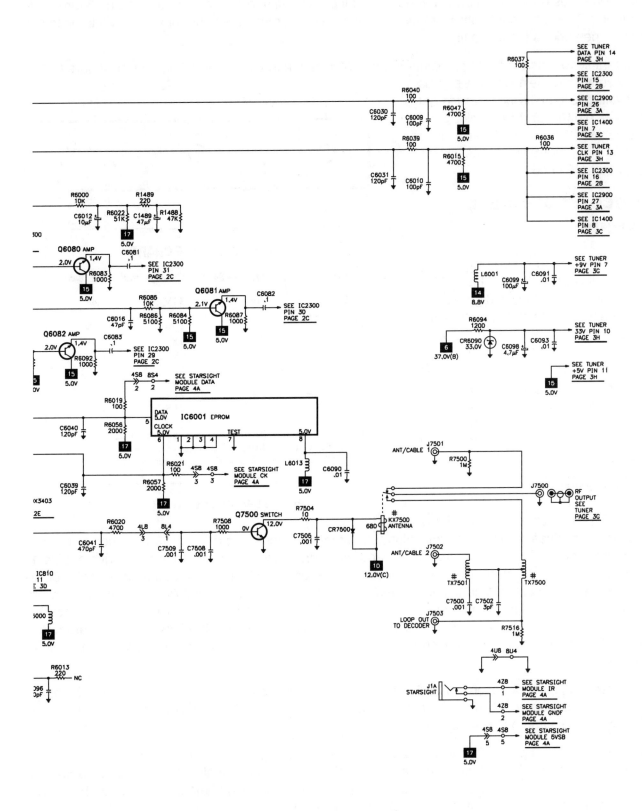

Figure 6-8 System Control Schematic (Continued)

chip, generates band-select and channel-select instructions and sends them out of pin 13 directly to the tuner. The tuner controller IC mounted inside the tuner interprets the information and uses it to control the various tuner circuits.

If you think you have a tuner problem, use the information in the "tuner voltage chart" (figure 6-9) to make sure it has the necessary signals and voltages to do its job. Don't forget to check pins 13 and 14 for clock and data activity. If it doesn't receive these signals from the microprocessor, the tuner cannot function.

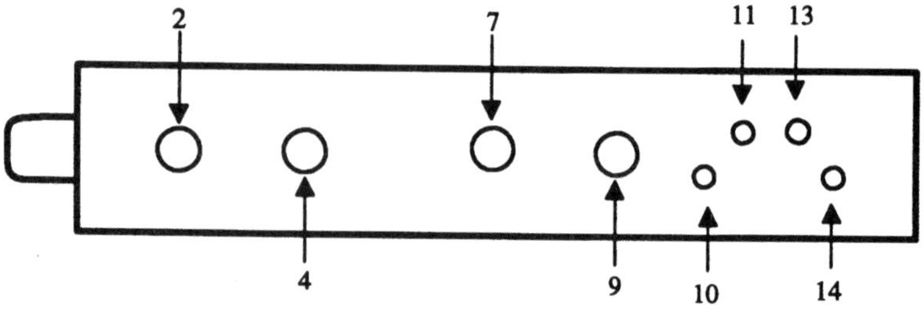

Video Processing Circuit

The video processing circuit is marvelously complicated. I am going to attempt to reproduce it (figure 6-10) from the Sams I'm using. I say attempt because it takes up four, 8 1/2 x 11 sheets of paper! Zenith reduces it to a block diagram (figure 6-11) that might be more helpful in the event you only have to do a bit of in-depth troubleshooting. Use it as a reference while I give you a brief description of the video processing system.

Video Processor

IC2300 is the centerpiece of the system. It receives chroma and luma information that it processes to develop vertical, horizontal, and RGB video drive. The chip controls these and other processes by commands that arrive at pins 15 and 16 from the microprocessor. The commands originate either from the customer menu or the service menu and are related to items contained in the service menu. You may have wondered why there are 30 to 35 items in the service menu, some of which aren't used. Engineers designed the system so that when there are circuit changes they can reset some of the service menu parameters rather than manufacture a new microcomputer loaded with new commands to implement the circuit changes.

TUNER VOLTAGE CHART

Pin	VHF Low Band	VHF High Band	UHF Band
2 (TUNING VOLTAGE)	.8V	3.7V	6.4V
4 (AGC)	4.2V	4.2V	4.4V
7 (+9V)	8.9V	8.8V	8.9V
9 (IF OUT)	0V	0V	0V
10 (33V)	32.1V	32.2V	32.2V
11 (+5V)	5.0V	5.0V	5.0V
13 (CLK)	2.4V	2.4V	2.4V
14 (DATA)	2.8V	2.8V	2.8V

NOTE: VHF Low Band voltages taken on channel 2.
NOTE: VHF High Band voltages taken on channel 7.
NOTE: UHF Band voltages taken on channel 14.

Figure 6-9 Tuner Voltage Chart

On-screen display information is mixed – or more properly, time-shared – with the main video information by switching circuits inside IC2300. As I have indicated, OSD originates inside IC6000 and comes to IC2300 as RGB and blanking signals.

RGB information (figure 6-11) exits IC2300 at pins 18 through 20 and goes to pins 2, 4, and 6 of IC2500, the RGB video output driver device. IC2500 also contains a circuit that provides automatic color tracking. In other words, don't waste your time looking for potentiometers or items in the service menu to set the gray scale because it is done automatically.

IF Module

If you paid attention to figures 6-2 and 6-3, you saw a small rectangular box mounted on the main module right next to the tuner. The box appears in figure 6-10 as a rectangle formed by dotted lines around IC1201 and its associated components. These figures identify the IF module. Since you need to know what goes in and what comes out, I'm including figure 6-12 to aid your troubleshooting.

The module receives the IF signal from the tuner at pin 3 of connector 1 and outputs composite video at pin 21 of IC1201 (figure 6-11). The composite video is buffered by Q1201M and sent to pin 3 of connector 2 and on to the source switch. Composite audio leaves pin 1 of the IC1201 and exits pin 1 of connector 1 on its way to the source switch.

The IF module contains several adjustable controls, L1205 (AFC), L1206 (VCO), L1201 (4.5 Mhz trap), AGC (R1207), and R1223 (video gain). The literature says that except for video gain and AGC delay these controls should be left alone because they are "factory adjustable" only. In other words, leave them alone unless you have the equipment and knowledge to adjust them.

Source Switch

Figure 6-11 identifies IC2900 as the source switch, an electronic device that receives audio/video input from the tuner and the various inputs available at the jack pack. It receives instructions from the microprocessor via the I2C and executes them by accepting audio and video from a particular source and routing it to the appropriate circuits for processing.

Picture-In-Picture Assembly

The PIP is also a fully shielded assembly soldered onto the main module. Its operation is remote controlled. In other words, the viewer uses the remote control to select any one of the available inputs as a signal source. Those input sources, depending on the model, include: ANT/CABLE 1, ANT/CABLE 2, VIDEO 1, VIDEO 2, and S-VIDEO 1.

The PIP module is in series (figure 6-11) with the chroma and luma lines to the video processor. This means the PIP can – and it will – cause chroma, luma, and sync problems. Zenith says that you need

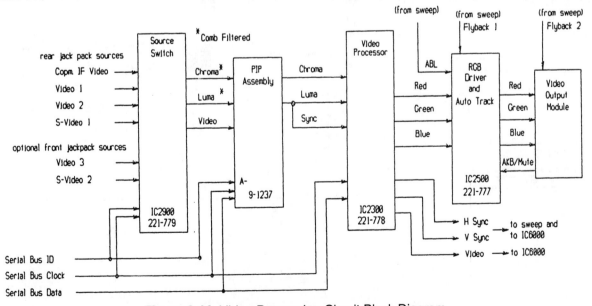

Figure 6-11 Video Processing Circuit Block Diagram

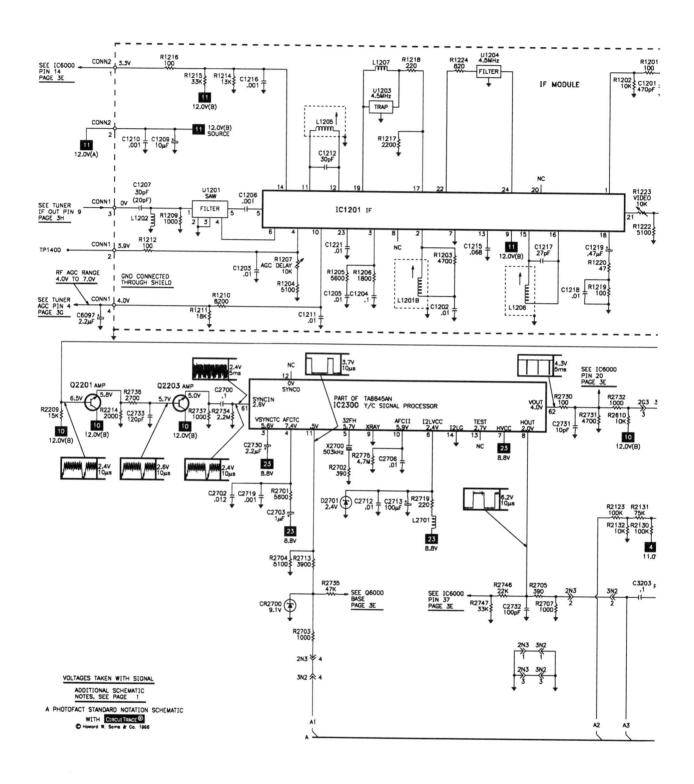

Figure 6-10 Television Schematic

The C-8 Chassis

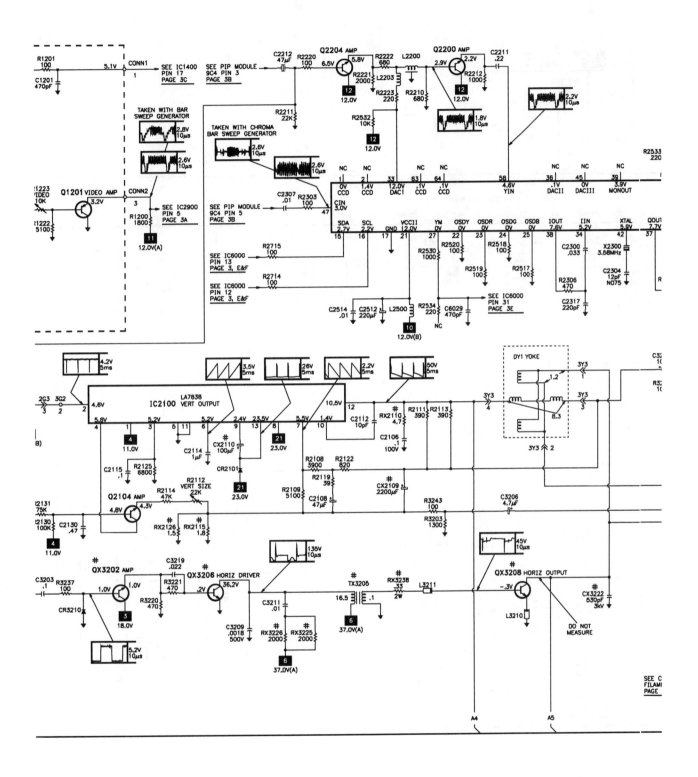

Figure 6-10 Television Schematic (Continued)

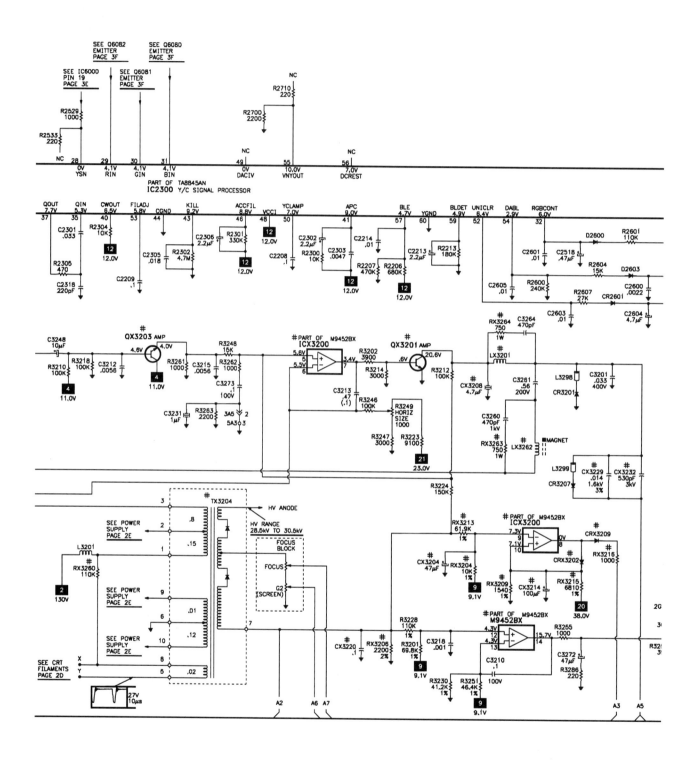

Figure 6-10 Television Schematic (Continued)

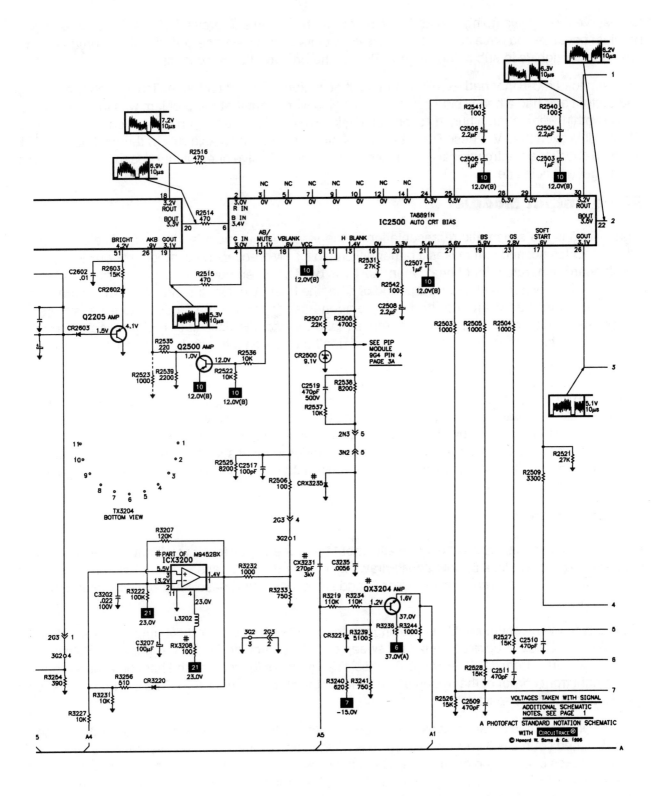

Figure 6-10 Television Schematic (Continued)

Servicing Zenith Televisions

to observe the signals going into and coming out of the module to prove that it is the source of your problem because you have no way to bypass its circuits. My experience confirms their comment. The best way, though, is to sub a known good PIP for the one you think is defective.

I just serviced a C-8 that had excellent audio but no picture. The raster had pulled in slightly at the top and bottom and had a heavy blue color to it. I originally thought that the problem was in the IF circuit, but on nothing more than a hunch I substituted a known good PIP for the one on the chassis. The new PIP module fixed the TV. These PIPs are kind of difficult to service. As a rule of thumb, I use an ESR meter to check the electrolytic capacitors and replace the defective ones. If that doesn't fix the module, I try to get a rebuilt one.

Video Output Module

Figure 6-13 is the schematic representation of the video output module. It is complicated as video driver modules go, and that makes it rather difficult to fix. Moreover, the components on video output circuit board generate a lot of heat when the TV is on. The video output transistors are mounted, for instance, on hefty heat sinks. Over a period of time, the heat discolors the circuit board and damages the circuit board traces around the major components, especially the video output transistors. A complicated circuit plus heat-damaged traces make the module difficult to repair reliably and often leads me just to replace the module. However, don't give up on it till you have at least tried.

A word of caution before I leave this topic.

A defective video module might cause you to think you have a bad picture tube. Let me give you an example. Assume that the TV in your shop changes color especially when you tap on the neck of the picture tube. That's a sure sign of a problem with the electron gun in the neck of the tube, isn't it? But before you write the TV off as a loss, check and resolder if necessary the leads of each of the video output transistors. The TV you save belongs to your customer, and the money for the repair goes into your bank account.

Sweep Circuits

Remember to consult figure 6-10 for details about the sweep circuits. If you want a peek at it from a different perspective, look at how Zenith draws it in figure 6-14. Because I spent time discussing the sweep circuits of the C-3 chassis (Chapter 2) and since these two chassis are similar, I intend to keep this discussion brief.

IC2100 is the vertical output IC and belongs to the LA78xx family of vertical output chips. QX3202, QX3206, and QX3208 develop horizontal sweep exactly as they do for the C-3 chassis. Both circuits receive their drive from IC2300. The C-8 chassis has controls for vertical and horizontal size mounted on the right front corner of the sweep and power supply module (figure 6-2). The controls are labeled R2112 and R3249 respectively.

The 16 volts switched initially biases the sweep circuits on. If you are following Sams notations, the 16 volts switched source is 3 and labeled "18 volts." When horizontal sweep comes up, the scan-derived +23 volts takes over the job of supplying B+ to the sweep circuits. Shift your attention now to figure 6-10 and locate TX3204. When you find it, look to its immediate right and locate R3245 and CR3234. These components form the 16-volt switched and 23-volt scan-derived crossover and blocking circuit. CR3234 does the job of keeping the two voltage sources apart. If it fails, the +16-volt source goes low, keeping the chassis from starting. When R3245 and CR3234 work properly, they permit the +23 volt source to assume responsibility for the +16-volt line. Incidentally, R2345 and CR3224 are located on

The C-8 Chassis

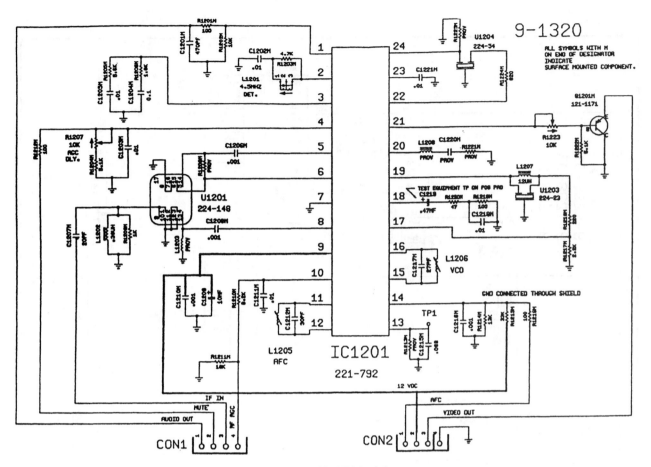

Figure 6-12 IF Modules

the sweep module. Besides developing the +23 volts, horizontal sweep also develops the filament voltage, +215 volts for the video output transistors, EHT, G-2, and focus voltages.

I said when I discussed the C-3 chassis (Chapter II) that a blown horizontal output transistor usually heralds additional trouble. I have seldom replaced an output transistor without having to replace either the flyback or the timing capacitor (CX3222) in the collector circuit of the horizontal output transistor. The flyback has been the problem most of the time. Whether you repair the module yourself, order a replacement module, or send the old one in to be rebuilt depends on several things. You could do it yourself and save in the neighborhood of $20. You might save even more if you have a good, used flyback on hand. But, you might not have time or the customer may want the set back the quickest possible way. Additionally, you might install the new or used part and discover you have additional problems hampering your troubleshooting attempts because of the physical layout of the board. So, the question is a toss-up, and you must decide what is best for you. Personally, I hate to have someone else do it for me, but that's my problem!

Just yesterday, I had a monster 35" set come in with the complaint, "Sound but no picture." The fuse in the +130 line was blown, indicating a problem with the horizontal output transistor, flyback, or worse. To save a bit of money, I decided to replace the output transistor with a Zenith original (TZ1148-01) and the fuse and see what happened. I really didn't expect the set to fire up. In almost all instances, the output transistor immediately shorts and blows the replacement fuse, but it fooled me.

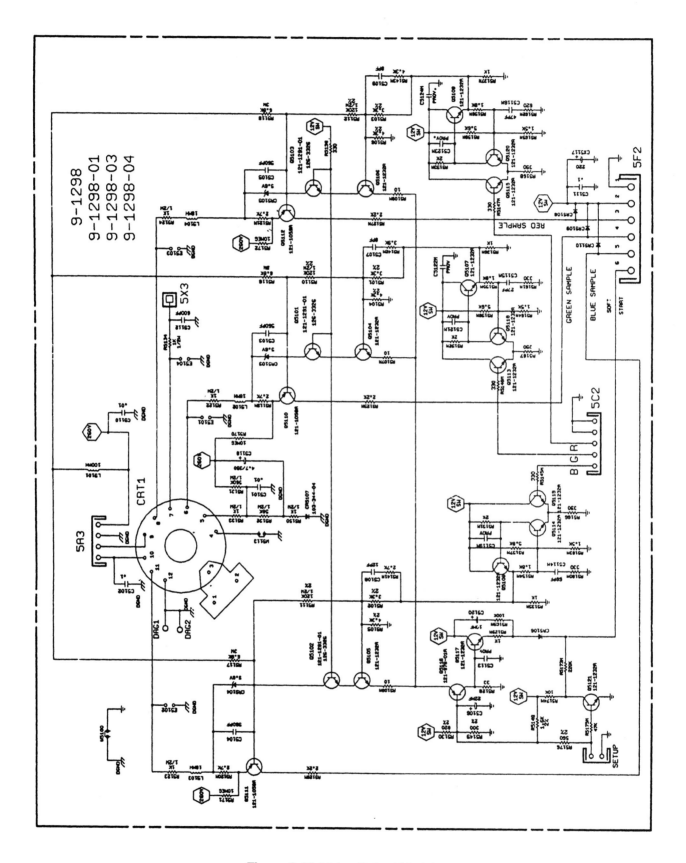

Figure 6-13 Video Output Module

The C-8 Chassis

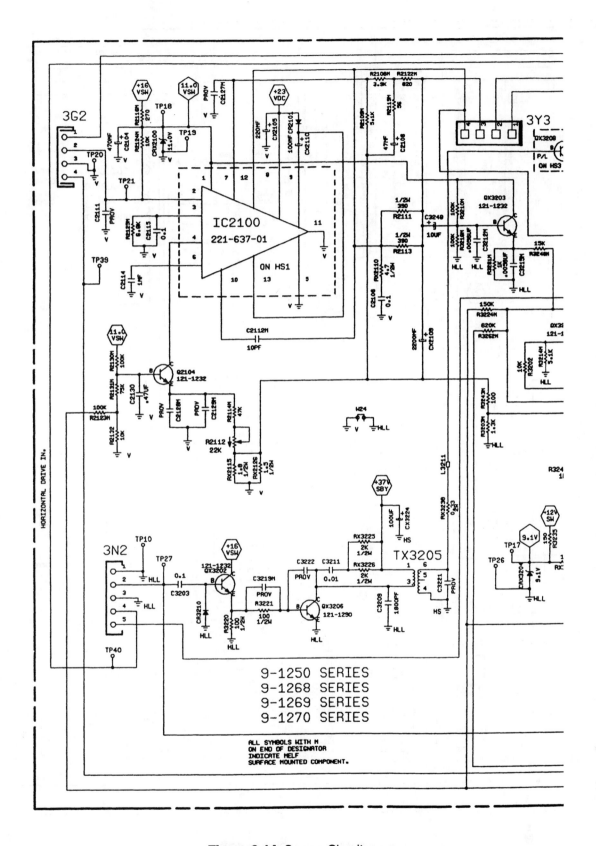

Figure 6-14 Sweep Circuits

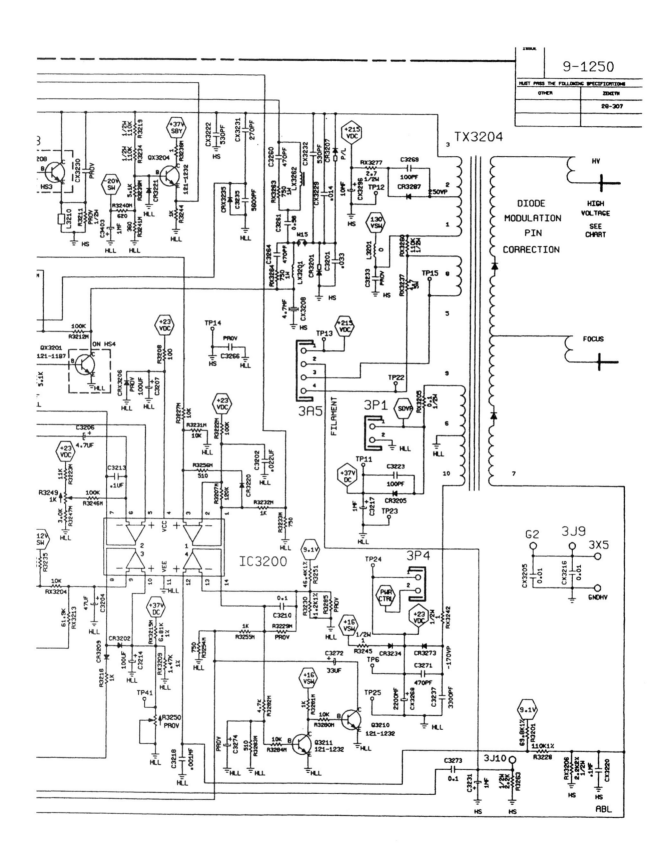

Figure 6-14 Sweep Circuits (Continued)

It came on immediately! I let it play for several minutes and felt the heat sink on which the output transistor was mounted, and it was very hot. I checked the amperage the set was pulling and found that it was using 1.5 amps which was well below its maximum rated current consumption of 2.3 amps. I checked the B+, and it was at +130 volts. I checked the peak-to-peak value and the width of the retrace pulse, and both were right on target.

Why was the heat sink getting hot, and why did the original transistor fail? I happened to notice a minute flash in one corner of the circuit board when the set came on. Checking further, I found a bad solder joint on one leg of the horizontal width coil. Was this the culprit? I resoldered the connection and turned the set back on. The heat sink got hot again. The loose connection explained the failure of the transistor but not the heat problem. I called Zenith technical assistance. He said the flyback or the picture tube could be failing. Since the current draw and waveforms were normal, he said I should play the set and see what happened. I did, and the blessed TV is still playing. All of this underscores one of the basic philosophies, "If a television can get you, it will."

Some final words as this section closes. Keep in mind that these words apply to service situations outside the contents of this book. They, in fact, apply almost every time you use factory service literature and a PHOTOFACT at the same time. For example, you have surely noticed that Zenith's voltage checks don't square with those given in the PHOTOFACTS. If you haven't picked up on it, just reread the last few paragraphs. I hope you are wondering about the disagreement. Before they produce a PHOTOFACT, the folks at Sams buy the TV, check all the voltages using the actual set, disassemble it, and then draw the schematic. Of course, the Sams engineers also use factory literature as a source of information along with their "autopsy" results. The voltages listed in PHOTOFACT 3679 have, therefore, been taken from an actual, working TV whereas the voltages given in Zenith's literature are expected or anticipated voltages based on engineering specifications. Don't think of this in terms of "right versus wrong" because that's not the case. You are savvy enough to know that voltages can vary from TV to TV. Just accept the difference and continue on your merry, troubleshooting way!

Audio Circuits

Figure 6-15, a typical C-8 audio configuration, should give you an idea about how the audio circuit works. Let's begin our tour of the audio system by taking a brief look at the center of activity, IC1400. It receives instructions from the system microprocessor over the I2C bus at pins 6 and 7, composite audio from the TV signal processor at pin 21 and auxiliary audio from the jack pack at pins 25 and 26. It has three audio outputs: (1) variable audio at pins 1 and 2; (2) a fixed output at pins 41 and 42; and (3) audio output to the speakers at pins 38 and 39.

Audio, on its way to the speakers, goes through part of IC1404 for buffering. The variable audio output goes to pins 2 and 6 of IC810. The main left audio output is pin 17 of IC810 while the main right output is pin 10. Each of these audio outputs goes to the jack pack assembly. Surround audio is matrixed by part of IC1404. The output of the matrix circuit goes to IC1410 for surround sound volume control. Part of IC810 is set aside for the surround channels.

Repair History

I have already given you some details about common problems and fixes, and I hope to extend the list in this section. Some of the problems I am about to list involve common-sense fixes, but I'm going to list them anyway because common-sense fixes are among the easiest to overlook and often cause us to waste bushels of time.

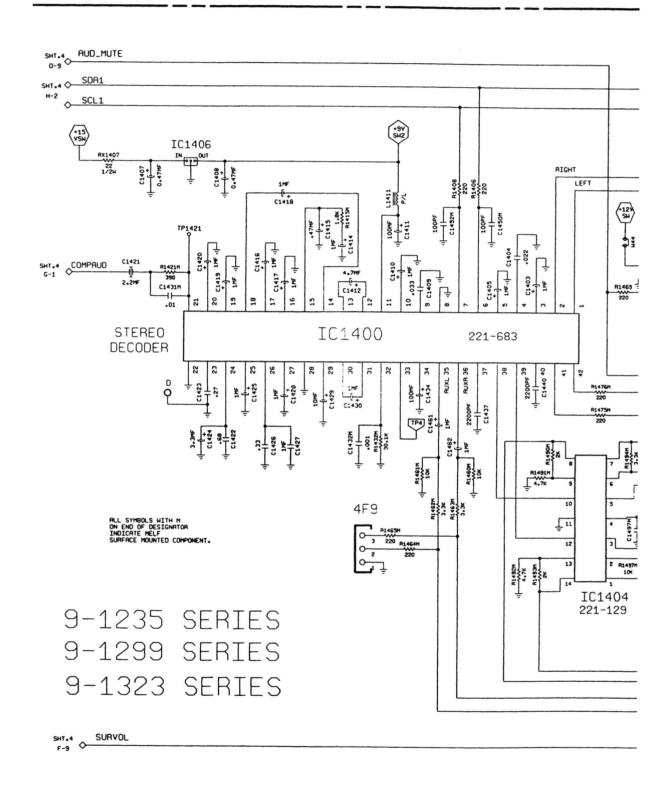

Figure 6-15 Audio Circuit

The C-8 Chassis

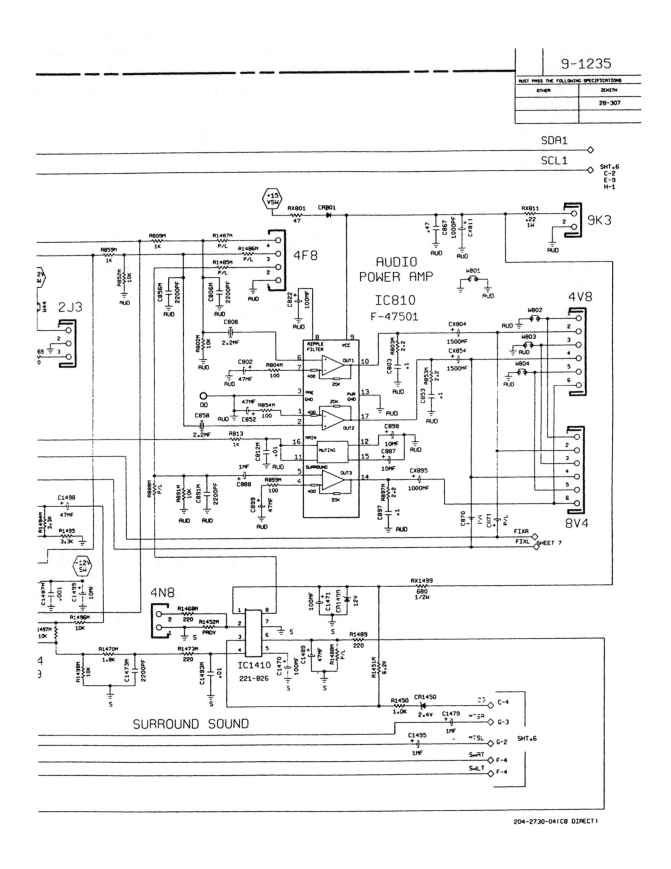

Figure 6-15 Audio Circuit (Continued)

(1) Dead set. Quite a few set problems are caused by poor solder connections. When you run across a C-8 that won't come on, check for poor solder around the bridge diodes, those big diodes mounted vertically, and the solder connections around the switching transistor and the switch mode transformer. While you are at it, take a few extra moments and look over the rest of the circuit board.

Of course the dead set symptom might be caused by a failed component. QX3407, FX3403, and ICX3405 head the list. If your inspection of the circuit board doesn't turn up defective connections, begin your troubleshooting efforts with these. Don't overlook the electrolytics in the startup and feedback circuits either.

(2) Pincushion problems. Pincushion problems cause the raster to be either too large or too small for the screen. My notes indicate that coil LX3201 causes more of these kinds of problems than any other component. If you recall, I mentioned finding a loose solder connection at this coil that led to failure of a horizontal output transistor. Two other failures are common to LX3201: shorted turns and a loose ferrite core.

(3) No picture and hiss in the audio. Check for poor solder around both ends of diode CR3412. If the fuse associated with the diode is open, immediately suspect a shorted horizontal output transistor.

(4) Loss of blue to the picture tube. Check for an open R5110. The same applies to the other colors. If, for example, green is missing, check the corresponding resistor for that color.

(5) No color or distorted color. IC810 and CRX3406 have been known to cause a no-color problem. CRX2300 might cause color distortions to appear in the picture.

(6) Dark screen that mimics a defective picture tube. Take a look at Q5116 on the video output module before you condemn the picture tube.

(7) No audio. Suspect an open RX801, a defective IC810, or a failed audio output IC.

(8) Replacing the horizontal output transistor. When you replace a shorted horizontal output transistor, check for cold solder connections around LX3201, the flyback, the horizontal driver transformer, CR3412, and the horizontal yoke before you fire the set up. Bad solder around any one of these components could cause the new transistor to fail. Use your scope to monitor the collector waveform as you fire up the set. If you detect ringing in the retrace pulse, immediately suspect the flyback.

CHAPTER 7
THE C-10 CHASSIS

C-10 chassis include, but aren't limited to, the following modules: 9-1332-01 (19"/20"), 9-1333-01 and 9-1334-01 (25"), and 9-1335-01 (27"). The 9-1407 and 9-1408 modules are often used as substitutes for the 9-1334 and 9-1335. An A-17815 jack pack accompanies those modules that have external audio/video inputs. If the module has stereo capability, the stereo jack pack comes as an integral part of the circuit board.

The C-10 is another single printed circuit board chassis like most of the others Zenith has manufactured in the last 10 to 12 years (figure 7-1). Five integrated circuits handle tuner, audio/video, sync, and sweep functions. The switching power supply uses a couple of additional chips, and the MTS stereo models employ yet another chip. IC6000, the microcomputer, handles all of the PLL tuning and band switching functions. It receives input from the keyboard and IR receiver and executes commands via the I2C bus. ICX1200 processes audio and video, sync, and sweep drive. IC2100 drives the vertical yoke. IC1400, the MTS decoder, and IC801, the audio amplifier, comprise the audio circuit. If the set is monophonic, IC820 handles both the processing and amplifying audio jobs. ICX3431 is, of course, the switch mode regulator. IC4121 develops the +9 volts switched supply from the +12 volt line.

Available Literature

I am depending on the Zenith booklet *Technical Training Program: C-10 Sentry 2* for most of my technical information. As a matter of fact, I have taken several of the schematics used in this chapter from it. However, I keep PHOTOFACT 3319 at my elbow in the shop and use it to supplement Zenith's material as I proceed.

Beginning with the C-10 series, I expect to supplement the bibliography I give you with reference to available sources for information about servicing commercial products. If you have a hotel, motel, or hospital that depends on the commercial C-10 chassis, I suggest you order *Technical Training Program: C-10 Commercial Products* because I believe you'll find it to be a valuable addition to your service library.

There is one other source of information on the C-10 chassis. The September 1999 edition of *Electronic Servicing And Technology* has a nice article on a couple of repair problems that concern the 9-1407 module. The article itself is worth the cost of the back issue (about $4). You may order it (or any back issue for that matter) by writing to:

> Nils Conrad Persson, Editor
> Electronic Servicing And Technology
> Post Office Box 12487
> Overland Park, Kansas 66212

The magazine lists the fax number as 630-584-9289. The editor is an excellent man who not only knows our business but also delights in helping us. He will be glad to send you the issue in question and even assist you in getting a subscription started. I have subscribed to it for years and find it is worth every penny.

Servicing Zenith Televisions

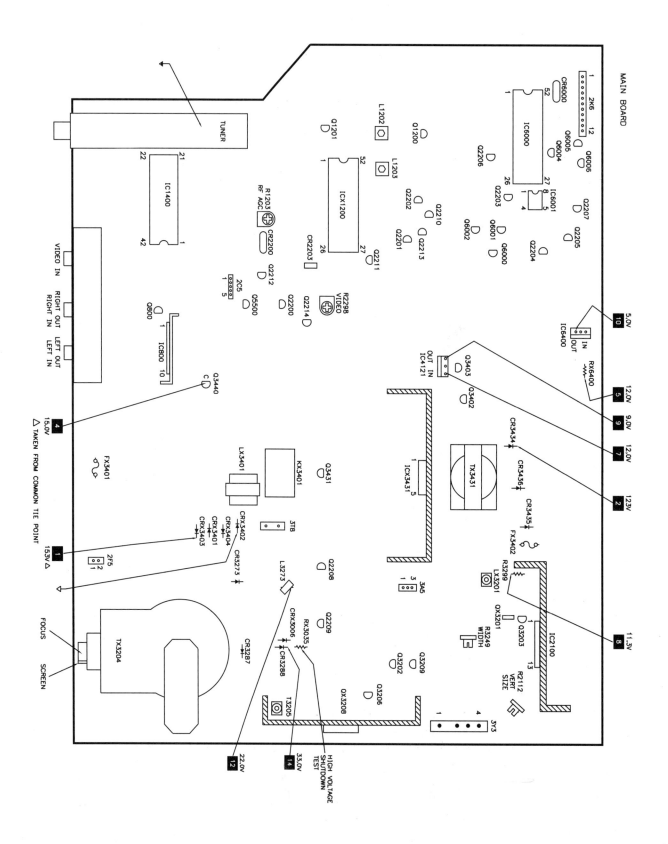

Figure 7-1 C-10 Chassis Printed Circuit Board

Customer Menu

There aren't many new items in the customer menu (figure 7-2), but I need to comment on some of the entries because they have changed slightly. "Channel Add/Del" used to be called "Favorite Channel" in other Zenith TV menus. Its function is self-explanatory. "Channel Labels" is a new addition, and the C-10 is one of the first Zenith lines to have it. As the name suggests, the customer can use it to assign labels such as NBC, Fox, TBS, and so forth to frequently watched channels. However, the customer is limited to the labels in the channel label list.

"Captions" is the next-to-last item in the setup menu and has five entries: off, caption 1, caption 2, text 1, or text 2. The user accesses the captions menu directly by using the "Caption" or "CC/Quit" buttons on the remote control. If "Captions" or "Text' is selected, the selected item appears in about 10 seconds. If captions or text data aren't present, captions or text won't appear, but the TV remains in the selected mode. If the text mode has been selected and no text is present, a black box appears right in the middle of the screen with video peeking around its sides. I am sure that you have encountered the situation where the customer accidentally turned the text mode on, saw the screen go black, and called you thinking the TV had blown up. Even though these features have been out for several years, we still get calls from customers who think their TVs have quit. I even get new ones shipped in for "store stock" warranty service for the same problem. Some people seem to have an aversion to reading an owner's manual. Some situations, it seems, don't change.

"Channel Background," which is available in selected models, permits the user to add a background to the on-screen channel display. The channel display appears over the video if no background has been selected.

If the C-10 model has external audio/video inputs, the customer switches between tuner and external audio/video signals by selecting the channel either below the lowest channel or above the highest channel in memory. Switching sources may also be accomplished by using the "Source" key on the remote control

CUSTOMER MENU

SETUP MENU
→ AUTO CH. SEARCH
CH. ADD/DEL
CH. LABELS
TUNING BAND
AUTO FINE TUNE
CLOCK SET
CAPTIONS
CH. BACKGROUND

AUDIO MENU
→ BASS
TREBLE
BALANCE
AUDIO
SEQ

VIDEO MENU
→ CONTRAST
BRIGHTNESS
COLOR
TINT
SHARPNESS
PICTURE PREF

Figure 7-2 Customer Menu

Service Menu

Early production C-10 service menus look like table 7-1 with its 25 items in two columns on one page. Late production C-10's have 16 items in two columns on one page as in table 7-2. The difference between the early and late-production models lies in the fact that the late-production runs eliminate references to PIP functions in the service menu even though the C-10 didn't come equipped with a PIP. "Why," you ask, "are PIP functions included in the service menu of a TV that never used a PIP?"

→ 860-1 - 1.13		AC ON	OFF
PRESET PX STORE		KEY DEFT	OFF
PIP PR STORE		A/V LOCK	OFF
BRIGHTNESS	16	AUTO SRCH	2
CAP PHASE	64	PIP TEST	0
VERT POS	5	PIP HPOS	4
HORZ POS	33	PIP VRAMW	8
PIP AVAIL	N/A	PIP VRAMR	4
ZEN/PL	ZENTH	PIP FR CT	45
CHAN LOCK	OFF	PIP GRAY	64
START CHAN	OFF	PIP SWTCH	10
MAX VOLUM	OFF	C10 AUX	ON
MIN VOLUM	OFF		
		04/15/93-0	

→ 802-2.17		C10 AUX	ON
PRESET PX STORE		A/V LOCK	OFF
BRIGHTNESS	18	SRCH DISP	ON
VERT POS	5		
HORZ POS	33		
ZEN/PL	ZENTH		
CHAN LOCK	OFF		
START CHAN	OFF		
MAX VOLUM	OFF		
MIN VOLUM	OFF		
AC ON	OFF		
KEY DEFT	OFF		
AUTO SRCH	2	04/28/93-0	

Table 7-1 and 7-2 Early Production (Top) and Late Production (Bottom) C-10 Service Menu

Good question. The microprocessor Zenith engineers used in the early production C-10 chassis is the same one used in the C-11 chassis that did have a PIP option. It was easier and cheaper to use the same microprocessor for both than to come up with a new one "on the spot." As a rule thumb, just disregard the reference to PIP in the C-10 service menu. At the very least, set them to "off" when you do stumble across them.

In the event that you have to make adjustments to the software, I suggest that you use tables 7-3 and 7-4 as your guide. These tables give you both the expected settings and range in which adjustments are made.

Power Supply

The C-10 is driven by a power supply that is pretty much the same one we have looked at several times. Figure 7-3 shows you how Zenith draws it while figure 7-4 illustrates how Sams engineers see it. Both are useful because, looked at together, they give you different perspectives on the same subject. At the risk of repeating myself, I intend to devote several paragraphs to this circuit in Chapter 10 because I am reserving discussion till we get to the really new stuff. However, I need to say a few words about it now.

When it comes on-line, the power supply outputs the voltages necessary to operate the set with the exception of scan-derived voltages. It is, therefore, in the standby mode. For the moment, let's focus on figure 7-4. When it is in the standby mode, the power supply produces +123 volts (source 2) and +12 volts (source 5). The +12-volt line is also the source for the standby by +5 volts (developed by IC6400) for the microcontroller and IR receiver. Did you see that the +12-volt source is protected from an overload by FX3402? The +123-volt source doesn't require a fuse because it is protected from an overload by overcurrent limiting that is integral to the power supply itself.

The C-10 Chassis

ITEM	RANGE	SETTING
860-1.13	ON-OFF	OFF
PRESET PX STORE	STORE/STORD	STORD
PIP PR STORE	STORE/STORD	STORD
BRIGHTNESS	00-31	16
CAP PHASE	0-254	64
VERT POS	0-31	5
HORZ POS	0-82	33
PIP AVAIL	N/A	N/A
ZEN/PL	ZENTH/PL	ZENTH
CHAN LOCK	OFF-CH	OFF
START CHAN	OFF CH	OFF
MAX VOLUM	OFF-0-62	OFF
MIN VOLUM	OFF-0-63	OFF

COLUMN 2

ITEM	RANGE	SETTING
AC ON	ON-OFF	OFF
KEY DEFT	ON-OFF	OFF
A/V LOCK	ON-OFF	OFF
AUTO SRCH	0-18	7
PIP TEST	0-15	0
PIP H POS	0-7	4
PIP VRAMW	0-15	8
PIP VRAMR	0-7	4
PIP FR CT	0-127	45
PIP GRAY	0-127	64
PIP SWTCH	O-15	10
C10 AUX	ON-OFF	*ON

ITEM	RANGE	SETTING
802-2.17	ON-OFF	OFF
PRESET PX STORE	STORE/STORD	STORD
BRIGHTNESS	00-31	18
VERT POS	0-31	5
HORZ POS	0-82	33
ZEN/PL	ZENTH/PL	ZENTH
CHAN LOCK	OFF-CH	OFF
START CHAN	OFF CH	OFF
MAX VOLUM	OFF-0-62	OFF
MIN VOLUM	OFF-0-62	OFF
AC ON	ON-OFF	OFF
KEY DEFT	ON-OFF	OFF
AUTO SRCH	0-18	2

COLUMN 2

ITEM	RANGE	SETTING
C10 AUX	ON-OFF	*ON
A/V LOCK	ON-OFF	OFF
SRCH DISP	ON-OFF	ON

*If the receiver has jacks, place in the on mode.
PIP is not used on the C10 chassis.

Table 7-3 & 7-4 Expected Settings and Ranges for Adjustments

The move from standby to full power begins when the microprocessor outputs a high on pin 35 that is coupled to the power supply from the emitter of Q6004 (figure 7-4). The high turns on Q3402, Q3403, and Q3440 bringing up the switched +12 and +15 volts. Since it is derived from the switched +12 volts, the +9 volts also comes on-line via IC4121. Horizontal sweep starts when the +9 volts becomes available, bringing up the scan-derived voltages that consist +23 volts, +35 volts, +215 volts, EHT, focus, and G-2.

The service literature lists the voltages in the order in which they should be checked. I really do urge you check the voltages in the order in which I list them. A successful troubleshooter always checks voltage in that order: standby, switched, and derived. Use figure 7-1b to help you locate the test points.

Begin by checking the standby voltages first. Please notice that I don't list them in any particular order.

+5 volts DC at pin 3 of IC6400

+12 volts DC at the emitter of Q3403

+15 volts DC at the emitter of Q3440

+150 volts at the junction of CRX3403 and CRX3404

If you are troubleshooting a dead set condition and have confirmed the presence of the standby voltages, check the power on request at pin 36 of IC6000. If it is present, check to see if pin 35 responds by going high. As you recall, the high goes to the various transistors that turn on the switched DC voltages. Simply follow the line of components until you find the one that doesn't respond to the on command.

Third, check for the presence of the switched voltages. Again, I'm not listing them in any particular order.

+9 volts DC at pin 3 of IC4121

+12 volts DC at the collector of Q3403

+15 volts DC at the collector of Q3440

+23 volts DC CR3273

+35 volts DC at CR3288

+123 volts DC at CR3434

+215 volts DC at CR3287

These tests ought to give you an excellent feel for what's going on with the C-10 on which you are working. I can't stress enough that you should check the voltages in the order given. After you have confirmed the presence, for instance, of the standby voltages, you may proceed with confidence to the next level of troubleshooting. If, however, you haven't confirmed the standby voltages, your efforts just might be wasted because nothing is going to work without them.

System Control

System control (figure 7-5) is built around a 52-pin CMOS processor in a DIP package and is similar to those used in previous chassis. Keyboard commands enter at pins 41 and 43-49 while IR commands arrive at pin 36. Brightness, color, contrast, tint, and sharpness commands exit at pins 30, 29, 24, 22, and 9 respectively on their way to the video processor.

The C-10 Chassis

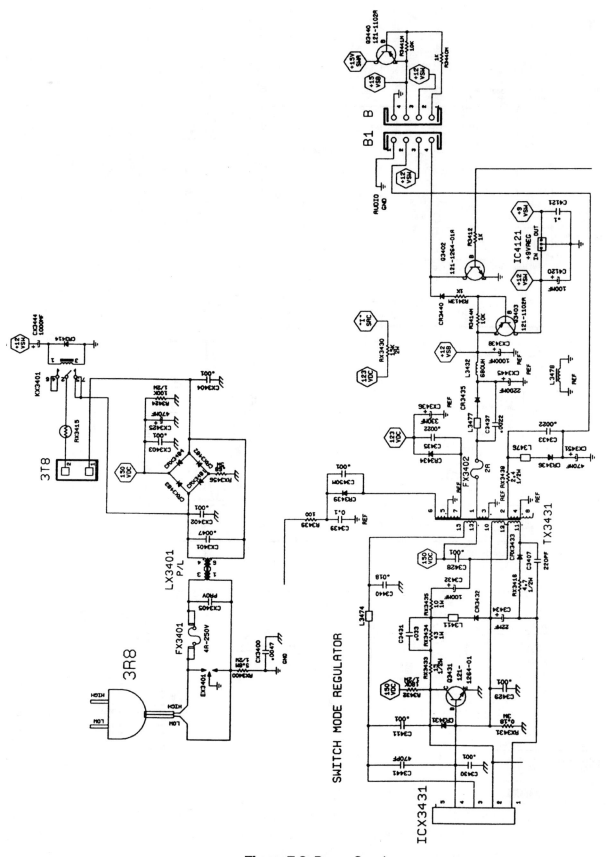

Figure 7-3 Power Supply

Servicing Zenith Televisions

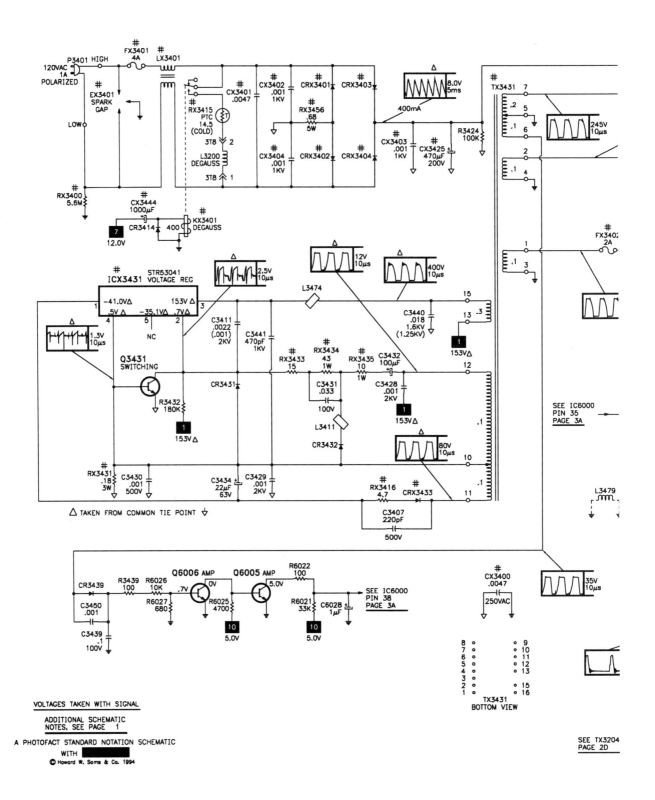

Figure 7-4 Power Supply Schematic

The C-10 Chassis

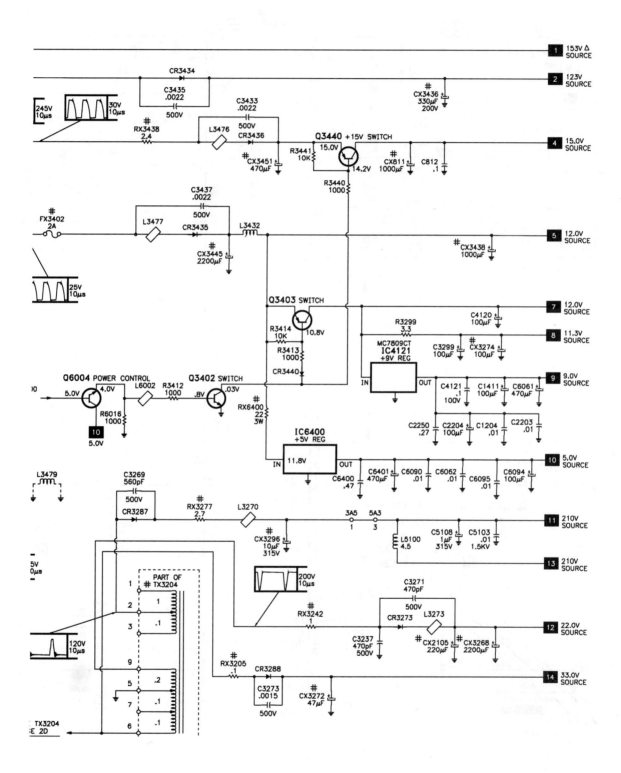

Figure 7-4 Power Supply Schematic (Continued)

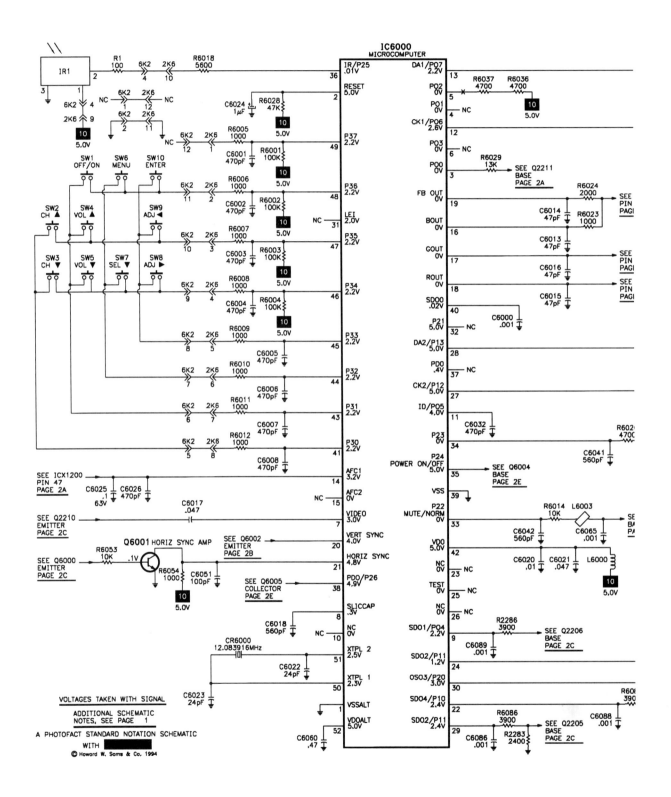

Figure 7-5 System Control Schematic

The C-10 Chassis

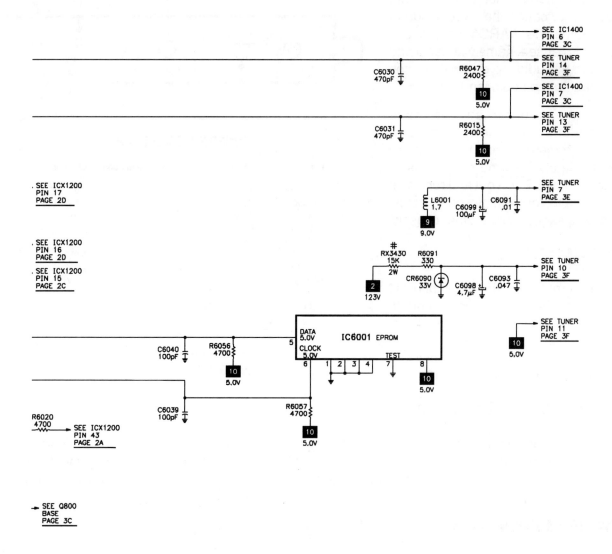

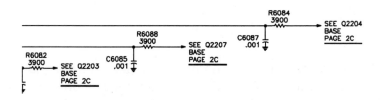

Figure 7-5 System Control Schematic (Continued)

IC6001 is the EAROM, acting as "external memory" for IC6000. It can, and does, cause the same kinds of problems the EAROM in the C-5 chassis caused, namely raster but no picture or audio, no channel control, picture but no audio, and so forth.

If you encounter any of these symptoms and find nothing that seems to cause them, I suggest you immediately become suspicious of the EAROM. Pull it out and replace it if you have, for example, one that you have salvaged from another chassis.

Tuner

Use figure 7-6 and the accompanying voltage chart when you need to troubleshoot possible tuner problems. Don't replace the tuner until you have confirmed that it has its necessary voltages and inputs from the I2C bus.

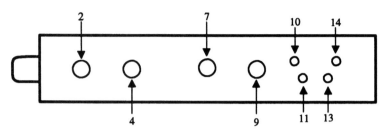

TUNER VOLTAGE CHART

Pin	VHF Low Band	VHF High Band	UHF Band
2 (Tuning)	1.0V	4.0V	6.6V
4 (AGC)	7.3V	7.3V	7.5V
7 (+9V)	9.0V	9.0V	9.0V
9 (IF Out)	0V	0V	0V
10 (33V)	31.8V	31.8V	31.8V
11 (+5V)	5.0V	5.0V	5.0V
13 (CLK)	2.6V	2.6V	2.6V
14 (Data)	2.2V	2.2V	2.2V

NOTE: VHF Low Band voltages taken on channel 2.
VHF High Band voltages taken on channel 7.
UHF Band voltages taken on channel 14.

Figure 7-6 Tuner Information

Deflection Circuits

The deflection circuits (figure 7-7) are, shall I say, "old hat." But I will give you enough information to help you go through them without having to flip back to previous discussions. However, if you need additional information, I suggest you read – or reread – some of the material we have already been over.

Vertical Deflection

Vertical drive leaves pin 28 of IC1200 and enters pin 2 of IC2100. Yoke drive exits on pins 3 and 4 of connector 3Y3. Check for the scan-derived +23 volts B+ at pin 8. If you need to replace the vertical output chip, use the correct one from the popular LA78xx series. Of course the correct part number depends on the module and that, in turn, depends on the size of the picture tube. Use R2112 to adjust vertical height.

Horizontal Deflection

Horizontal drive exits pin 23 of IC1200 and is coupled to the horizontal output transistor via Q3209, Q3202, Q3206, and T2305. The horizontal output transistor, QX3208, responds by developing drive to the flyback and the yoke. The yoke connectors are at pins 1 and 2 of 3Y3. QX3208 (Zenith part number, 121-1148) crosses to an ECG2302, NTE2302, and a SK9422.

Be aware that the flyback, TX3204, does have a history of failure. Its Zenith part number is 95-4372. I don't know if ECG or NTE makes a substitute, though I suspect they do. I usually keep junk modules

The C-10 Chassis

and sometimes don't return duds to Zenith because they are worth a small fortune as a source for used parts. So, I have "previously owned" flybacks to replace the defective ones I find.

Horizontal Holddown. Pin 24 of IC1200 controls the XRP circuit (figure 7-8). I thought about reproducing the schematic from Sams, but I realized it is virtually identical to several XRP circuits we have already examined, for example the XRP circuit in the C-5 chassis. Let me simply note that if you suspect you have a problem with it, lift CRX3006 while you monitor the voltage levels at the collector of the horizontal output transistor. If those voltages appear to be correct and the set doesn't shut down, proceed to troubleshoot the XRP circuit paying particular attention to CRX3006 and CRX3004.

Pincushion Correction. I see no need to reproduce the pin correction circuit because, like the XRP circuit, it is almost exactly identical to the circuits we have already been over.

Video Processor

The video processor, IC1200 (figure 7-8), does a host of jobs. It processes audio and video IF, sync, horizontal and vertical drive, develops RGB signals and drive for the video output transistors, and generates the RGB signals for on-screen display. It is quite similar to the chip used in the C-5 and C-6 chassis. Since I really can't tell you anything new about IC1200, I suggest you read what I have written about them in previous chapters.

Video Output Module

The video output module causes a problem or two now and then (figure 7-9). For instance, you might encounter the TV that has a bright screen with heavy retrace lines. Before you condemn the picture tube, check the luminance driver transistor, Q5106, on the video output module. An emitter-to-collector leak or short is its most common failure. Use an ECG159 if you have to replace it.

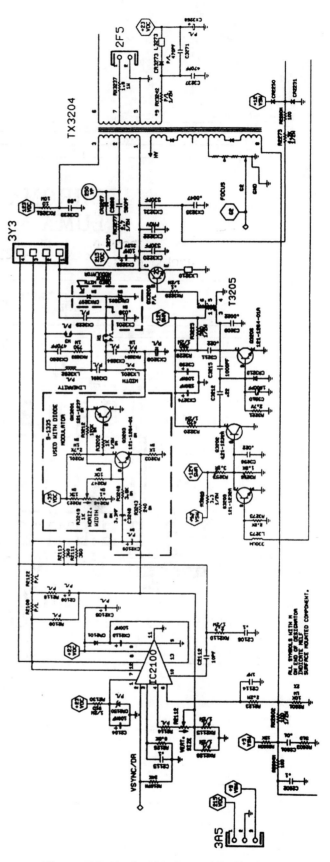

Figure 7-7 Vertical/Horizontal Deflection

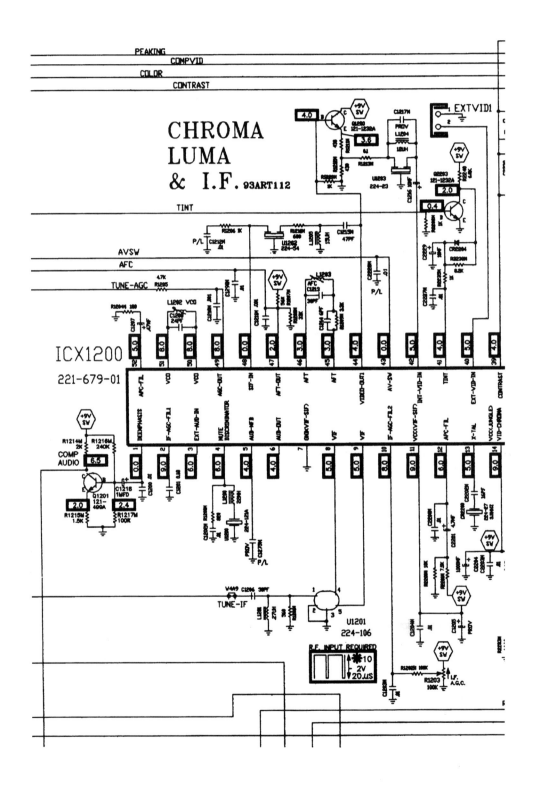

Figure 7-8 Video Processor Schematic

The C-10 Chassis

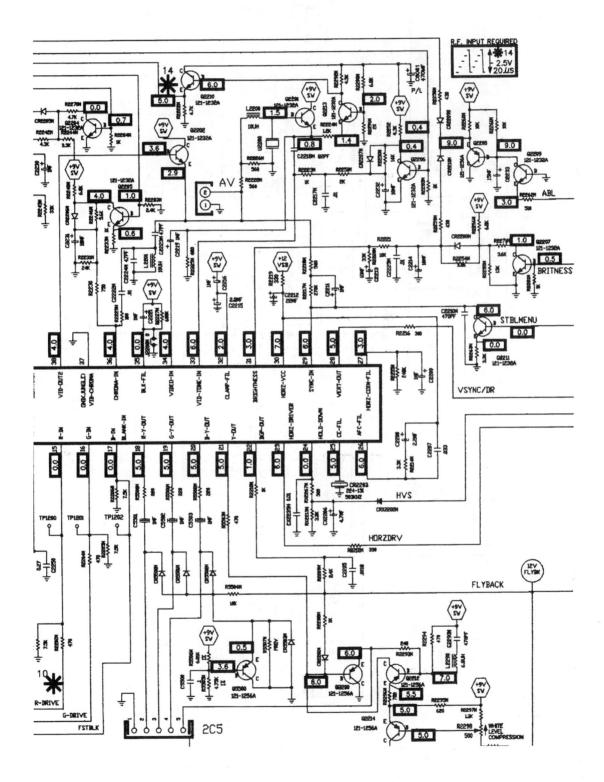

Figure 7-8 Video Processor Schematic (Continued)

Servicing Zenith Televisions

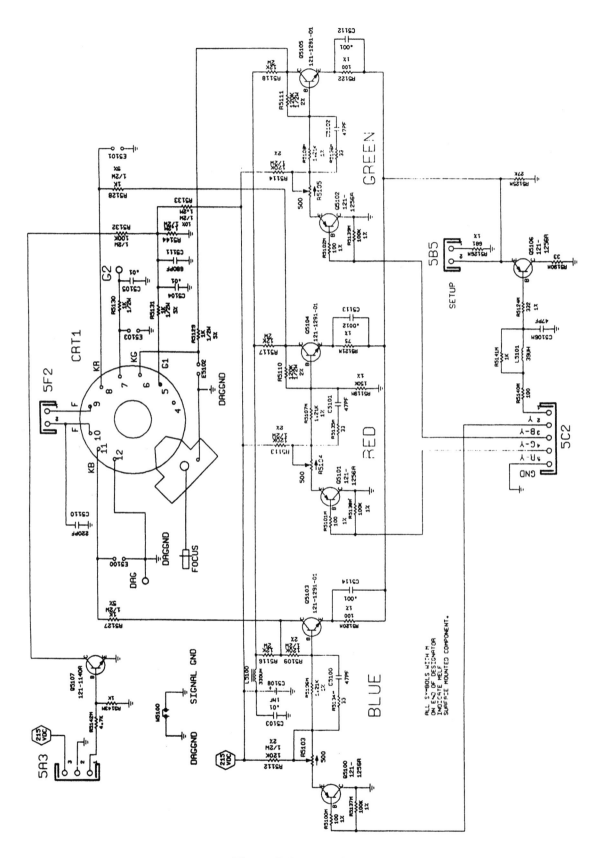

Figure 7-9 Video Output

The C-10 Chassis

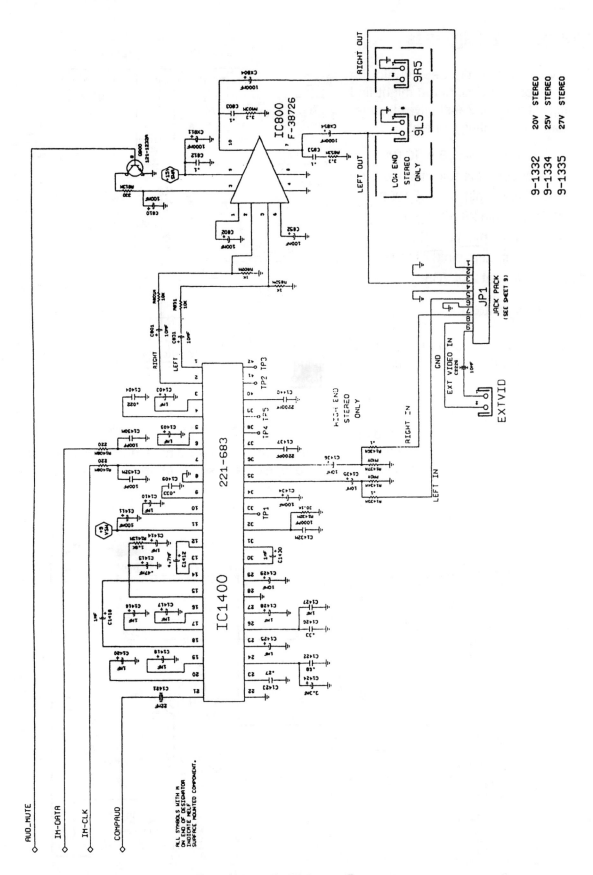

Figure 7-10 Stereo/Audio System

Servicing Zenith Televisions

Operating Signals

When you need to dig deeply into the video processing circuit, you might find this little list helpful. Zenith gives it under the rubric of operating signals:

composite video at pin 38 of IC1200
composite audio at the collector of Q1201
horizontal drive at pin 23 of IC1200
vertical drive at pin 28 of IC1200
blue video output pin 4 of 2C5
green video output at pin 3 of 2C5
red video output at 2 of 2C5

Audio System

IC1400 (figure 7-10) is the heart of the MTS decoder-audio processing circuit in those models equipped with MTS capabilities. Composite audio from IC1200 enters it at pin 21. External audio makes its appearance at pins 35 and 36.

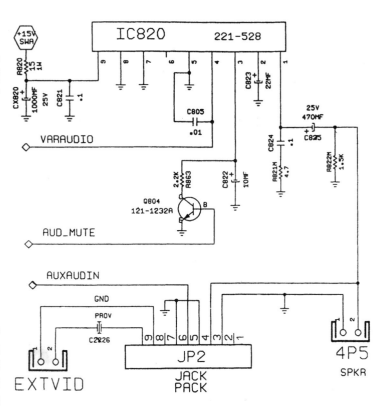

Figure 7-11 Mono Audio System

Left and right channel audio exits at pins 1 and 2 and goes to pins 1 and 2 of the audio power amplifier, IC1800. The audio amplifier boosts the signals and sends them out to the speakers at pins 7 and 10.

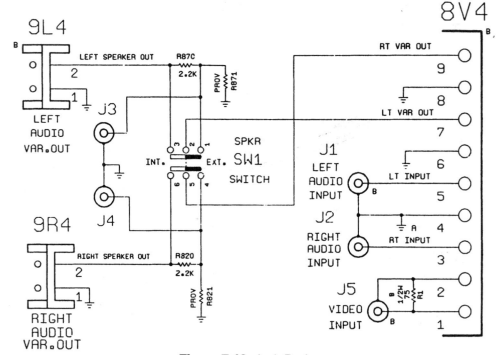

Figure 7-12 Jack Pack

Of course, monophonic audio systems use a different circuit (figure 7-11). The composite audio enters IC1820 at pin 4 and the recovered and amplified signal exits at pin 1 where it goes to the speaker.

Jack Pack

A stereo jack pack is a feature of certain models. It is, of course, attached to the main printed circuit board (figure 7-12). The diagram should be self-explanatory. Cold solder around the pins that connect it to the main chassis has been the only problem the jack pack has ever given me.

Repair History

(1) Power supply. Repair of the power supply follows suggestions I have already given. When you encounter a blown AC fuse, check the bridge diodes for leaks or shorts. Replace defective ones using an ECG125. Then replace the fuse, Q3421 (121-1264-01), and ICX3431 (STR53041). Check the resistors in the power supply and the remaining diodes, replacing any defective components you find. It might be necessary to replace C3432 and C3434. Apply AC, and listen for the familiar chirp. If you hear it, you have fixed the power supply. If you don't hear it, you have missed a defective part.

(2) Dead set. If the components in the power supply seem to be okay, check R3434 and R3436M (a surface mount resistor) for open conditions. If these are good, proceed with the checks to confirm the presence of the switched voltages.

(3) Bright screen with retrace lines. Check Q5101 on the video output module. Replace it with an ECG159 if it is defective.

(4) Jail bars in the picture. Replace C2230 off pin 39 of IC1200.

(5) No vertical deflection. Before you do anything, check the solder around the vertical deflection IC. If the solder connections are good, check for B+. If it is missing, go directly to the fusible resistor in the +23-volt line because it has a history of failure. Replace it with a flameproof resistor only.

(6) No vertical sync. IC4121 may have failed.

(7) Horizontal lines and sync problems. Replace C2212.

(8) Squeal at plug-in. Check the horizontal output transistor. If it is defective, replace it with a Zenith original or one of the generics I mentioned in the text. Be aware that a defective flyback might have caused the output transistor to fail.

CHAPTER 8
THE C-11 CHASSIS

The C-11 chassis is another single printed circuit board chassis that fits into the "System 3" line of televisions (figure 8-2) and is used in 25" and 27" sets only. Since we servicers keep up with Zenith products by module rather than chassis, you'll want to know which modules make up the family. The non-PIP modules include 9-1444-01, 9-1445-01, 9-1447-01, and 9-1449-01. The PIP-equipped ones are 9-1446-01 and 9-1448-01. All have MTS stereo with SAP and surround sound features. The PIP-equipped chassis have the PIP as a soldered-in feature at 4D9, 4G9, and 4C9. As far as I can tell, 9-1551, 9-1552, 9-1553, and 9-1559 also belong to the family.

Ten integrated circuits handle the tuner, audio-video, sync and sweep functions. An additional circuit in the power supply brings the total to eleven. In addition to its duties as the

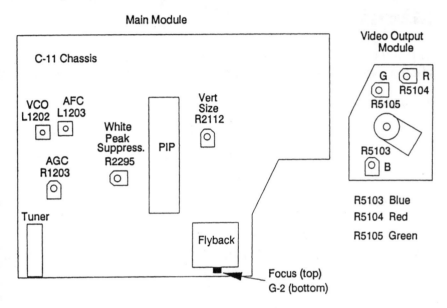

Figure 8-2 C-11 Chassis Printed Circuit Board

system control microprocessor, IC6000 also controls the tuner by developing the PLL tuning and band switching data. IC6001 interfaces the tuner bus line and IC6000. IC6003 is the system reset. IC1200, IC2201, and IC2202 handle audio and video, sync, and sweep processing. IC2100 develops vertical drive for the yoke. IC1400 and IC801 receive, decode, and amplify the audio signals, with IC2200 serving as the audio-video interface. ICX3431 is the now-familiar switch mode regulator. Finally, IC4121 develops the +9 volts from the +12 volts supply.

Chassis Service Controls

Several service controls are mounted directly on the main chassis (figure 8-2). Four are self-explanatory: G-2, focus, vertical size, and AGC. The remaining three should for the most part be left alone because they require special signals and setup procedures if they are to be set correctly. In the unlikely event that you do have to adjust them, consult the appropriate PHOTOFACT for the TV on which you are working for instructions. The Zenith literature doesn't include all the information the servicer needs to make those adjustments. But, you still need to know what they are and what they do.

L1202 is a variable coil used to adjust the VCO (voltage controlled oscillator) that controls the oscillator responsible for detecting the IF video. If it is misadjusted, it causes the circuit to output a signal that creates a scrambled picture with scratchy audio. I have lucked out a time or two and been able to set L1202 for stable picture and audio by turning it perhaps an eighth of a turn, but I held my breath when I did. L1203 sets the automatic frequency control (AFC) of the IF video detector. Third, R2295 adjusts white peak compression. When properly adjusted, the white peak compression circuit prevents saturated white signals from overdriving the video and CRT circuits.

Available Literature

I know of just one publication covering the C-11, *Technical Training Program: C-11 System 3*. It has some useful circuit information and readable schematics and is worth the price particularly if you take many C-11s into your shop for repair. If you know of additional service-related material, please let me know because I would like to add it to the bibliography.

Menus

The C-11 chassis has, of course, two menus. Since I have dealt extensively with both in previous chapters and because there is nothing really new here, my discussion will be brief.

Customer Menus

The customer menu is six pages long, devoting a page each to these topics: Source, Setup, Audio, Video, PIP Video, and PIP Source. See table 8-1 for an explanation of the items contained in each menu page.

Service Menu

The service menu contains entries that we have seen and discussed in previous chapters. When you get into it, you find that the display corresponds to the one in the diagram in table 8-2 (26 items in two columns on one page). I include in table 8-2 a brief explanation of each item and its typical setting. Disregard the last item at the bottom of column two because it is for factory use only.

Zenith service menus provide the servicer and customer with interesting choices which is one of the reasons I spend a lot of time talking about these menus. For example, consider "Start Chan." If a customer uses a cable box and wants the TV to come on to channel 3 when the set is turned on, merely turn the set on and tune in channel 3, access the service menu, and turn "Start Chan" on. The TV now automatically tunes in channel 3 every time it is turned on regardless of the last channel used.

Power Supply

Guess what? This is the power supply Zenith has been using for quite some time. I intend, therefore, to limit my discussion just to the details necessary to keep you from having to flip pages looking for information given in previous discussions.

Figure 8-3 lays out the AC input and bridge circuit responsible for developing the raw B+ (usually about +150 volts) while figure 8-4 displays schematically the rest of the power supply. When AC is applied, the power supply comes alive and makes available +123 volts for horizontal deflection and +12 volts DC standby (CR3435). IC6003, which is not shown, develops the +5 volts standby for the system control and IR receiver.

An on command from the front panel controls or the remote control causes IC6000 to output a high at pin 35 that is coupled to the power supply from the emitter of Q6004 (figure 8-5). The signal from

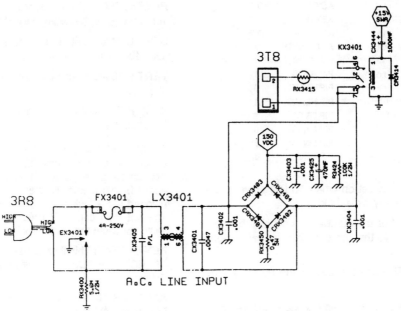

Figure 8-3 AC Input and Bridge Circuit

SOURCE MENU

Allows the selection of CABLE, ANTENNA or AUXILIARY input.

SETUP MENU

AUTO CH. SEARCH: Finds the available channels and stores them in memory for access by using CHANNEL (CH) Up/Down.

CH ADD/DEL: Changes the list of active channels selected by using channel (CH) Up/Down.

CH LABELS: Allows the addition of identification such as CBS, NBC, HBO etc... to the on screen channel read out.

TUNING BAND: Determines the operation of the channel tuner inside the TV.

AUTO FINE TUNE: Lets your TV compensate for variations in broadcast and cable TV frequencies

CLOCK SET: Sets the clock in the receiver to the correct time.

CAPTIONS: Displays closed captions (CC) or informational text when available.

CH BACKGROUND: A background can be added to the on screen channel read out. Normally the read out is over the video.

AUDIO MENU

BASS: Adjusts the low frequencies.

TREBLE: Adjusts the high frequencies.

BALANCE: Adjusts the balance between the left and right channel audio output.

AUDIO: Allows the selection of Stereo, Mono, or 2nd Audio/Sap.

SEQ: The selections are on or off. In the "ON" mode the Stereo signal is enhanced to create a full rich room filling sound.

VIDEO MENU

CONTRAST: Adjusts the overall contrast level of the picture.

BRIGHTNESS: Adjusts the brightness level of black areas of the picture.

COLOR: Adjusts the intensity of the colors in the picture.

TINT: Adjusts the color of the flesh tones.

SHARPNESS: Adjusts the video response for the best picture.

PICTURE PREF: Lets the viewer decide if he wants to use his own CUSTOM video settings, or the factory PRESET video settings.

PIP VIDEO MENU

CONTRAST: Adjusts the contrast of the insert picture.

COLOR: Adjusts the intensity of the color in the insert picture.

TINT: Adjust the color of the flesh tones in the insert picture.

PICTURE PREF: Lets the viewer decide if he wants to use his own CUSTOM video settings, or the factory PRESET video settings for the insert picture.

PIP SOURCES MENU

ANT/CABLE, VIDEO 1: When the source key is pressed in the PIP mode, the PIP sources menu will appear. Pressing the source key again will toggle the source between ANT/CABLE, and VIDEO 1. The Left/Right ADJUST and SELECT Up/Down keys will also do the same function.

Table 8-1 Customer Menu Explanation

Servicing Series: Zenith Televisions

→913-1-1.04		AC ON	OFF
PRESET PX	STORD	KEY DEFT	OFF
BRIGHTNESS	15	A/V LOCK	OFF
CAP PHASE	64	AUTO PRGM	2
VERT POS	12*	PIP VPOS	7
HORZ POS	20**	PIP HPOS	11
PIP AVAIL	ON	PIP VSPAC	8
ZEN/PL	ZENITH	PIP HSPAC	32
CHAN LOCK	OFF	PIP FR CT	8
START CHAN	OFF	PIP MAC W	0
MAX VOLUM	OFF	PIP SWITCH	5
MIN VOLUM	OFF	PIP L DLY	16

AC ON - Turns the set on when AC line power is applied. The power key is disabled.

AUTO PRGM - Set the number of channels between channels 14 and 48 that are searched for cable signal during auto search.

AUX CHAN - Enables the use of the auxiliary input jacks.

A/V LOCK - Disables all audio and video inputs - gives blue screen and menus only.

BRIGHTNESS - Limits brightness control range.

CAP PHASE - Sets up the operation point of the closed caption decoder.

CHAN LOCK - Locks the set to one channel use only.

HORZ POS - Sets the horizontal position of the on screen displays.

KEY DEFT - Disables menu use keyboard keys: MENU, ENTER, ADJUST, and SELECT.

MAX VOLUM - Sets the maximum volume limit.

MIN VOLUM - Sets the minimum volume limit.

TEXT MODE - Disables the closed caption text mode operation.

PIP AVAIL - Turns the PIP feature ON or OFF.

PIP VPOS - Upper left PIP frame vertical position relative to the upper left corner.

PIP HPOS - Upper left PIP frame horizontal position relative to the upper left corner.

PIP VSPAC - Other PIP frames vertical position relation to the upper left corner frame.

PIP HSPAC - Other PIP frames horizontal position relative to the upper left corner frame.

PIP FR CT = Sets PIP frame contrast - 0 = BLK ; 15 = WHT - non-service setting only.

PIP MAC W - PIP macro width - non-service setting only.

PIP SWITCH - PIP switching - non-service setting only.

PIP L DLY - PIP luma delay - non-service setting only.

PRESET PX - Used to store changes to the customer video menu as the settings for "PRESET".

STRT CHAN - Sets the tuned start channel when the set is turned on.

VERT POS - Sets the vertical position of the on screen displays.

ZENITH/PL - Sets the IR receiver decode to Zenith or Private Label.

AC ON: (AC Power On) This is the first item at the top of the service menu column 2. In the "on" mode, the set can be turned on automatically when AC power is applied. This feature is used at times when the set is plugged into a switched AC power outlet or cable box. The set than can be turned on and off with the AC switch or when the cable box is turned on.

KEY DEFEAT: This item is used to disable the MENU, SELECT, and ADJUST keys on the front panel of the TV set.

A/V LOCK: In the "on" mode, all video and audio (Tuner or Aux) is blocked and a blue screen is substituted. If MENU is activated, it will be displayed on a Black background. This item is for commercial use. For normal TV, leave this in the "off" mode.

AUTO SRCH: (Auto Search) This item is used to improve the action of the auto search feature. The range of 0 to 18 is the number of channels above 14 that the system will look at, to determine if it is broadcast or cable. Set the auto search between 2 and 7 for most cable systems. Zero is cable and 18 is broadcast.

PIP TEST: This is for factory use only. The range of this adjustment is from 0 to l5. Set this to 0.

PIP H POS: This adjustment moves the insert picture horizontally. The range is 0 to 7. Zero moves the picture to the left and 7 to the right. Set this to 4.

PIP VRAMW: (Picture In Picture Video RAM Write) This is for factory use only. The range is 0 to 15. Set this to 8.

PIP VRAMR: (Picture In Picture Video RAM Read) This is for factory use only. The range is 0 to 7. Set this to 4.

PIP FR CT: (Picture In Picture Frame Contrast) This changes the frame around the insert picture from white to black. The range is 0 to 127. Set this at 45, which gives a gray frame around the insert picture. Do not use 127 or white, as this may damage the phosphor of the crt if left on for long periods of time.

PIP GRAY: This is for factory use only. The range o this adjustment is from 0 to 127. Set this to 64

PIP SWITCH: (Picture In Picture Switch) This is for factory use only. The range is from 0 to 15. Set this to 10.

C10 AUX: This is not available on the C11. Set this to off.

Table 8-2 Service Menu Explanation

The C-11 Chassis

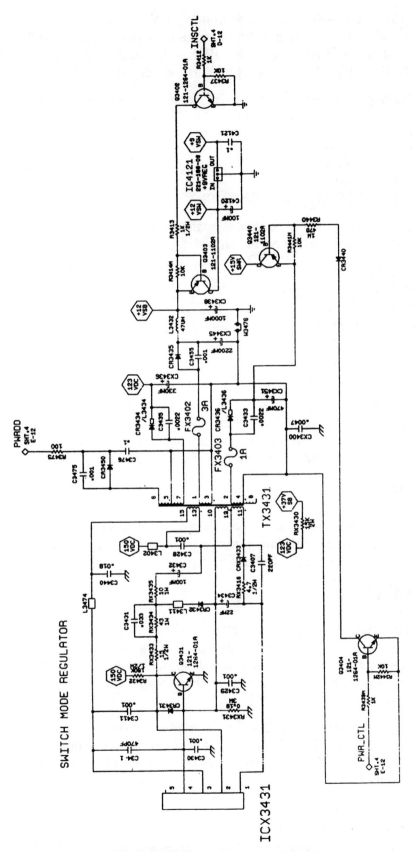

Figure 8-4 Power Supply Schematic

Q6004 turns on Q3404 and Q3440 (figure 8-4) bringing on-line the switched voltages +5 volts, +9 volts, +12, volts, and +15 volts. The remaining voltages necessary to operate the TV come up with horizontal deflection, namely +23 volts, +35 volts, +245 volts, focus, G-2, and EHT. The switched +15 volts also turns on the degauss relay (KX3401) to start the degauss cycle.

To aid your troubleshooting, check the following points for standby voltages. I do not list these voltages in any particular order.

+5 volts at pin 7 of IC6003
+12 volts at the emitter of Q3403
+15 volts at the emitter of Q3440
+150 volts at the junction of CRX3403 and CRX3404

Check the following points for switched voltages. Again, I am providing you with a list of voltages to check without listing the voltages in any particular order.

+5 volts at pin 6 of IC6003
+9 volts at pin 3 of IC4121
+12 volts at the collector of Q3403
+15 volts at the collector of Q3440
+23 volts at the cathode of CR3273
+123 volts at the cathode of CR3434
+245 volts at the cathode of CR3287

Do I need to remind you that these voltages probably vary slightly from chassis to chassis? So, look for approximate values when you check them. However, get suspicious if the voltages you check are very far off. Use your judgment as you move through the list. For instance, the +5 volts standby should be at +5 volts, but I'd give the others no more than a five percent margin of error.

Microprocessor

IC6000 (figure 8-5) is the now familiar 52-pin CMOS processor in a DIP package that we have seen on several occasions. If you need additional information, check the last chapter because early C-10 chassis use the same processor the C-11 chassis uses.

Deflection Circuits

I suspect the deflection circuits (figure 8-6) are also familiar to you by now. If they aren't, they should be! Vertical drive exits pin 28 of IC1200 and enters pin 2 of the vertical deflection IC (IC2100). It is powered by +23 volts at pin 8 and outputs vertical drive at pins 1 and 2 of connector 2Y2. R2112 controls vertical size. As you have seen in previous chassis, IC2100 belongs to the LA78xx series of vertical output ICs.

Horizontal drive comes out of pin 23 of IC1200 and is coupled to horizontal output transistor (QX3208) by Q3209, Q3202, Q3206, and T3205. Pins 1 and 3 of connector 3Y3 couple drive to the yoke. The horizontal output transistor is a TZ1148-01 and crosses to the now familiar ECG2302 and NTE2302 group of generic transistors.

Video Circuit

IC1200 (figures 8-7a and b) is the workhorse of the video circuit, doing what IC1200 typically does in a Zenith television. It has a great deal in common with the processor used in the C-10 chassis. The microprocessor receives commands via the front controls or the remote control and sends the customer information to it using pulse-width modulated signals.

The horizontal section employs the usual dual, PLL loop system with a horizontal coincidence detector to control the gain of the phase detector in order to optimize performance of the horizontal section. The coincidence detector also permits fast acquisition of the signal during channel change or loss of signal conditions.

The vertical section utilizes a narrow range countdown circuit. Its operating range is within +/10 percent at 60 hertz. The operating parameters should sound familiar since Zenith uses the same circuit in other chassis, but there is a difference. This system defaults to 60 hertz in the absence of a signal. The difference in circuit design results in minimal change of vertical size and on-screen menu position in absence of signal conditions.

The luminance processing section doesn't markedly differ from other Zenith designs. Video from the tuner enters IC1200 at pins 8 and 9. The processed video exits at pin 44. It passes through an emitter follower configuration (Q1200) and reenters the chip at pin 3 (figure 8-7a). The video exits again at pin 14 and goes through another emitter follower (Q2212). If the chassis doesn't have a comb filter, the signal passes through Q2203, Q2204, Q2205, and Q2210 before reentering IC1200 at pin 34. If the chassis has a comb filter and PIP, the output of Q2212 goes to the delay line (DL2200) before it is applied to pins 1 and 3 of 4D9 on the PIP module (figure 8-7b). The PIP processes the video via transistors Q2206 and Q2207 and sends it on to pin 6 of 4D9. The main and PIP luminance signal exits the PIP at pin 3 of 4C9 and makes its way to pin 34 of IC1200. The main and PIP chroma signal exits the PIP module on pin 5 of 4C9 on its way to pin 39 of IC1200.

The Y signal goes out of pin 21 while the R-Y, G-Y, and B-Y signals go out at pins 18, 19, and 20. Sound IF enters IC1200 at pin 48. The composite audio output exits at pin 1. As you may have guessed, IC1200 handles video and audio switching by sensing the voltage applied to pin 43. That is, the voltage on pin 43 determines whether IC1200 processes the tuner signal or signals from the jack pack.

Operating Signals

When you have to troubleshoot the circuit, look for the following signals at the designated points.

composite video out at pin 21 of IC1200
composite audio out at the collector of Q1201
horizontal drive at pin 23 of IC1200
vertical drive at pin 28 of IC1200
blue drive out at pin 4 of 2C5
green drive out at pin 3 of 2C5
red drive out at pin 2 of 2C5

Video Output Module

The video output module is a breakaway portion of the main printed circuit board (figure 8-8). Luminance information enters at pin 1 of 5C2 while chroma information enters at pins 2, 3, and 4. Q5106 amplifies the luma signal before it is applied to the emitters of the three video driver transistors. Because it is a key player in the circuit, Q5106 not only can, but also does, cause problems. When the screen turns white and has heavy retrace lines in it, be sure to check it for an emitter to collector leak. Think in terms of replacing it with an ECG159 or NTE159.

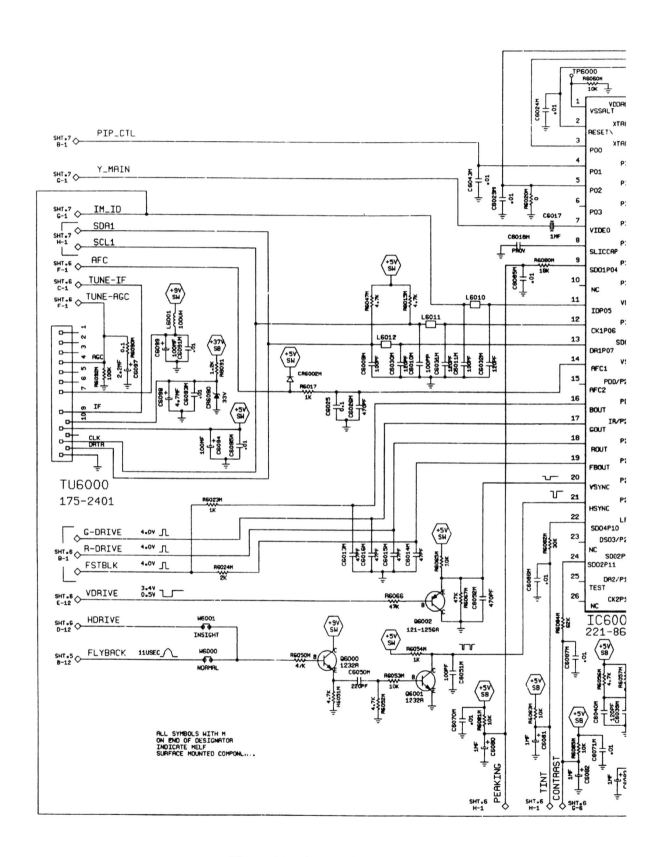

Figure 8-5 IC6000 Microprocessor

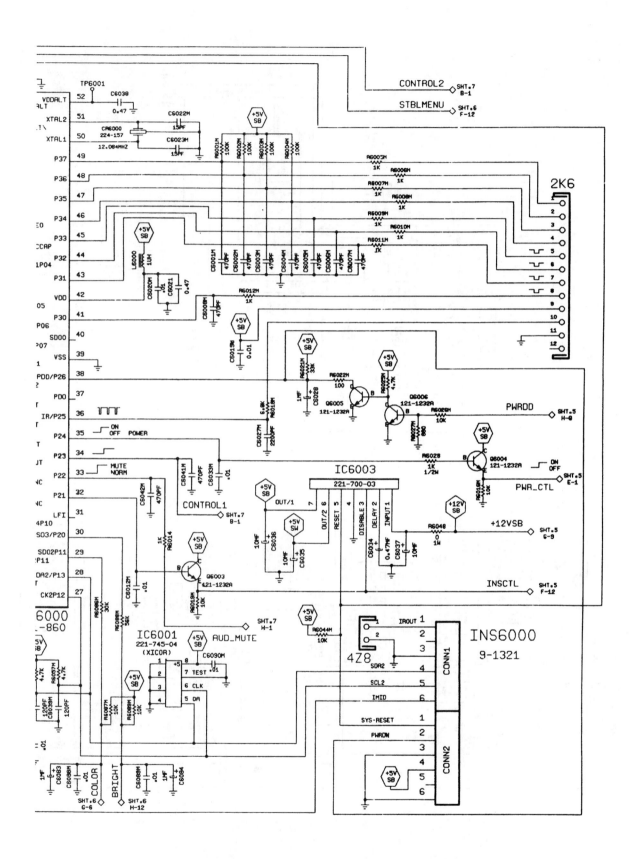

Figure 8-5 IC6000 Microprocessor (Continued)

Servicing Series: Zenith Televisions

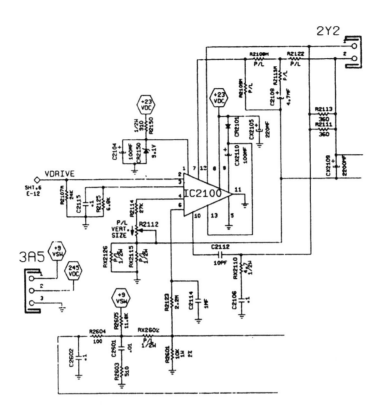

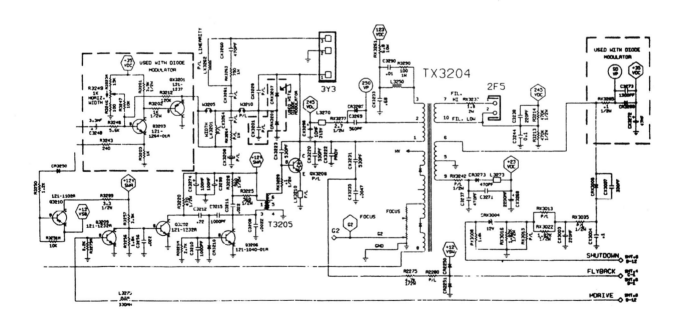

Figure 8-6 Vertical (Top) and Horizontal (Bottom) Deflection Circuits

The module needs several voltages to work – all of which you may check at the following pins of connector 5A3: +9 volts at pin 2; +245 volts at pin 3; and filament voltage at pin 3.

Audio Circuits

IC1400 (figure 8-7b) processes the baseband audio. The audio enters at pin 21. External audio inputs are available at pins 35 and 36. Left and right channel audio exits at pins 1 and 2 on its way to IC1801 which is the audio amplifier that amplifies the signals and sends them out at pins 7 and 10 to the left and right speakers.

Jack Pack

The jack pack resembles those we have seen time after time. It features a stereo jack pack for processing two external audio and video signals. The Internal/External switch (SW1) directs left and right audio outputs to jacks J1 and J2 for connection to external speakers or the internal speakers. Audio enters the system from external sources at jacks J3, J4, J5, and J6. Jack J9 is the audio output for the surround sound system. Jacks J7 and J8 are for external video. The C11 chassis does not have provisions for video out.

Repair History

Even though the repair history for the C-11 pretty much follows that of the modules we have discussed in previous chapters, I have a few things I would like to pass along.

(1) Dead power supply. Use any one of a number of previous discussions to help you troubleshoot a dead power supply.

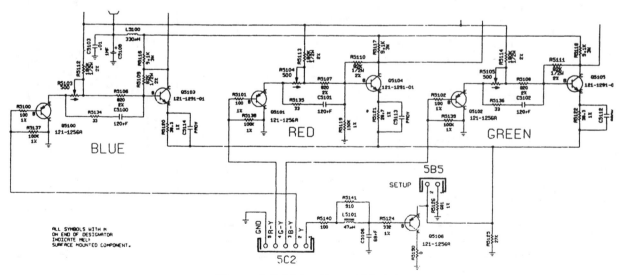

Figure 8-8 Video Output Module

Servicing Series: Zenith Televisions

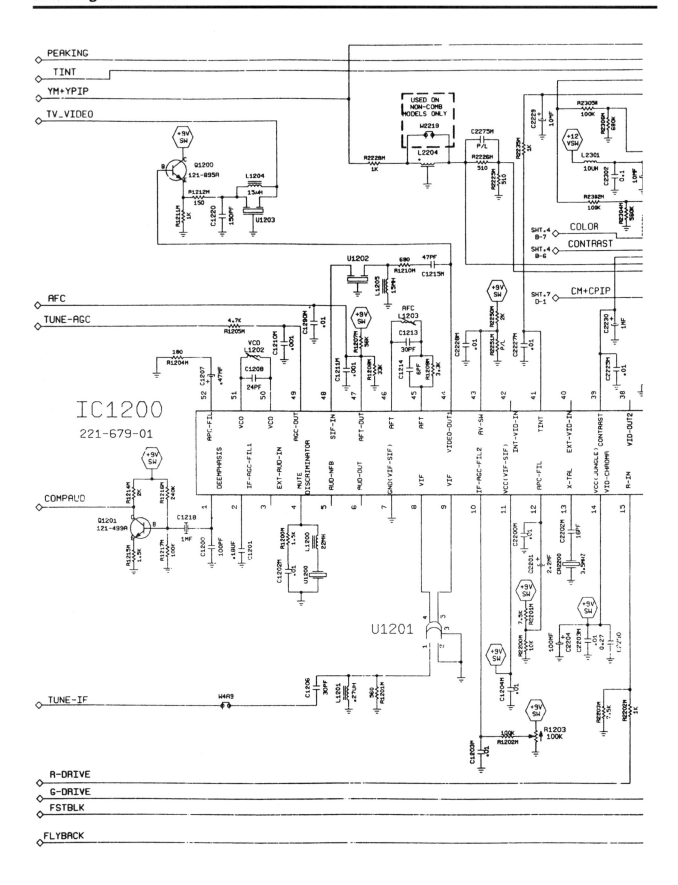

Figure 8-7a Video Circuit

The C-11 Chassis

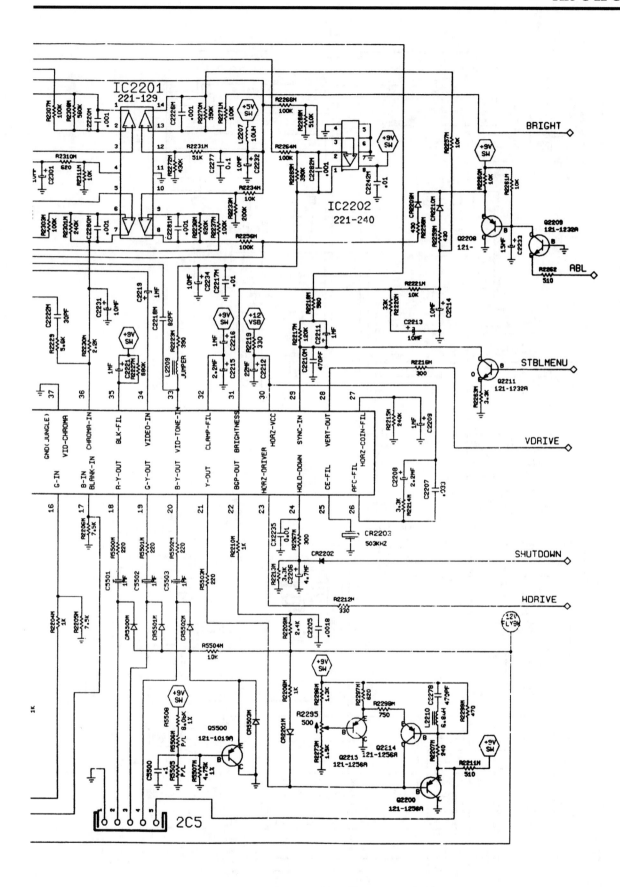

Figure 8-7a Video Circuit (Continued)

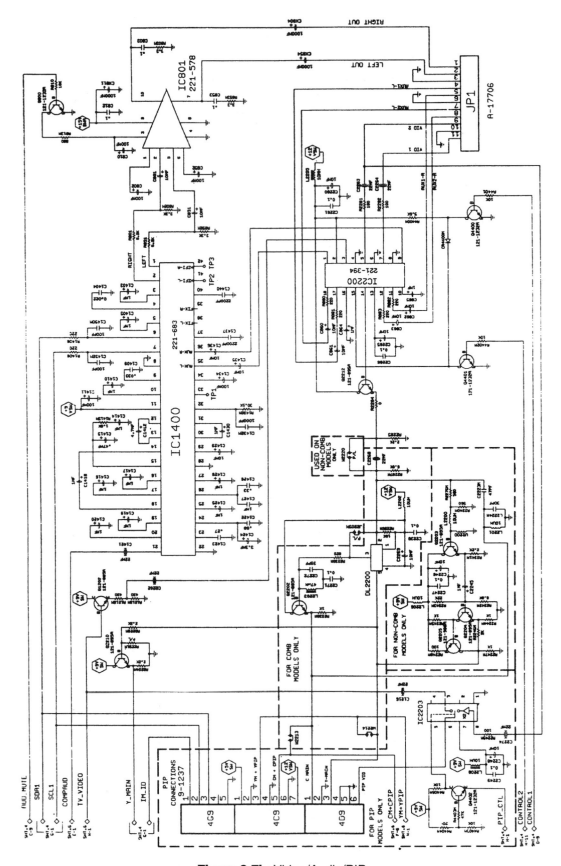

Figure 8-7b Video/Audio/PIP

The C-11 Chassis

(2) No vertical deflection. Check for loose connections at IC2100 and the vertical yoke plug. Make sure to check the diode and capacitor in the B+ line for the vertical output IC before you begin to change parts. Don't overlook the possibility of an open fusible resistor in the +23-volt B+ line. It is designated RX3242 and is located near the flyback. If the IC is bad, use an LA78xx as a replacement. The exact number depends on the screen size.

(3) Bright screen with retrace lines. Check Q5106 on the video output module. If it is defective, replace with the generics I have already mentioned.

(4) No audio and no video. Check the solder connections around the PIP module. They have a modest history of failure.

(5) Intermittent tuner problems. If tapping on the tuner makes the signal come and go, remove the tuner shields and check for broken solder connections where the printed circuit board solders to the frame. The problem will most likely be broken ground connections at one or more of these points.

(6) Great audio but no video. The soldered-in PIP is notorious for causing the C-11 to lose video. Since it is soldered onto the main chassis, check for video in and no video out. You might luck out and repair the PIP if it has failed because of defective electrolytic capacitors. Use an ESR meter to locate them.

CHAPTER 9
THE C-12 CHASSIS

The C-12 first came out as a part of Zenith's R-Line and is like most of the chassis Zenith has manufactured in the last several years. Built on a single circuit board, it depends on a few large-scale integrated circuits to do its work. Figure 9-1 will give you an idea about how the module looks even though it is a part of a PHOTOFACT Gridtrace, while figure 9-2 will assist you in locating key components and test points.

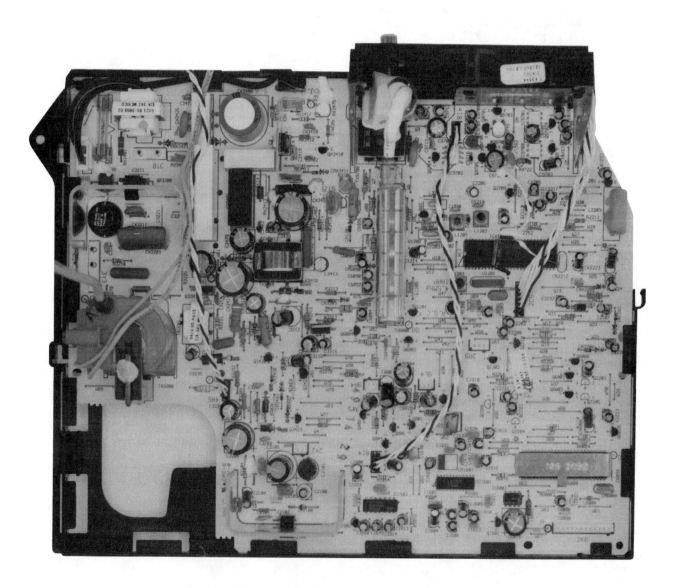

Figure 9-1 PHOTOFACT Gridtrace of C-12 Chassis

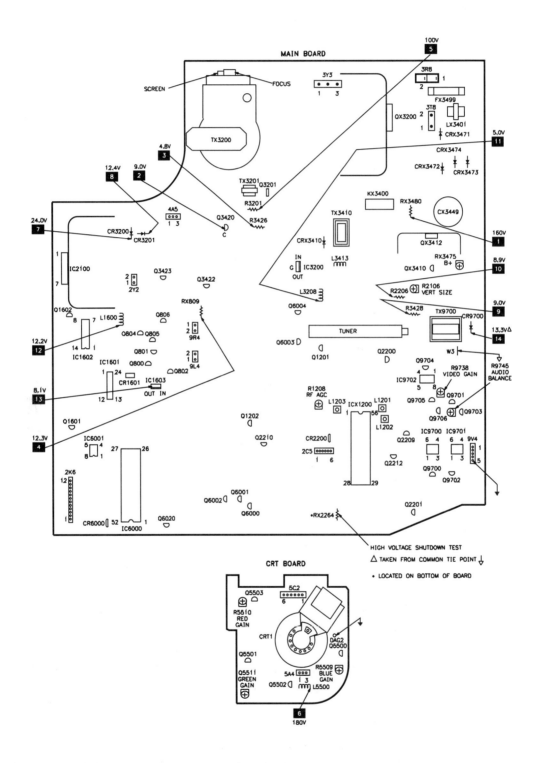

Figure 9-2 Placement Chart and CRT Board

The C-12 Chassis

The following modules make up the family of chassis:

9-1555 for 19/20-inch mono sets without a jack pack
9-1555 for 19/20-inch stereo sets with a jack pack
9-1557 for 25-inch mono sets without a jack pack
9-1558 for 25-inch stereo sets with a jack pack
9-1558-02 for 25-inch mono sets without a jack pack.

The C-12 is the last chassis I intend to discuss before getting to the Y-Line and the beginning of the "newest of the new." Since a great deal of its circuits are kind of a carryover from previous designs and have been discussed almost ad nauseam in previous chapters, you may anticipate a relatively brief discussion, accompanied by sufficient data to help you with the repairs you need to make.

Available Literature

I have not been able to locate any technical training literature for the C-12 chassis. I am not saying there isn't any. I just haven't been able to locate it. I do have access to Zenith's service literature however, and parts of it are reproduced here. In the event you are interested in the factory service literature, you may locate it or order it by using the number, CM-147/C-12(A). Since it is not as detailed as I personally like technical literature to be, I have supplemented it with PHOTOFACT 3448 for model SMS1943S. PHOTOFACT 3448 also covers models SMS1943SM, SMS1943S7, SMS2053S, SMS2053SM, SMS2053S7, and SMS2053W.

Service Menu And Chassis Adjustments

Figure 9-3, taken from Zenith's factory service literature, serves several functions. It tells you how to access the service menu, what you find when there, a list of the items in the menu, and how to adjust those items. Note that the literature makes reference to "early production" and "late production" service menus. The difference between the two is that the one has more/different items in it that the other doesn't – a phenomenon we have witnessed on several occasions. It also lists the controls that you may have to adjust from time to time and tells you how to make the adjustments. If you require additional information about the service menu, you may consult *Technical Training Program: Service Menus D Line Through R Line*.

Power Supply

You have seen the power supply before, haven't you (figure 9-4)? In case you don't remember or began your reading with Chapter 9, flip back a few pages to Chapter 5.

R3426, CR3493 (5.1 volt zener), C6019, and C6098 develop standby B+ for the microcomputer and IR receiver (PHOTOFACT source 3). Q3420 is a switch that turns on the +9-volt source (PHOTOFACT source 2) that powers the horizontal oscillator and initiates horizontal scan. When horizontal deflection becomes available, the +12.4 scan-derived voltage (source 12) becomes the source for the +4.8 volts and the +9 volts because the standby voltage source cannot provide the necessary current to operate these circuits by itself. CR3488 isolates the two sources. If it becomes leaky, expect the +4.8 volts to drop below the level the set needs to operate.

The main power supply becomes active when AC is applied and produces +100 volts (source 5) for the horizontal deflection circuit and a switched +12.3 volts (source 4) for the audio circuit. The C-12 makes use of a "buck regulator" type of switching power supply. The term "buck regulator" usually

Servicing Zenith Televisions

SERVICE MENU:

NOTE! SERVICE MENU DISPLAYS ONE ITEM AT A TIME, IN SEQUENTIAL ORDER. AND MOST ITEMS ARE NOT INTENDED TO BE FIELD ADJUSTMENTS. IF SET INCORRECTLY, MAY CAUSE WHAT APPEARS TO BE A DEFECTIVE MODULE., ALWAYS CHECK THE SERVICE MENU SETUP, (IF THERE IS VIDEO,). BEFORE MAKING ADJUSTMENTS, OR CHANGING MODULES, READ AND UNDERSTAND THE SERVICE MENU TERMS LISTED BELOW

MAKE SPECIAL NOTE OF THE FOLLOWING (CONTROLS) AND WHAT THEY DO WHEN TURNED ON.

| AC PWR ON. | A/V LOCK. | MAX VOLUM. | MIN VOLUM. |
| KEY DEFT. | CHAN LOCK. | STR CHAN. | ZEN/PL. |

MENU TERMS GLOSSARY

AC PWR ON TURNS THE SET ON WHEN A.C. LINE POWER IS APPLIED. THE POWER KEY IS DISABLED.

AUX CHAN ENABLES THE USE OF THE AUXILIARY INPUT JACKS.

A/V LOCK DISABLES THE USE OF THE AUXILIARY INPUT JACKS.

BAND/AFC ALLOWS MANUAL SETTING OF THE TUNING BAND, AND A.F.C. SELECTION FOR SPECIAL SIGNAL CONDITIONS, WHERE AUTO SEARCH DOES NOT SELECT THE REQUIRED BAND OR A.F.C. SETTING. THERE ARE EIGHT POSSIBLE SETTINGS FOR THIS CONTROL.

0 - BROADCAST BAND, A.F.C. OFF   4 - BROADCAST BAND, A.F.C. ON
1 - CATV BAND, A.F.C. ON   5 - CATV BAND, A.F.C. OFF
2 - HRC CATV BAND, A.F.C. ON   6 - HRC CRTV BAND, A.F.C. OFF
3 - IRC CATV BAND, A.F.C. ON   7 - IRC CATV BAND, A.F.C. OFF

BRIGHTNESS LIMITS BRIGHTNESS CONTROL RANGE.

CAP PHASE SETS THE OPERATION POINT OF THE CLOSED CAPTION DECODER

CHAN LOCK LOCKS THE SET TO ONE CHANNEL ONLY

HORZ POS SET THE HORIZONTAL POSITION OF ON SCREEN DISPLAY

KEY DEFT DISABLES MENU USE KEYBOARD KEYS: MENU, ENTER, ADJUST, AND SELECT.

MAX VOLUM SETS THE MAXIMUM VOLUME LIMIT.

MIN VOLUM SETS THE MINIMUM VOLUME LIMIT.

TEXT MODE DISABLES CLOSED CAPTION TEXT MODE OPERATION.

PRESET PX STORES CHANGES TO THE CUSTOMER VIDEO MENU AS PRESET, SETTINGS

STRT CHAN SETS THE START CHANNEL, WHEN THE SET IS TURNED ON

VERT POS SETS THE VERTICAL POSITION OF THE ON SCREEN DISPLAYS.

ZEN/PL SETS THE I.R. RECEIVER DECODER TO ZENITH OR PRIVATE LABEL.

ENTERING THE SERVICE MENU USING THE REMOTE:
PRESS AND HOLD THE MENU KEY UNTIL THE CUSTOMER MENU DISAPPEARS THEN, PRESS KEYS 9, 8, 7, 6, AND ENTER QUICKLY. THE SELECT KEY MOVES THE ARROW AND HIGH-LITE TO THE SERVICE ITEM (CONTROL) AND THE ADJUST KEY IS USED TO MAKE THE CONTROL CHANGE.

ENTERING THE SERVICE MENU USING THE KEYBOARD:
PRESS AND HOLD THE MENU KEY UNTIL THE CUSTOMER MENU DISAPPEARS THEN, PRESS AND HOLD THE ADJUST RIGHT AND CHANNEL UP KEYS AT THE SAME TIME.
THE SELECT KEY MOVES THE ARROW AND HIGH-LITE TO THE SERVICE ITEM (CONTROL) AND THE ADJUST KEY IS USED TO MAKE THE CONTROL CHANGE.

SERVICE MENU DISPLAY
LATER PRODUCTION (221-949-01)

```
5 HORZ POS  34
xx/xx/xx-xx
```

SERVICE MENU ITEMS (CONTROLS) AND SETTINGS:

949-1-1.1	0	MAX VOLUM	0
PRESET PX	1	MIN VOLUM	0
BRIGHTNESS	10	AC PWR ON	0
CAP PHASE	95	KEY DEFT	0
VERT POS	14	A/V LOCK	0
HORZ POS	34	BAND/AFC	0
ZEN/PL	0	AUX CHAN	1
STRT CHAN	0	TEXT MODE	1

SERVICE MENU DISPLAY
EARLY PRODUCTION (221-898)

```
→ 898-1.XX             AUX      ON
  PRESET PX    STORD
  BRIGHTNESS   11
  VERT POS     17
  HORZ POS     31
  ZEN/PL       ZENITH
  CHAN LOCK    OFF
  STRT CHAN    OFF
  MAX VOLUM    OFF
  MIN VOLUM    OFF
  AC ON        OFF
  KEY DEFT     OFF
  AUTO PRGM    2
```

ADJUSTMENTS: 94ART063

REFER TO THE EASY ACCESS COMPONENT LAYOUT
(ALLOW FOR A MINIMUM OF 15 MINUTES WARM UP)

G2 (MAY BE REQUIRED WHEN REPLACING PARTS)
OFF AIR OR CROSS HATCH, (STRONG, AND NOISE FREE.) REDUCE TO MINIMUM, (VIDEO MENU CUSTOMER CONTROLS), CONTRAST, BRIGHTNESS, COLOR, AND G2 THEN, RAISE G2, TO SEE RETRACE LINES, THEN REDUCE G2 UNTIL PICTURE WHITE HIGHLIGHTS ARE JUST BARELY VISIBLE. AT THE MENU, GO TO PRESET, AND, ADJUST (VIDEO MENU CUSTOMER CONTROLS) (CONTRAST, BRIGHTNESS, COLOR.) TO MATCH THAT OF PRESET.

RED, GREEN, BLUE, GAIN (BIAS)   (COLOR SETUP)
(MAY BE REQUIRED WHEN REPLACING PARTS OR ADJUSTING G2)
OFF AIR OR CROSS HATCH, (STRONG, AND NOISE FREE.) REDUCE TO MINIMUM, RED, GREEN, BLUE, VIDEO OUTPUT CONTROLS. AND COLOR LEVEL, (MENU CUSTOMER CONTROLS). ADJUST ONLY THE TWO NON-PREDOMINANT CONTROLS, TO OBTAIN COLOR FREE, (NO TINT) WHITE AND GREY/BLACK. THEN, RESET (MENU CUSTOMER CONTROLS). TO MATCH THAT OF PRESET.

FOCUS. (MAY BE REQUIRED WHEN REPLACING PARTS)
OFF AIR OR CROSS HATCH, (STRONG AND NOISE FREE.) VERIFY, OR SET MENU TO (PRE SET.) ADJUST FOR SHARPEST PICTURE AT C.R.T. CENTER.

VERTICAL SIZE.
(MAY BE REQUIRED WHEN REPLACING PARTS)
ADJUST T.V. BRIGHTNESS TO MAXIMUM, A.C. LINE AT 108. VOLTS (LOW LINE) ADJUST VERTICAL SIZE TO (SLIGHT) OVER SCAN, TOP AND BOTTOM.

AUDIO DET. (ADJUSTMENT IS NOT NORMALLY REQUIRED)
CARRIER INPUT ONLY. LEVEL JUST ADEQUATE FOR NO NOISE, WITH NO SOUND. ADJUST FOR A MEASURED 4.0 VOLTS D.C. AT PIN 54 OF (ICX1200). NOTE! METER MUST HAVE A 1.0 MEG OHM, OR HIGHER INPUT IMPEDANCE.

AUDIO BALANCE.
(ADJUSTMENT IS NOT NORMALLY REQUIRED)
1.Khz SIGNAL, ABOUT 50. MILLI VOLTS INPUT, APPLY TO BOTH LEFT AND RIGHT, AUX AUDIO IN JACKS, ADJUST T.V. TO MAX VOLUME, MEASURE EACH PAIR OF SPEAKER LEADS. NOTE! (SPEAKERS MAY BE DISCONNECTED FOR ADJUSTMENT PROCESS AND YOU BE IN AUX MODE) ADJUST AUDIO LEVEL INPUT FOR ABOUT 1.0 VOLT OUTPUT, ADJUST BALANCE CONTROL FOR EQUAL LEVEL AT L AND R OUTPUTS.

VIDEO GAIN.
(ADJUSTMENT IS NOT NORMALLY REQUIRED)
1.0 VOLT PEAK TO PEAK VIDEO, INPUT TO AUX VIDEO IN JACK, ADJUST FOR 1.0 VOLT PEAK TO PEAK AT PIN 42 OF (ICX1200).

B+ ADJ. (ADJUSTMENT IS NOT NORMALLY REQUIRED)
ADJUST T.V. TO MINIMUM BRIGHTNESS AND VOLUME, A.C. LINE AT 120. VOLTS ADJUST B+ FOR 100. VOLTS D.C. AT VOLTAGE SOURCE LOCATION AS NOTED.

A.G.C. (ADJUSTMENT IS NOT NORMALLY REQUIRED)
USE ANY SIGNAL ADEQUATE FOR SNOW FREE PICTURE, ADJUST, FIRST INTO SNOW, THEN BACK, TO JUST SNOW FREE.

A.F.C. (ADJUSTMENT IS NOT NORMALLY REQUIRED)
(REQUIRES EXACT FREQUENCY ACCURATE SIGNAL.) OFF AIR, STANDARD BROADCAST CHANNELS ARE ACCEPTABLE. MEASURE PIN 52 OF (ICX1200) ADJUST FOR 3.3 VOLTS WITHIN THE RAPID VOLTAGE CHANGE PORTION OF THE ADJUSTMENT.

V.C.O. (ADJUSTMENT IS NOT NORMALLY REQUIRED)
COLOR BAR OR ANY TEST PATTERN, ADEQUATE FOR NOISE FREE VIDEO. ADJUST FOR MINIMUM BURST AMPLITUDE AT PIN 45 OF (ICX1200), THEN FINE TUNE THE ADJUSTMENT, FOR BEST TRANSIENT RESPONSE.

Figure 9-3 Service Menu Items and Adjustments

The C-12 Chassis

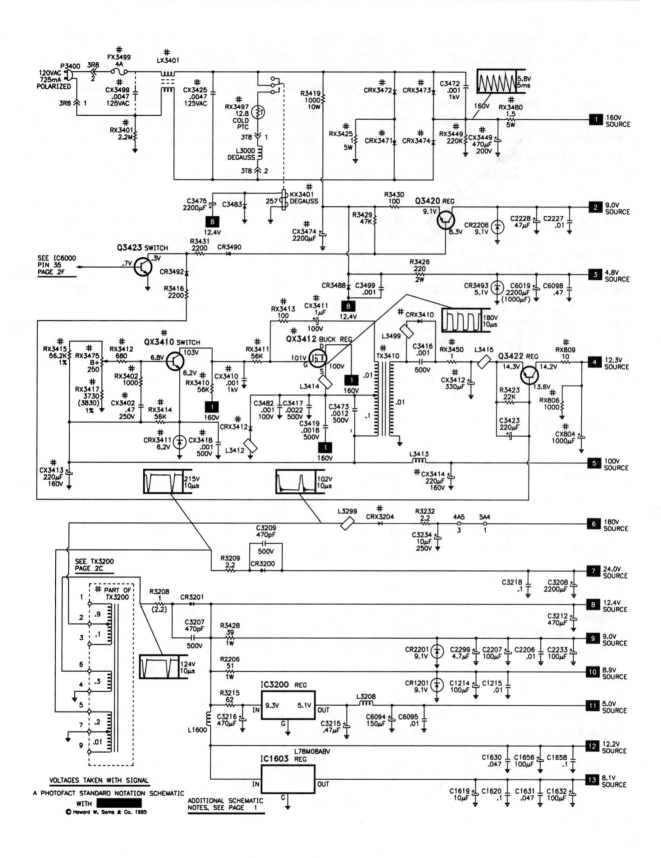

Figure 9-4 Power Supply Schematic

means that the output of a switching power supply is less than its DC input. In this instance, the DC input is in the neighborhood of +150 volts while its output is about +100 volts.

QX3412 is the regulator. It is a 2SK640, Zenith part number 121-1299, but an ECG2385 makes an excellent replacement. QX3410 controls the on/off time of the regulator by monitoring the +100 volt line. Its part number is 121-1146A. I have been told that an ECG123AP makes a perfectly acceptable substitute, but I haven't confirmed it. Given the voltages present in the circuit, I see no reason to doubt the information. If I needed a substitute and didn't have one, I wouldn't hesitate to stick a 123AP in the circuit. Keep these part numbers in mind folks, because these transistors do fail, especially QX3412.

While we are on the subject of substitute parts, let me say a word or two about generic parts versus OEM products. I am sure you have noticed that most of the components in the primary of the power supply have been printed in a dark shade, meaning they are critical, safety components. If I were doing warranty work, I would never use a generic or off-the-shelf part to replace a manufacturer's original. That's stupid. Always use Zenith's original, and do it for two reasons: (1) Zenith guarantees their parts, whereas an off-the-shelf part might or might not meet their specification, and (2) using non-Zenith parts in a warranty situation violates the contract you signed when you accepted your status as a warranty repair station. Out-of-warranty products are different, but even then don't violate the precautions in the service literature. What do I mean? Don't replace critical components (those in shaded areas in a PHOTOFACT or that have an "X" as a part of their designation in Zenith TV's, like RX3413 or CX3411 in figure 9-4) with just any old part. Use flameproof resistors and high quality capacitors. In other words, be the professional you are.

System Control

Figure 9-5 gives you a Sams depiction of the system control circuit, including information about the tuner and the voltages that it needs to operate. IC6000 receives its instructions via the IR receiver at pin 36 and the front panel controls at pins 41 through 49. When it receives an on command, the microprocessor outputs a high on pin 35 that turns on Q3423 (figure 9-4). When it turns on, Q3423 pulls the bases of PNP transistors Q3420 and Q3422 low, turning them on. The +9 volts for horizontal B+ (Q3420) and +12.3 volts for the audio circuit (Q34222) become available, and the chassis comes to life.

If you suspect a microprocessor-related problem, begin by checking all VSS and VDD connections. If they are good, check the oscillator at pins 50 and 51 for a clean, stable sine wave of about 5 volts peak-to-peak and at a frequency of 12 megahertz. If the oscillator is working, check for reset voltage at pin 2. You know by now that I also recommend checking data-in and data-out lines for abnormalities before changing a microprocessor. If these parameters are as they should be, then and only then change the chip.

The EAROM (IC6001) completes the system control setup. As you know, it stores vital information that the system control microprocessor retrieves and uses to set the operating parameters of the TV. If its information becomes corrupt, then the microprocessor cannot issue correct instructions to the TV processor, tuner, and audio processor, leading to symptoms like no audio, no picture, dim raster, and/or horizontal frequency and phase problems. When you run across these symptoms and really can't find a cause for them, you might profitably consider replacing the EAROM especially if you have one you have just extracted from a dud.

The C-12 Chassis

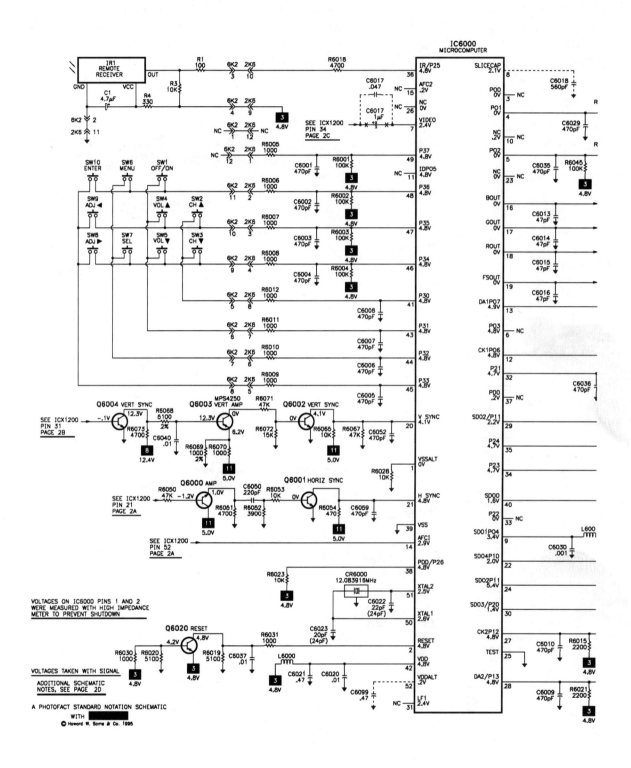

Figure 9-5 System Control Schematic

Servicing Zenith Televisions

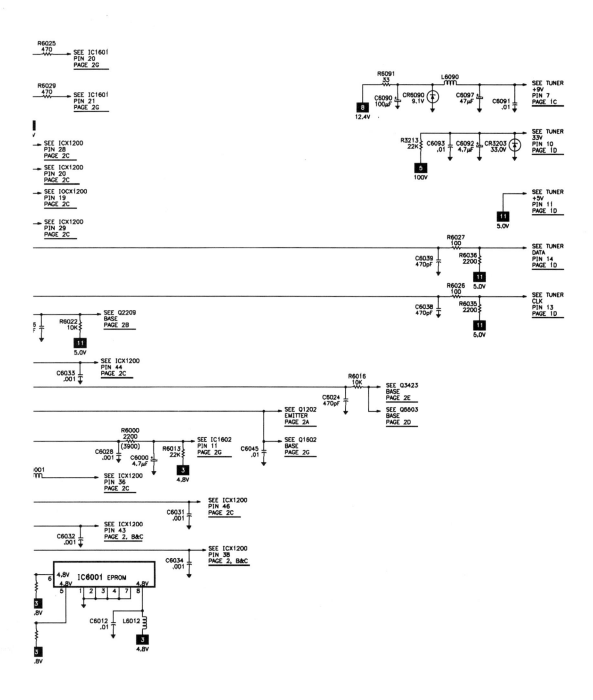

Figure 9-5 System Control Schematic (Continued)

Deflection Circuits

Horizontal drive makes it way out of pin 23 of ICX1200 (a TA8879). Unlike most other Zenith chassis, horizontal drive goes directly to the horizontal driver, Q3201, where it is amplified and coupled by means of TX3201 to the base of the horizontal output transistor to provide drive for the horizontal output transistor.

The circuit is reliable, but it does occasionally fail. My records indicate that I have replaced the horizontal driver on occasion. I recommend using an OEM if possible, Zenith part number 121-1290. One customer was in such a hurry that I used a generic, but I can't remember which one! I suppose it held up because she never did call back. If you have to replace the horizontal output transistor, use Zenith part number 121-1148 or good friends whom we know as ECG2302 and NTE2302.

IC1200, though not prone to problems, does occasionally fail to produce horizontal drive, especially if it has been subjected to a lightning bolt. On those occasions when I have replaced it, I have used a TA8879N that I order from a parts house like MCM merely because I got it a buck or two cheaper.

Vertical drive exits pin 27 of ICX1200 and goes directly to pin 4 of IC2100 which is an LA7830 for the 25-inch sets. When you encounter a "no vertical deflection" problem, check two things before you begin to replace parts. First, check the solder around the pins of IC2100. If the solder connections are good, check for about +24 volts at pin 7. If it is missing, check resistor R3209 off pin 5 of the flyback (figure 9-4) because it has a history of opening for no apparent reason. If I remember correctly, Zenith even issued a technical bulletin calling servicers' attention to its propensity for failure. Replace it with a flameproof resistor only.

Video Circuit

Figure 9-6 lays out the video circuitry, the heart of which is ICX1200. Figure 9-7 gives you a peek inside the chip and ought to be a valuable addition to your troubleshooting arsenal. Since I discussed in some detail the TA8879N processor in the chapter on the C-6 chassis (9-1244 module), I won't repeat the material here.

There is one component in the video circuit that is prone to failure. I am, of course, talking about our old friend the video (Y) amplifier, Q2210. As we have seen time and time again in past chassis, its failure causes the video driver transistors to bias fully on causing the picture tube to become excessively bright, leading you

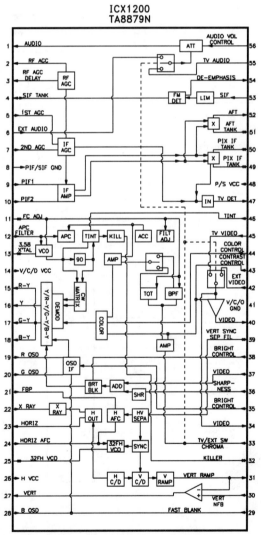

Figure 9-7 TA8879N (ICX1200)

to think you might have a shorted electron gun in the CRT. Don't condemn a picture tube until you have checked the video amplifier. Also, be aware of the fact that a shorted Q2210 might cause the picture tube to draw so much current that it damages the power supply. I almost condemned a C-2 set the other day because I thought the CRT had "checked out," but playing a hunch I checked Q2212 and found it to be bad. Hopefully now you don't make the same mistake!

Audio Circuit

The audio circuit (figure 9-8) differs from most of the audio circuits we have studied. One IC decodes the baseband audio (IC1601), another serves as a switch and volume control (IC1602), and a series of discreet components rather than an IC powers the speakers. Even though the "piece count" has increased, the audio circuit is still straightforward and ought to present you with no special servicing challenges.

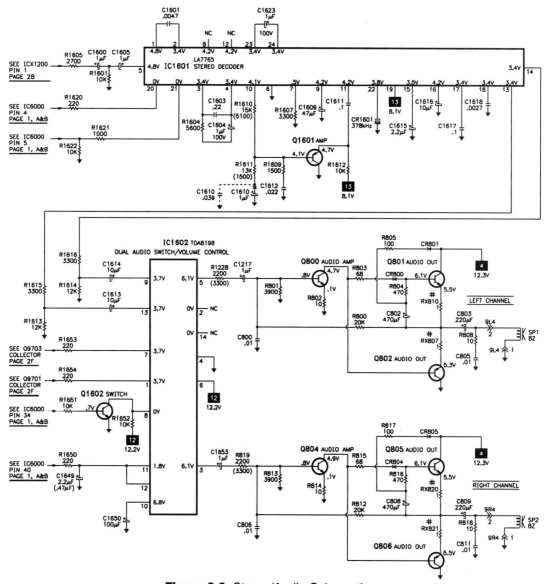

Figure 9-8 Stereo/Audio Schematic

Audio/Video Input Schematic

You might want to take a look at the A/V jack pack circuit (figure 9-9) because it is different. Three optoisolators separate the world of the consumer from the hot chassis, a bit of a departure from the way Zenith usually does things. Voltage to operate the jack pack is developed by TX9700, a transformer that serves to isolate the main chassis from the jack pack. A flyback pulse excites the primary and is coupled to the secondary where it is rectified by CR9700 and filtered by C9700 to provide the +13.3 volts to power the electronics on the jack pack. Magnavox used a similar circuit when it came out with its Y-1 and associated models of televisions a few years ago. The transformer doesn't do anything except to keep the main chassis isolated from the electronics on the jack pack.

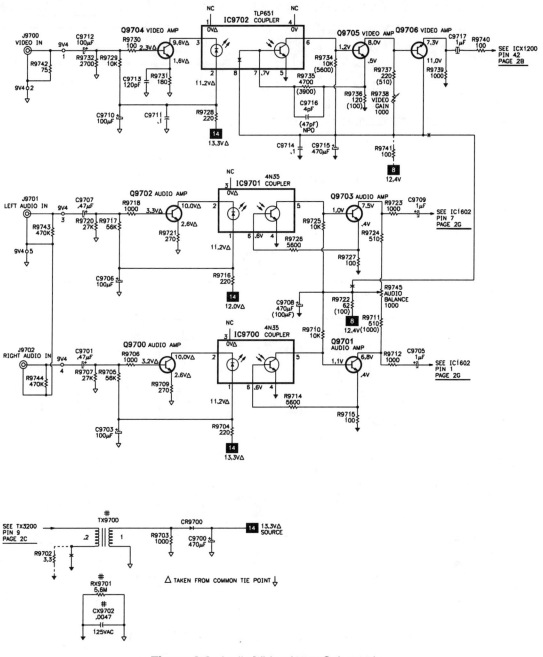

Figure 9-9 Audio/Video Input Schematic

Servicing Zenith Televisions

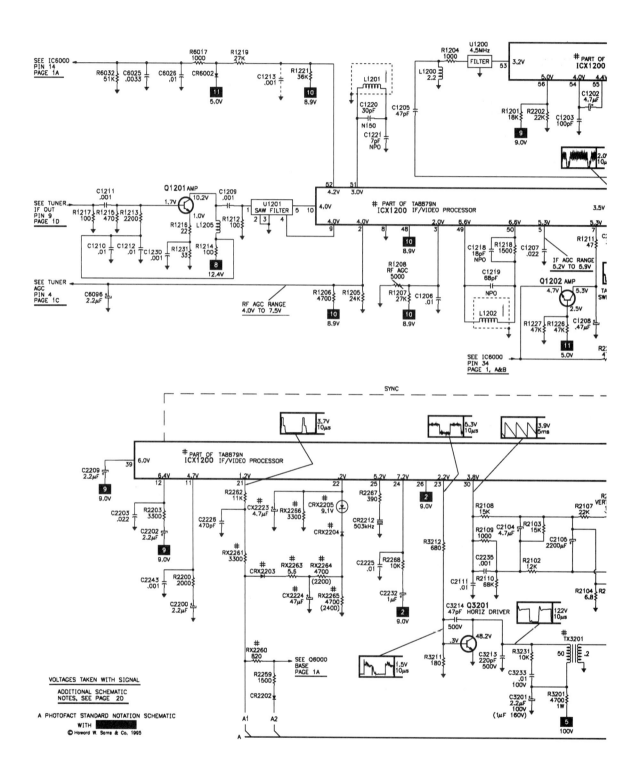

Figure 9-6 Television Schematic

The C-12 Chassis

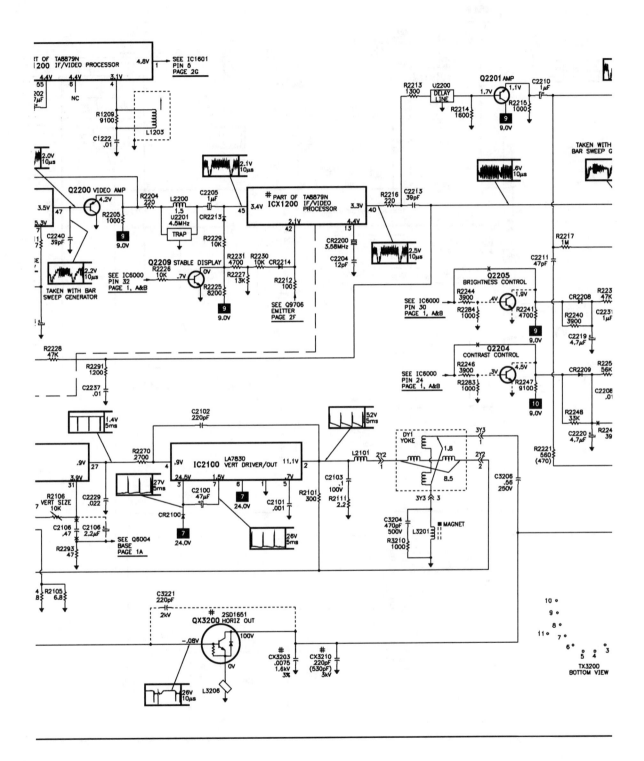

Figure 9-6 Television Schematic (Continued)

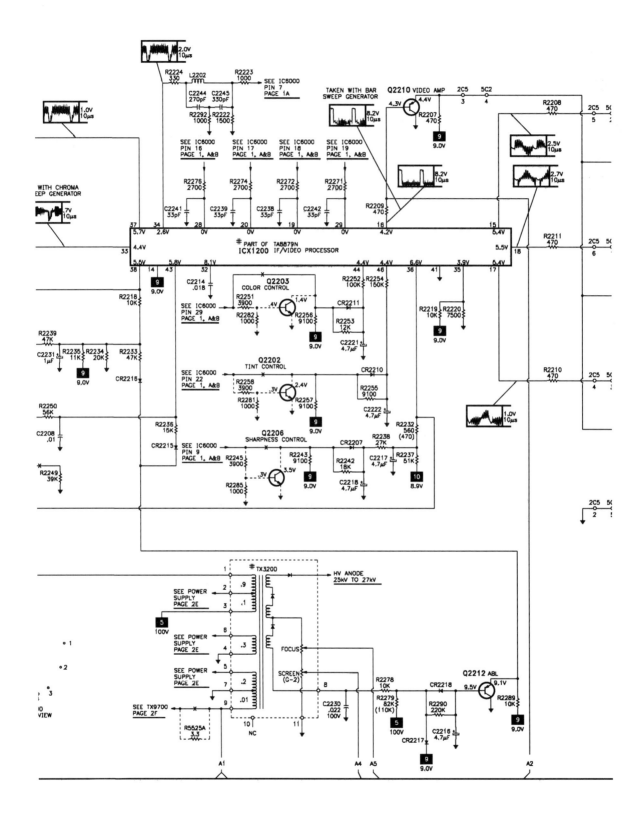

Figure 9-6 Television Schematic (Continued)

The C-12 Chassis

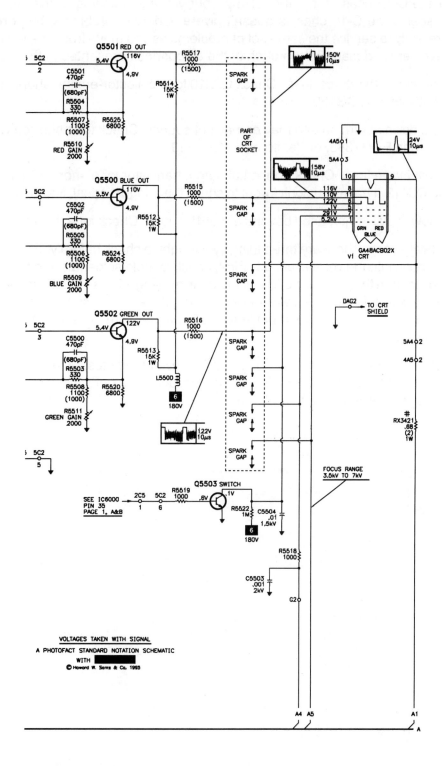

Figure 9-6 Television Schematic (Continued)

Servicing Zenith Televisions

Repair History

My database for the C-12 chassis is rather skimpy, but you are welcome to the information I have accumulated. It seems the C-12 chassis doesn't give a variety of problems, but it repeats similar problems. You are likely to service the same set of problems as you move from one set to the next. To put it simply, I have serviced many TVs in each of the categories I presented in this section.

(1) Bright screen with retrace lines. Check Q2210 for a collector-emitter short. If it is defective, replace it with an ECG159.

(2) TV shuts down at turn on and may or may not squeal. Check for a shorted Q3412. Use an 2SK640, and ECG2385, or Zenith part 121-1299.

(3) Dead with blown fuse. Check horizontal output transistor for a short. You may also have to replace Q3412. I hope you don't forget to check the bridge diodes! Never forget the obvious.

(4) Shuts down if AC is increased beyond about 100 volts. Check for a defective buck regulator.

(5) No vertical sweep. There are three things you ought to check before you roll up your sleeves and get deep into the vertical circuit. First, check for cold solder around the pins of IC2100. Second, confirm the presence of B+. If the B+ is missing, check R3209 for an open condition. Replace it with a flameproof resistor only. Third, if the solder connections are good and if there is B+ and vertical drive, replace IC2100 with the correct part for the screen size you are servicing.

(6) Video fades out after the set has been on for several minutes. Check Q2210 because it may be opening under load. Don't overlook the fact that a defective picture tube or a failing G-2 voltage causes the same problem.

CHAPTER 10

THE Y-LINE

The Y-Line appeared in 1996 and for a variety of reasons represented a radical departure from the way Zenith had been doing consumer televisions. Oh, the chassis still consisted of a one-piece printed circuit board with the video output module constructed as a breakaway piece of the main board. But the similarity ends there. For example, all of the components were located in exactly the same position on the circuit board regardless of the module number (figure 10-1). Seven modules belong to the GX family, one of the two member chassis of the Y-Line, and the seven modules are identical with respect to component location! That makes servicing a bit easier.

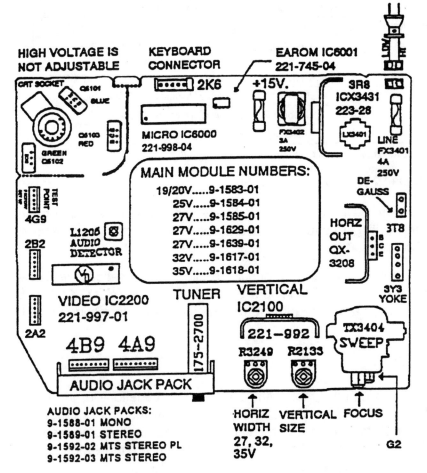

Figure 10-1 Y-Line Chassis Circuit Board

Moreover, Zenith designed the Y-Line sets to be serviced to the component level and informed its authorized service centers that we were expected to fix the modules for 27" and smaller sets. Zenith continued to support the large screen and projection sets at the module level. In fact, we were told NOT to try to fix the large screen and PTV modules without prior authorization unless the repair was an emergency. Since repair shifted from module support to component support, Zenith's technical training and literature took one of those giant leaps forward and is now an indispensable tool for the tech who does Zenith repair.

The GX and GZ chassis make up the Y-Line. The GZ chassis is a world unto itself and is found in projection televisions and certain direct-view sets that bear the "INTEQ" logo. Table 10-1 should give you an idea about the features incorporated into the GZ chassis. If that doesn't cause you to marvel, then you should know that the operating instructions are in a booklet that is 55 pages long! If I remember correctly, the field engineer at the service meeting said sets with the INTEQ logo would be serviced by select service centers only. It is, to say the least,

GENERAL FEATURES	AUDIO FEATURES
❏ Optional StarSight® Program Guide	❏ 2-Level Audio Mute
❏ Automatic Demonstration of TV Features	❏ MTS Stereo
❏ On-Screen Menus Accessed by Remote or TV Control Panel	❏ Variable Audio Output Jacks
❏ Multi-Brand Programmable Remote Control	❏ External Speaker Output Jacks
❏ Z-Trak Trackball (Programmable Remote Control	❏ Surround Sound Speaker Output Jacks
❏ Closed Captions and Text Modes	❏ Preset Surround Sound
❏ 181 Channel Tuning	❏ Front and Rear Surround Sound Speaker Controls
❏ Automatic Channel Programming	❏ On-Screen Audio Adjustments
❏ Channel Lockout Parental Control	❏ SoundRite Auto Volume Limiter
❏ Channel Labels (ABC, NBC, MAX, SHO, etc.)	❏ On-Screen Captions when Audio Muted
❏ Input Source Identification (VCR, Cable Box, etc.)	**PIP (Picture-In-Picture) FEATURES**
❏ Sleep Timer	❏ 2-Tuner PIP
❏ On/Off Timer (Vacation Timer)	❏ Picture / Main Screen Picture Video Swap
❏ On-Screen Menu Icon Identifiers	❏ PIP Picture / Main Screen Picture Audio Swap
❏ Antenna/Cable Connection Jacks	❏ 1-Minute Surf-to-PIP Channel Scanning
❏ Stereo Audio/Video (A/V) Input and Output Jacks	❏ PIP Window Freeze
❏ Front Panel Stereo Audio/Video (A/V) Input Jacks (Available on some models only)	❏ Placement of PIP Window Anywhere on TV Screen
❏ On-Screen Video Adjustments	**ADDITIONAL PROJECTION TV ONLY FEATURES**
❏ Favorite (Surf) Channel Scan	❏ Built-in Screen Protector (Available on Some Models)
❏ Unique Icon Menu for Z-Trak Remote Control	❏ Optional Screen Protector (Purchased Separately)
❏ Adjustable On-screen Pointer Speed Control	❏ VCR Shelf

(Design and specifications are subject to change without prior notice.)

Table 10-1 GZ Chassis Features

the high end of the Y-Line. I haven't seen even a single direct-view unit, though they have been available for almost four years as of this writing. Because of that and space limitations, I have decided to devote my time and effort to the GX chassis.

However, if you are interested in the GZ family, I suggest you consult the following publications: *Technical Training Program: GZ Direct-View and Projection TV* (number 923-3288TRM), *INTEQ GZ Chassis Service Manual* (number CM148/GZ). Both are thorough and provide sufficient information to work on these high-end products.

Overview of GX Chassis

The GX series was developed for 19" through 35" mono and stereo sets. Variations of the basic chassis were also manufactured for use in certain commercial products. At the risk of repeating myself, Zenith produced one board for all models in a particular screen size and programmed the EAROM for the features that the model incorporated. Additional plug-in boards like the various audio modules and the PIP support those features (figure 10-1).

The GX series utilizes four ICs for signal, sync, and sweep processing. ICX2200 handles audio-video, sync, and sweep drive processing. IC6000 is the microprocessor, and IC6001 is the EAROM. IC2100 develops vertical sweep. ICX3431, a fifth IC located in the power supply circuit, develops most of the voltages for the operation of the television.

Seven modules comprise the GX family of chassis:

9-1583-01	19" and 20" sets
9-1584-01	25" sets without comb filter
9-1629-01	27" sets with comb filter
9-1639-01	27" sets with comb filter and flat CRT
9-1617-01	32" sets
9-1618-01	35" sets

Table 10-2 is, I believe, a fairly complete list of Y-Line models. I don't think Zenith has added new models since the publication of this list, but I could be wrong.

You need to keep in mind the plug-in modules (daughter boards) that may belong to the set on which you are working, especially the audio modules. For example, certain mono models have the 9-1588-01; the MTS stereo models have the 9-1592-01; and the non-MTS stereo models have the 9-1589-01. Each jack pack contains not only the jacks themselves but also the support circuitry for the features they support.

Available Literature

Fortunately, I am able to recommend a couple of really good booklets to help you learn about the GX chassis. The first is *Technical Training Program: Y-Line Color Television GX Chassis* (number 923-3270TRM). The second is the service manual for the Y-Line direct-view television, Zenith number CM-148GX. As you know, I usually keep a PHOTOFACT handy as a supplement to factory service material. To that end, I use PHOTOFACT 3743 in conjunction with the last booklet I mentioned. Incidentally, the service manual has four sections and contains just about everything the servicer needs to know – circuit descriptions, schematics, menu settings, parts list, location guides, and so forth.

If you do hotel and motel service work, you might want to invest in two additional books. One of which is the *Technical Training Program: Y-Line GX Commercial Products* (923-3280TRM). There is also a fairly complete service manual dedicated specifically to the commercial products available.

Customer Menu

The Y-Line customer menus have a new look. The menus are displayed over the video and on the left side of the screen. The color of the menu varies depending on the "level" of the TV in

SY1931SG
SY1949Y
SY1949YM
SY1951Y
SY1953Y
SY1953YM
SY2031S
SY2731SM
SY2049DT
SY2049DTM
SY2049X
SY2053S
SY2053SM
SY2068DT
SY2068DTM
SY2500RK
SY2504BW
SY2518RK
SY2549S
SY2551S
SY2551SM
SY2568S
SY2569S
SY2572DT

SY2572DTM
SY2722RK
SY2723BW
SY2728RKM
SY2738RK
SY2738RKM
SY2739RK
SY2751Y
SY2765S
SY2765SM
SY2768S
SY2768SM
SY2772DT
SY2772DTM
SY3272DT
SY3272DTM
SY3572DT
SY3572DTM
SY7249DT
SY7568S
SY7768S
SY7772DT
SY7949Y

Table 10-2 Y-Line Models

question. Private labels are level 0 while Zenith sets are either level 1 or 2. Incidentally, the level also determines the features the set offers the consumer. More about "level" later on. For the present, just remember private level sets have a dark red background, Zenith level 1 has a blue background, and Zenith level 2 has a gray background.

Each press of the menu key toggles among Setup Menu, Video Menu, Source Menu, and PIP Menu if the set has a PIP. Move from one menu item to the next by using the "select up/down" keys. Use the "adjust left/right" to change the selected item. Press "enter" or "quit" to exit the menu. If you don't press a key, the on-screen display times out in about fifteen seconds.

Here are the menu items under each category:

Setup Menu	**Video Menu**	**Audio Menu**	**Pip Menu**
Auto Program	Contrast	Bass	Color
Add/ Del/ Surf	Brightness	Treble	Tint
Ch Labels	Color	Balance	Size
Clock Set	Tint	Audio Mode	
Timer Setup	Sharpness	Surround	Source Menu
Parental Ctl	Picture Pref	SoundRite	Main Source
Caption/Text			Pip Source
Audio Mode			
Language			
Auto Demo			

Since you are familiar with most of these items, I'll limit my comments to Surf, Parental Control, and Soundrite which are the brand-new items in the menu.

(1) "Surf" has been added to the "Add/Del" item usually found in Zenith's customer menus. The surf feature permits the customer to save a group of his or her most-watched channels so they can be scanned using the Surf key on the remote control. After you bring up this menu, use the adjust key to select "Surf." Then use the numeric buttons or channel up/down keys to add channels to the surf list. When you press the Surf button, the on-screen display shows one of the following, "Surf's Up" or "No Surfing." Press the Surf button again to toggle to the other selections.

(2) "Parental Control" has four subcategories. "Block Channel" permits the user to block one or more channels. "Block Video" allows the user to block the video inputs. "Set Hours" is used for setting the amount of time the channel(s) or video input is to be blocked. The time is from a minimum of one hour to a maximum of ninety-nine hours. "Set Password" lets the user to set a password that consists of four (4) digits that are set using the numerical keypad. "Lock On/Off" turns the parental control on or off.

NOTE: If the user forgets the password, he or she must wait the prescribed time before the channel(s) or video ports are unblocked. Do not unplug the set. If you do, it only delays the time-out function. *However, a technician can remove the block simply by entering the service menu and making the proper adjustments.*

(3) "Soundrite" is what Philips calls "Smart Sound." When it is turned on, Soundrite keeps the audio in the commercials at the same level as the audio in the program. You may think Soundrite is just a sales gimmick, but the feature really works.

The Y-Line Chassis

Service Menu

Enter the service mode the usual way. When it pops up, the display looks like the one in figure 10-2. The black bar at the top contains the part number of the software the set uses while the black bar at the bottom contains the date the module went through the factory and a number that indicates the module's test status. "Hpos" is the third item in the service menu. Make a note that the GX service menu always comes up on this item.

Unless you set Factory Mode from off to on, you have access only to the first seven items in the service menu. If you want access to all 38 entries, use the select up key to highlight Factory Mode and the adjust key to turn it on (00 to 01). When factory mode has been turned on, you may proceed to manipulate any of the items you choose. Table 10-3 lists all of the items in the menu, their adjustment range, and default settings. Consult table 10-4 if you need additional information about any of them.

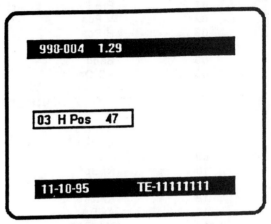

Figure 10-2 Service Menu

Let's begin exploration of the service menu by looking at item 04, "Level." As you see, there are three different levels, and each level has its own particular features (table 10-4). If you have to set the level, like when you change the module, begin by placing a jumper between pins 3 and 4 of connector 4G9 (figure 10-3). Don't short-circuit any of the five pins except pins 3 and 4 because, if you do, you will most likely damage the module! Once you have shorted pin 3 to 4, use the adjust key to change the level. Then remove the jumper and exit the service mode. Incidentally, the level for the set is listed on the same label that has the model and serial number.

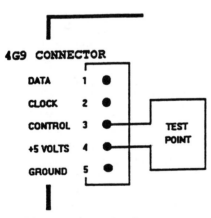

Figure 10-3 4G9 Connector

You may be wondering what happens if "level" has been incorrectly set. Well, the TV will work fine EXCEPT that it won't have the features it should have, meaning you should brace yourself for a call from the customer complaining that the set is missing certain features "it used to have."

Now let's take a look at items 29 through 38. Note that each item has an asterisk beside it referring you to an important footnote at the bottom of the list. When you change a PIP or an MTS module, you must adjust these items so that they correspond to the bar code on the PIP or the MTS module. More about those adjustments and the bar code later on.

Never overlook the importance of the service menu when you work with Zenith televisions. An erroneous setting might lead you to believe the module is defective. Or worse, it might cause you to waste hours of time and many dollars trying to fix a set when the problem could have been solved by a simple adjustment to an item in the service menu. Don't forget a point I have made several times, that lightning or a power surge might do no damage to the TV other than reset one or more service menu items. For example, a customer calls and says, "My TV isn't working the way it used to after the storm came through last night." You check the complaint, and sure enough, the set has lost

GENERAL SETTINGS

	ITEM	RANGE	DEFAULT SETTINGS
00	Fact Mode	01	0
01	Pre Px	01	1
02	V Pos	031	10
03	H Pos	075	44
04	Level	02	1*
05	Band	07	1**
06	AC On	01	0

VIDEO PROCESSOR SETTINGS

	ITEM	RANGE	DEFAULT SETTINGS
07	C.Phas	0254	80
08	C.Srch	01	1
09	C.Line	032	17
10	Rf Bpf	01	0
11	3.58T	01	0
12	RF Brt	063	20
13	Ax Brt	063	20
14	MaxCon	063	63
15	VPhase	07	0
16	HPhase	0-31	16
17	Aud Lvl	0-63	46
18	Aud Adj	0-63	63***
19	RF Agc	0-63	40
20	H Afc	0-1	1
21	WhComp	0	1
22	60hzSw	0-3	0
23	PifVco	0-127	63
24	R Cut	0-254	0
25	G Cut	0-254	0
26	B Cut	0-254	0
27	G Gain	0-254	75
28	B Gain	0-254	75

PIP SETTINGS

	ITEM	RANGE	DEFAULT SETTINGS
29	PipRas	0-254	1
30	PipSw	0-15	11
31	PipLuD	0-7	3

AUDIO SETTINGS

	ITEM	RANGE	DEFAULT SETTINGS
32	In Lev	0-15	9
33	St Vco	0-63	24
34	SapVco	0-15	6
35	SapLpf	0-15	8
36	St Lpf	0-63	24
37	Spectr	0-63	42
38	WideBa	0-63	30

*Set this to the level (Bar Code Data
**This setting depends upon input signal.
***This item is not to be field adjusted.

Table 10-3 Service Menu Items: Adjustments and Defaults

Level 0 is used for Private Label sets. In the private label mode the IR code is 21 or 121.

Level 1 and 2 is used for Zenith. In the Zenith label mode the IR code is 01 or 101.

Feature Level Summary

"LEVEL 0" Private Label
- Red Menus (with borders, italics)
- Auto Channel Set
- Set Channels
- Clock Set
- Caption/Text
- Languages (Trilingual)
- Sleep Timer

"LEVEL 1" LowEnd Zenith
- Blue Menus (with borders, italics)
- Auto Program
- Ch Add/Del
- Clock Set
- On/Off Timer
- Caption/Text
- Languages (Trilingual)
- Sleep Timer

"LEVEL 2" HighEnd Zenith
- Gray Menus (with borders, italics)
- ICONs
- Auto Program
- Ch Add/Del/Surf
- Channel Labels
- Clock Set
- On/Off Timer
- Parental Control
- Caption/Text
- Languages (Trilingual)
- AutoDemo
- Channel Surf
- Sleep Timer

BE SURE TO REMOVE THE JUMPER AFTER THE LEVEL HAS BEEN SET

05 Band There are eight positions. 0 is Broadcast fixed, 1 is CATV afc, 2 is HRC afc, 3 is ICC afc, 4 is Broadcast afc, 5 is CATV fixed, 6 is HRC fixed, 7 is ICC fixed.

06 AC On There are two positions; 0 is off and 1 is AC On. In the on position the set will turn on and off when AC power is applied.

07 C.Phas (Caption Phase) It determines whether or not captioning will be received. It has a range of 00 to 254. Default setting is 80.

08 C.Srch (Caption Search) The range is 0 to 1. Default setting is 1.

09 C.LINE (Caption Line) The range is 0 to 32. Default setting is 17.

10 Rf Bpf (Rf Bandpass) The range is 0 to I. Default setting is 0.

11 3.58 T This is the 3.58 MHz trap. The range is 0 to 1. Default setting is 0.

12 RF Brt (RF Brightness) This sets the subbrightness of the customer control for brightness in the RF mode. The range is from 0 through 63. Default setting is 20.

13 Ax Brt (Auxiliary Brightness) This sets the subbrightness of the customer control for brightness in the AUX mode. The range is from 0 through 63. Default setting is 20.

14 MaxCon (Maximum Contrast) This sets the adjustment range of the customer control for contrast. The range is from 0 to 63. Default setting is 63.

15 VPhase (Vertical Phase) The range is from 0 to 7. Default setting is 0.

16 Hphase (Horizontal Phase) The range is from 0 to 31. Default setting is 16.

17 AudLvl (Audio Level) This sets the gain for the Low Cost stereo and MTS Stereo. The range is from 0 to 63. Default setting is 46.

18 AudAdj (Audio Adjust) This set the balance between the right and left channel. The range is from 0 to 63. Default setting is 63. This item is not to be field adjusted.

19 RF Agc The range is from 0 to 63. Default setting is 40. Tune in the weakest channel and adjust for a snow free picture.

20 H Afc There are two settings 0 and 1. Default setting is usually 0.

21 WhComp (White Compression) there are two settings 0 and I. Default setting is 1.

22 60hzSw (60 Hertz Switched) The range is 0 to 3. Default setting is generally set to 2.

23 PifVco (PIF Voltage Controlled Oscillator) The range is 0 to 127. Default setting is 63.

24 R Cut The range is 0 to 254. Default setting is 0.

25 G Cut The range is 0 to 254. Default setting is 0.

27 G Gain The range is 0 to 254. Default setting is 75.

28 B Gain The range is 0 to 254. Default setting is 75.

29 PipRas* (Picture in Picture Raster) The range is 0 to 254. Default setting is 1.

30 Pip Sw* (Picture in Picture Switch Delay) The range is 0 to 15. Default setting is 11.

31 PipLuD* (Picture in Picture Luminance Delay) The range is 0 to 7. Default setting is 3.

32 In LEV* (Input Level) The range is 0 to 15. Default setting is 9.

33 St Vco* (Stereo Voltage Controlled Oscillator) The range is 0 to 63. Default setting is 24.

34 SapVco* (Second Audio Program Voltage Controlled Oscillator) The range is from 0 to 15. Default setting is 6.

35 SapLpf* (Second Audio Program Low Pass Filter) The range is 0 to 15. Default setting is 8.

36 St Lpf* (Stereo Low Pass Filter) The range is from 0 to 63. Default setting is 24.

37 Spectr* This is high frequency separation. The range is from 0 to 63. Default setting is 42.

38 WideBa* This is low frequency separation. The range is from 0 to 63. Default setting is 30.

* When changing PIP module or MTS modules in the field, these items must be adjusted.

Table 10-4 Feature Level Summary

some of its features! But, you are a knowledgeable repairperson. You access the service menu, and while you look through each item, you note that the setting for "level" is incorrect. You reset it, and lo and behold, you have fixed the TV. Fantastic? Absolutely not! You can expect those kinds of problems to crop up. The moral of this little lecture is, "Take the time to check the service menu before you begin extensive troubleshooting or attempt to replace expensive parts."

Now let's get to the electronics of the GX chassis. Or as they say in our modern world, "Let's move from the software to the hardware."

Power Supply

The power supply for the GX chassis should look familiar (figure 10-4a) because it is virtually the same one we have been seeing since the first chapter. There are, of course, some differences incorporated to accommodate different features and voltage and current demands. For example, a

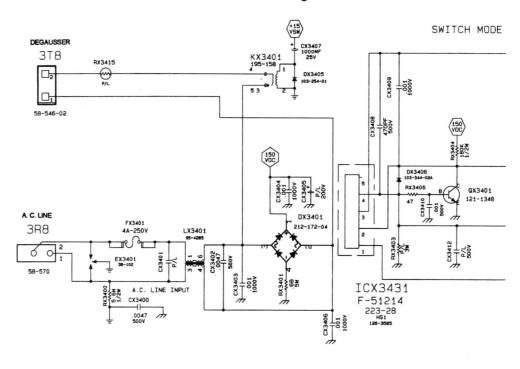

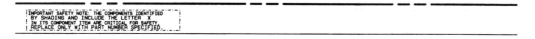

Figure 10-4a Power Supply Schematic

couple of the components have been "beefed up" to handle the larger current demands. I have worked on lots of these Zeniths, and I have to confess that this power supply is about as trouble-free as any I have ever seen. They even stand up well in the face of lightning storms and power surges. I saw one recently that had been damaged by lightning; the fuse looked like a piece of charcoal.

After checking all of the components and finding nothing amiss, I replaced the fuse, applied AC, and listened to the power supply come to life. Believe me, that's not unusual. I am in the process of reading about RCA's brand-new CTC203 chassis and have noted that Thompson has come up with yet another power supply scheme. My earlier book, Servicing RCA/GE Televisions, deals with their products from the CTC167 through the CTC195/197, a total of nine chassis. However, I had to deal with six – count them, six – totally different power supplies. The CTC203 uses a radically new power supply. Thomson is even going back to the use of discreet components as opposed to the use of integrated circuits! Therefore, the parts count in the CT203 power supply has gone up considerably. Zenith, on the other hand, has stayed with the same design for a long time.

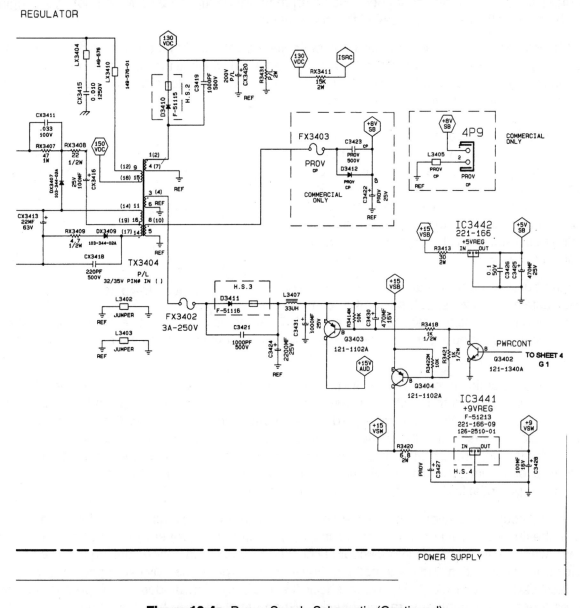

Figure 10-4a Power Supply Schematic (Continued)

If you have read this book from the beginning, you know that I have dealt with four different power supplies for the middle-of-the-road products and two for the dual-module, high-end products. And, with the exception of the high-end products, those four power supplies have not only been very reliable, but also very serviceable. I do occasionally fuss about Zenith products, but I tip my hat to them with respect to their power supply technology.

How It Works

As I promised, let's dig into it to see how it works. Figure 10-4a shows a full schematic of the power supply. Zenith employs a traditional circuit to develop a DC voltage from the applied 120 volts AC. AC enters at 3R8, the line cord connection, and goes through FX3401 (4-amp, 250-volt fuse) to the bridge rectifier, DX3401, and emerges at CX3405 as approximately 150 volts DC. Zenith has made a slight change in the AC input circuit by using a GJV fuse (a fuse with pigtails) instead of the usual AGC fuse that snaps into a fuse holder. I guess they made the change to save a few cents per set. But, don't make light of those few cents because when they are multiplied by a few million, you begin to realize how much their engineers saved the corporation.

The +150 volts goes to two components in the switching power supply. It goes first to pin 15 of TX3401 and exits at pin 9 on its way to pin 3 of ICX3431 (figure 10-4b) which is the collector connection for the switching transistor inside the IC. The +150 volts also goes to resistor RX3404 that draws an initial turn on current into pin 2, the base connection for the switching transistor inside ICX3431. When the base receives this initial pulse, the transistor turns on and switching action begins. You can hear the power supply when it starts up because it emits a very audible "chirp," a sound that never ceases to make me a just a little happy because then I know the power supply has started. After the initial chirp, the power supply settles down to a switching frequency that falls somewhere between 28 and 38 kilohertz.

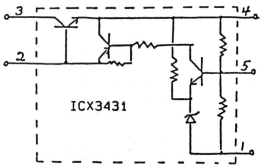

Figure 10-4b ICX3431

The initial burst of current starts the switching action. Drive to sustain the power supply's operation comes from a primary winding of the transformer. Diode DX3407 serves as a half wave rectifier to rectify the AC waveform and to develop a square wave the positive value of which drives the base of the switching transistor.

Now shift your attention to pin 4 of ICX3431. Pin 2 is the base of the transistor inside the IC; pin 3 is the collector; and pin 4 is the emitter. RX3403 connects between pin 4 and ground and serves as a current sense device. When the voltage drop across it reaches 0.7 volts, QX3401 turns on and grounds base drive, shutting the power supply down. This little transistor provides the switching action necessary to turn ICX3431 ona dn off and also protects the main power supply from damage due to an overcurrent situation. However, a gross, sustained overcurrent condition, like a shorted CRT, has the potential to destroy both it and the switching IC.

I have had a couple of unpleasant experiences with this shutdown circuit and can testify that it does work. In the first instance, I encountered a supply that seemingly never started. That is, I did not hear the chirp when I applied AC. Using a scope, I checked for switching action thinking that my ears had perhaps gone on vacation and found nothing. You learn by experience. My experience has taught me to replace QX3401 and ICX3431 and to check the resistors and diodes in the primary circuit. I replaced those components and ran my checks. When I applied AC, I heard nothing.

After a lot of head scratching and chin-rubbing, I thought I ought to check if perhaps something had shorted in one of the secondary circuits. Guess what? I found that D3411 in the 15-volt line had shorted. I know the line is fused, and even though it should have opened, the fuse was good! The overcurrent protection circuit had acted so quickly at plug-in that I hadn't heard the usual chirp. I have also had these sets to shut down because of a shorted diode in the +200-volt scan-derived circuit and a shorted vertical output IC. However, I have never had it to shut down because of a shorted horizontal output transistor. If the horizontal output transistor shorts, the power supply chirps, groans, wheezes, and squeals as long as it stays plugged into the AC outlet. I can't explain it. I know it doesn't make sense. I am only reporting it.

A sustained, excessive current drain often causes catastrophic failure of several components in the power supply. If you have worked on some of the older sets, you know that a shorted picture tube often takes out ICX3431, QX3401, RX3407, and RX3408. I have also seen DX3407 fail under stress. Fortunately, these parts are inexpensive and repair time is short. If you work on the older sets, be sure to watch the screen for signs of a picture tube short and pull the plug immediately when you see it happen. If you don't, prepare to see your money and time go up in smoke. Perhaps I should add a word of caution: a leaky luminance driver transistor mimics a shorted CRT. You can quickly tell where the problem lies by removing the CRT socket, turning on the TV, and checking the voltages at the collectors of the video output transistors. If these voltages are low, check the luminance driver transistor. If they are normal, you probably have a bad picture tube.

ICX3431 is the familiar STR53041. Order it from Zenith by using part number 221-997-01 which includes the IC mounted on a heat sink, or order it from any number of parts jobbers using the STR53041 number. The part number for QX3401 is 121-1348. There are generic substitutes for it that probably work, but I stick with Zenith on this one. RX3407 is a 47-ohm, 1-watt resistor; RX3408 is a 22-ohm, 1/2-watt resistor. If you pull them out of your stock, make sure they are flameproof as opposed to carbon composition because they are also safety devices (Note the "X" in "RX," and refer to Chapter 9 if you have forgotten what it means.). The diodes in the primary circuit are run-of-the-mill silicon types. I usually replace them with an ECG125. If you are repairing the set under warranty, you should always use Zenith original parts, and there is no excuse for not doing so.

You have probably noticed the designation "P/L" by certain parts in the schematic, for example RX3403 in figure 10-4a. The P/L stands for "parts list" and is Zenith's way of telling you that the part designated varies from screen size to screen size. When you need to replace a part that has the "P/L" beside it, check the parts list to see what the part should be for the screen size on which you are working.

Now, back to the "how it works" scenario.

Look now at the components tied to pin 9 of TX3404. Capacitor CX3415 is what engineers call a "snubber." It limits the voltage peak of the waveform that appears at the collector of the switching transistor when it turns off. LX3410 is really just a ferrite bead that limits the ringing on the collector line caused by CX3415 and the flyback voltage. If these components weren't in the circuit, transients caused by the switching action would destroy the switching transistor in short order.

I haven't yet commented on pins 1 and 5 of ICX3431. If you have worked on these power supplies in the past, you know that pin 5 is not used. That leaves pin 1, which is the control input for the IC. The regulator works best when it is at -41 volts. A winding of the switching transformer and DX3409, RX3409, and CX3413 develops the feedback. If the output voltage increases, the resulting increase is coupled to pin 1 causing the voltage on pin 1 to rise. The IC responds by adjusting the duty cycle and/or frequency to bring the voltage at pin 1 back down to -41 volts. If you have a regulation problem such as B+ that is too high or too low, begin by checking the components tied to pin 1.

You have probably noticed several ceramic capacitors in the primary of the power supply. They have been strategically placed to suppress electromagnetic interference (EMI) and to serve as high frequency bypass control.

And that, as they say, is that. It is a simple design that has proved itself over several years of service and still holds its own in a world that sees constant change in consumer electronics. If you don't believe me, do a little research into the engineering efforts of other manufacturers like the RCA product I mentioned earlier in this chapter.

Power Supply Output

The time has come to shift attention from the primary to the secondary. Each secondary winding has its own half wave rectifier to develop the necessary voltages for the TV. All of the power supply's voltages come up as soon as switching action starts. The +130 volts goes to horizontal deflection. The +15 volts does a sort of double duty. Part of it drives IC3442 to develop +5 volts for system control and the IR receiver. Part of it is used to develop the +9 volts switched supply via IC3441, and part of it is used in other places in the TV as the switched +15 volts. The commercial GX chassis use a winding on the switching transformer to develop a +8 volts, but that doesn't concern us because the commercial products are outside the scope of discussion.

In other words, some of the voltages are available at all times while others are switched. PNP and NPN transistors do the switching, and work as follows. The PNP transistors remain off as long as standby voltages alone are needed. When switched voltages are needed, a high signal from the microcomputer turns the NPNs on. Their collectors go low, pulling the bases of the PNP transistors low and turning them on, making the switched voltages the control available at their collectors.

When the Television Turns On

Let's turn theory into a bit of reality by looking at the power-on sequence in a typical Y-Line television. You have one on your bench to which you have just given an on command. The microprocessor responds to the on command by outputting a high on pin 52. The high turns on Q3402, causing it to saturate and pull its collector voltage low. The low on the collector is coupled to the bases of Q3404 and Q3403. These transistors are PNP types, meaning a low on their bases causes them to turn fully on. When they turn on, the voltages they switch become available to the rest of the chassis.

Two other actions take place as soon as the switched voltages come on-line. First, the degauss relay energizes to start degauss action. The relay remains energized until capacitor CX3407 fully charges, at which time it de-energizes, and the degauss cycle ends. Second, the horizontal sweep circuit starts up, producing sweep-derived voltages of +23 volts for vertical deflection, +245 volts for the video output transistors, G-2 and focus voltage, and EHT. The set is now fully operational, producing picture and sound.

Troubleshooting Primer

If you ever need to do a quick check of the power supply, I suggest you use this information. Please remember that I haven't put the list in any kind of order:

standby voltages
150 volts at RX3402
130 volts at CX3420
15 volts at the emitter of Q3403

switched voltages
9 volts at pin 3 of IC3441
15 volts at the collector of Q3404
5 volts at pin 3 of IC3442

Cold Start-Up Problems

You may be called upon to fix a cold start-up condition in the 9-1583 series, a condition in which the TV sometimes won't start up after it has been unplugged for a while. According to a service bulletin, you should remove the 62-ohm 1/2-watt resistor at location RX3299 and put in its place a 47-ohm 1/2-watt resistor.

Concluding Service Note

As I was editing this chapter for the last time, I remembered servicing three Y-Line products for unusual power supply problems. Since I expect to see these problems occur again, I thought you ought to know about them. Two of the sets failed to respond to an on command. The degauss relay clicked on and off, but horizontal deflection wouldn't start. To compound the symptoms, the TVs would eventually come on and work fine and even turn off and back on without a moment's hesitation. The other set was just dead.

Let's begin by dealing with the first problem. Obviously, several things have the ability to keep the set from turning on – low standby voltages, the XRP circuit, failure of one of the switched voltages, etc. I ruled out several circuits simply because I didn't have to unplug the TV to get it to respond to a second on command. I could press the power button from opening till closing of the shop and get a response each time. Therefore, rule out XRP or CRT protection. I also ruled out a low standby voltage and failure of one of the switched voltages to come on-line.

Goodness, what is left? I put my scope probe on the collector of the horizontal output transistor to monitor the voltage as the set came on. I noticed that the voltage was about +125 volts but dropped to about +95 volts when horizontal drive was applied to the transistor. Not enough to start horizontal scan, don't you see.

The problem was in the power supply, but where? Having worked on these little jewels for several years, I suspected a regulation problem and looked immediately to CX3213 and CX3416 (figure 10-4a). CX3413 checked fine in circuit with an ESR meter, but I replaced it anyway. CX3416 failed the test. That sucker had not only developed a high ESR but also had lost half of its capacitance. After replacing it, the TV set has "lived happily ever after," as they say in fairy tales!

The third set was just dead. Some techs approach a dead Zenith by first replacing a few parts. I don't. I make some voltage and resistance checks and follow the leads I turn up. I couldn't turn up any leads with this one because every component checked as if it were brand new. Having just struggled with the two Y-Lines I previously discussed, I wondered if CX3416 was the culprit, and it was.

System Control

IC6000 is a 52-pin, eight bit, CMOS processor in a DIP package designed to run at 12 MHz. It has the following characteristics and capabilities:

an on-screen display (OSD) generator
two I2C communication buses
infrared input hardware
multiple eight-bit A/D converters
multiple D/A converters (that is, pulse-width modulators)
multiple general purpose input/output (I/O) ports
a built-in closed caption decoder

Incidentally, make a note that figure 10-5 is a complete pinout diagram for the microprocessor. I think you'll find a use for it when you need to troubleshoot the circuit because it offers you a far simpler view of the function of each pin than the full schematic does.

The microprocessor (figure 10-5) communicates with the other ICs on the chassis via two communication buses. Bus 1 (pins 33 and 34) controls or supports the tuner, video processor, PIP and MTS module while bus 2 (30 and 31) handles communication with the EAROM. Let's take a look at what happens on bus 1 to illustrate how the system works. The microcontroller receives information about brightness, color, contrast, and sharpness settings from the user and transmits those instructions over the bus to ICX2200. It also generates on-screen information and sends it to ICX2200 over the same communication lines. The microcontroller, you see, not only responds to your input but also displays its compliance on the screen.

You probably have already picked up on some differences between this processor and previous ones Zenith has used. It should come as no surprise then that it really is new. How it receives data from the front panel controls merely underscores the fact that it is indeed new. It makes use of a new keyboard that is really an A/D (analog to digital) device capable of producing a distinct DC level for each key pressed (figure 10-6). The keyboard produces these varying DC voltage levels by employing two resistor-ladder circuits, the outputs of which go into pins 19 and 20. Pin 22, labeled "AVDD," is the analog voltage reference for the input. So, when you press a certain key on the front of the TV, you cause the keypad to generate a specific voltage that corresponds to the key you just pressed. The microprocessor receives the voltage, interprets it, and executes the command. Interesting, isn't it?

Troubleshooting System Control

I cannot overemphasize the fact that the newer consumer electronic devices require the technician to adopt new ways of thinking and troubleshooting. The days have passed when a tech can find a problem using a trusty Simpson 260 and virtually nothing else. He or she has to have up-to-date test equipment and a pretty thorough knowledge of how the new gadgets work. To a certain extent troubleshooting is troubleshooting, and that's a fact. You begin by dividing the circuit into half, eliminating one of the halves, and dividing the remaining half into another half. You proceed like this until you find the problem circuit and finally the problem component. But you have to have the proper equipment and a reasonable knowledge of the device on which you are working to do the job adequately.

Just a little sidebar comment on troubleshooting before getting on with the discussion. Assuming you want to brush up your skills and don't know how to begin, take a look at Samuel Goldwasser's article, "Troubleshooting Techniques," in the December 1999 issue of *Electronic Servicing And Technology*.

Now let's get back to Zenith.

For example, suppose you have a GX chassis in your shop with this complaint written on the work order, "TV comes on but has black screen, no picture, and no sound. However, the menu comes up on request." Where on earth do you begin? I am not a genius, but I do know a thing or two. The problem could be in one of three areas, couldn't it? It could be a power supply problem, a video path problem, or more precisely a problem involving ICX2200 since there is no picture and no audio. It could also be a system control problem. Where do you start?

Well, let's use a DMM to start with because your Simpson 260's low input impedance will load down some of these sensitive circuits and give you a faulty reading. Moreover, it won't act fast enough to read the changes that occur in these circuits. So, you take a few voltage readings around ICX2200 and find that every voltage you measure is correct. Is the chip defective? But before you change it, ask yourself, "Is this chip receiving instructions from system control?"

Remember how communication takes place in this chassis? Of course, it's by clock and data lines from the microprocessor to the other circuits. But you can't check them with a DMM. So you turn on your scope, make a couple of adjustments, and scope pins 27 (SCL) and 28 (SDA) of ICX2200. You see something, but what you see is far below the value the literature calls for. You move your probe to bus one (pins 33 and 34) of IC6000 to check the signals there and see the same thing. But you are a bit savvy. Before you start replacing parts, you "wick out" pins 33 and 34 of the microprocessor to see if the signals change. Sure enough, the waveforms jump up to about 5 volts peak-to-peak. Something has obviously pulled these pins low.

Okay, what's attached to bus one? Well, ICX2200, the tuner, the PIP and MTS module are attached. One or more of these should be the culprit. The next logical step is to unsolder ("wick out") the clock and data lines going to each circuit and check the results. It doesn't matter where you start as long as you cover your bases. When you unsolder the clock and data lines going to the tuner, apply AC, and turn the TV on, you get raster and white noise coming out of the speakers. You check clock and data signals and find them normal!

You see, the tuner was the culprit. The controller chip inside that little tin box had shorted pulling these lines low. Those peculiar symptoms noted on the work order were the results of a fault on the communication bus. This is what I mean when I say you have to keep up with what's going on and have to have decent equipment to deal with the new generation of televisions. Therefore, learn all you can!

Use the following table to make a few quick checks because these measurements tell you a great deal about the health of the microprocessor and each of its associated components.

the IR input at pin 2
power control at pin 52 (It outputs a high to turn the set on.)
+5 volts at pins 22 and 39 using the ground of the microprocessor as reference
oscillator at pins 36 and 37 (check for peak-to-peak value as well as frequency)
reset at pin 35
horizontal sync at pin 28
vertical sync at pin 29
R, G, B output at pins 24, 25, and 26

Tuner

Since Zenith usually discusses the tuner as an integral part of system control, let's talk about the tuner. Zenith initially used two tuners in the GX chassis, 175-2411 with the RF connection at the top edge and 175-2721 with the RF connector near the middle. The two tuners are electrically identical but mechanically different. Some 17 modules were manufactured using the 175-2411 tuner while replacement modules came equipped with the 175-2721. It was not possible to mount the original jack pack module on the replacement module. So, Zenith instructed its authorized service centers to order Zenith part number 12-10681-10, the audio jack pack housing, to match the 175-2721 tuner in the event the main module had to be replaced.

The tuner that gives problems like the one I just discussed belongs to this group. It gives so much trouble I usually order them in batches of three to six. I am talking about the 175-2721 tuner. Fortunately,

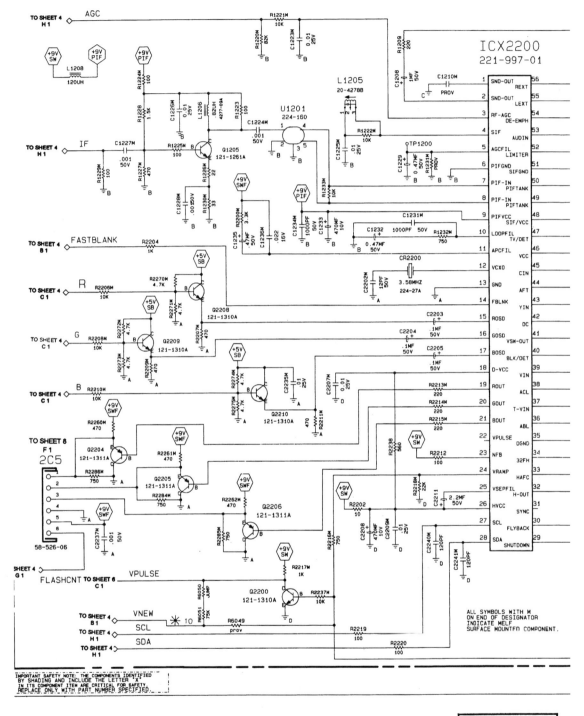

Figure 10-5 Microprocessor Schematic

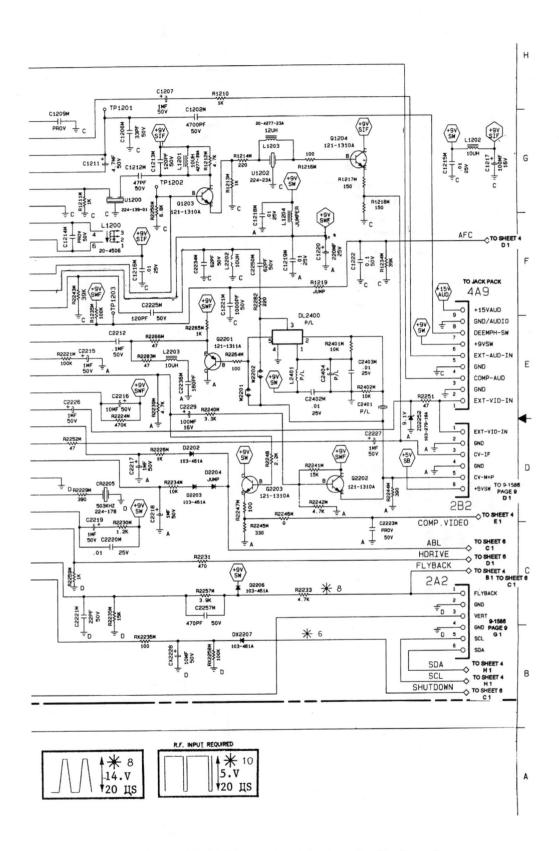

Figure 10-5 Microprocessor Schematic (Continued)

Servicing Zenith Televisions

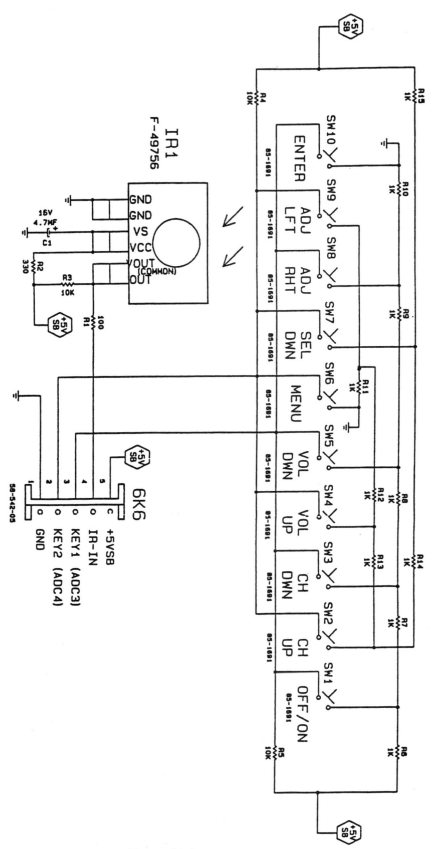

Figure 10-6 Keyboard with IR

The Y-Line Chassis

it is one of the least expensive tuners I have ever bought, selling for about $28. Even PTS sells them cheap! If you do much Zenith work, be sure to keep one or two handy.

The service manual is a bit vague about the tuner's pinout arrangement. It is there, but you have to dig for it. So, let me give you the information in a format that ought to make your servicing a bit easier:

pin 1	AGC	pin 7	5 volts
pin 2	TU	pin 8	Lock
pin 3	EN/AS	pin 9	33 volts
pin 4	Clock	pin 10	IF 2/NC
pin 5	Data	pin 11	IF 1
pin 6	9 volts		

When you troubleshoot a tuner problem, pay particular attention to all of its voltage needs (5 volts, 9 volts, and 33 volts), the AGC requirements, and especially clock and data information. Use pin 1 to inject an IF signal into the system.

Deflection Circuits

The deflection circuits have the feel of "something old" and "something new" about them, especially the vertical deflection circuit. Since they are not trouble free and are prone to give more problems than some of the other circuits, I have decided to discuss both of them in some depth.

Horizontal Deflection

Horizontal drive leaves ICX2200 at pin 32 and goes to the base of Q3202, a transistor that serves as a horizontal predriver (figure 10-7). It is configured as an emitter follower to increase the current gain of the horizontal drive signal and provide impedance matching. Q3206, the next stage along the way, is the horizontal driver transistor and is used to drive the horizontal driver transformer, T3205. Pay attention to R3299 in its collector circuit because this resistor has been selected to control base current to ensure that the transistor is neither underdriven nor overdriven. The "P/L" designation beside the location number alerts you to look at the parts list and select the appropriate value for the module on which you are working. The service manual says – and you should pay attention to this – "Its value will vary for each screen size because the best drive is a function of collector current, which is chassis dependent" *(Technical Training Program: Y-Line Color Television GX Chassis*, p. 14).

The horizontal driver transformer couples the drive signal to the base of the horizontal output transistor that excites the flyback and the horizontal winding of the deflection yoke. I once again call your attention to the "P/L" designation in the parts designation for the horizontal output transistor and flyback. The exact part number depends on the screen size.

It might be a good idea to note that the high voltage also varies according to screen size. Expect to find about 28 KV at the second anode of screen sizes 19" through 27" and about 30 KV for the 31" and 35" sets.

The flyback – Zenith calls it "the sweep transformer" – develops the usual voltages in the usual way. DX3273 and CX3268 generate +23 volts for vertical deflection. DX3287 and CX3296 develop the voltage for the video driver transistors which is +215 volts for screen sizes up to 27" and +245 volts for the larger screen sizes.

Pincushion Correction. Small-size televisions don't require external pincushion correction circuits because they have the necessary correction built into their yokes. But the 27" and larger ones do.

Servicing Zenith Televisions

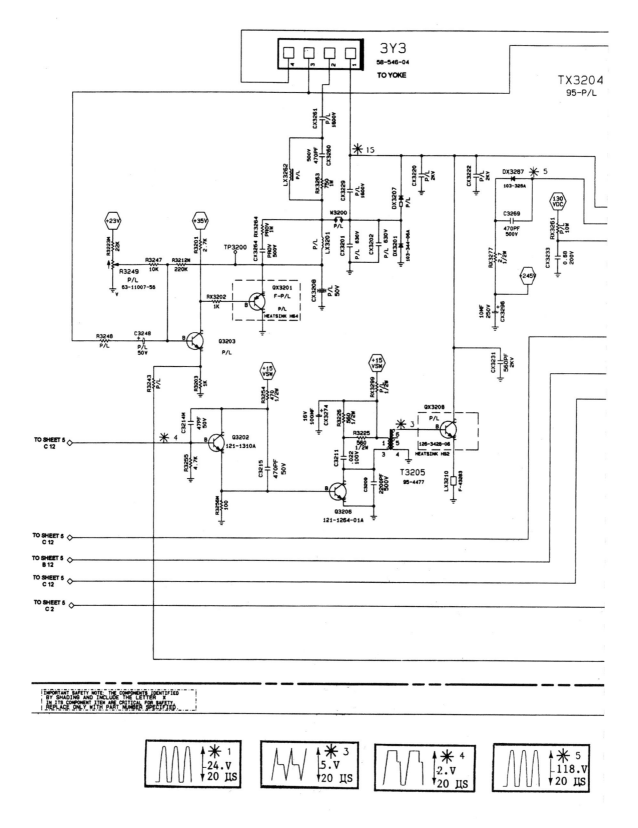

Figure 10-7 Deflection Circuits

The Y-Line Chassis

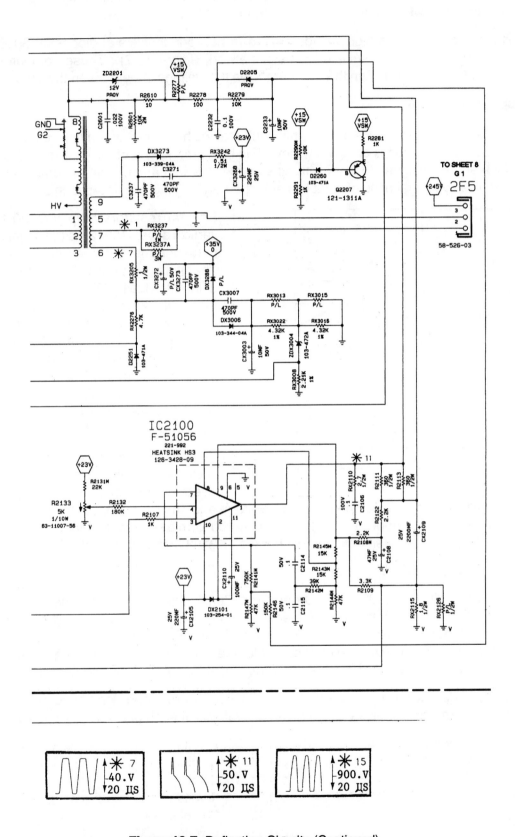

Figure 10-7 Deflection Circuits (Continued)

Zenith uses a diode modulator (QX3201 and associated components) to make the necessary correction. R3249, the horizontal width control, is also a part of the circuit.

High-Voltage Shutdown. The high-voltage shutdown network consists of diodes DX3006, DX3004, capacitor CX3003, and the associated resistor network (figure 10-7). These components feed a voltage to pin 29 of ICX2200. Zenith's literature says this voltage should be on the order of 2.0 volts.

Vertical Deflection

The horizontal deflection circuit is basic Zenith stuff that we have seen in previous chassis. However, the vertical deflection circuit is completely new. It is built around one of the newer vertical deflection ICs (IC2100) and is designed to work with all screen sizes with a minimum of component changes. Figure 10-8 gives you a glimpse of what is inside the new 11-single-in-line-pin chip.

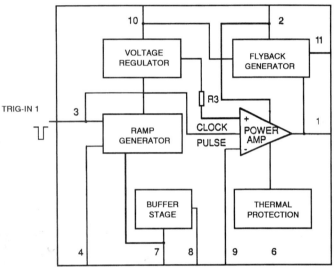

Figure 10-8 11-Single-In-Line-Pin Chip

Several resistors help IC2100 to develop vertical drive. RX2126 is the major component that varies from screen size to screen size. The literature says that RX2122 also has a different value in 27" sets. I am assuming it is true, but I haven't confirmed it. RX2115 in parallel with RX2126 serves to sample vertical yoke current. The value of RX2126 trims the total resistance so that the proper yoke current produces a signal at RX2126 that is about 1.5 volts peak to peak. The vertical circuit always works with the same voltages no matter what the screen size is if this value is maintained.

Vertical drive originates inside ICX2200, exits at pin 24, and goes through Q2200 where it is inverted and applied to pin 3 of the IC2100. C2114 and C2115 form the reference sawtooth voltage and provide a point at their junction to which linearity correction can be applied. DX2101 and CX2100 form a voltage doubler circuit to provide the necessary voltage to ensure smooth vertical retrace. R2133 sets the vertical size.

Now just a couple of concluding notes.

The part number for IC2100 is 221-992. Order it from Zenith using that part number or use an ECG or NTE substitute. I have encountered no problems with either generic. If you are called upon to service a poor vertical linearity problem, suspect C2114 and C2115. When you to change any parts, be absolutely certain that you select the correct part for the chassis you are servicing. Look up the part number if you aren't sure what to use because you'd best be safe rather than sorry later on.

In the event you do have to service a vertical deflection problem and don't have quick access to a service manual, I'm going to give you the expected voltage at each of the 11 pins of IC2100:

pin 1	10 volts	pin 6	0 volts
pin 2	23 volts	pin 7	4 volts
pin 3	8 volts	pin 8	5 volts
pin 4	7 volts	pin 9	4 volts
pin 5	n.c.	pin 10	23 volts
		pin 11	0 volts

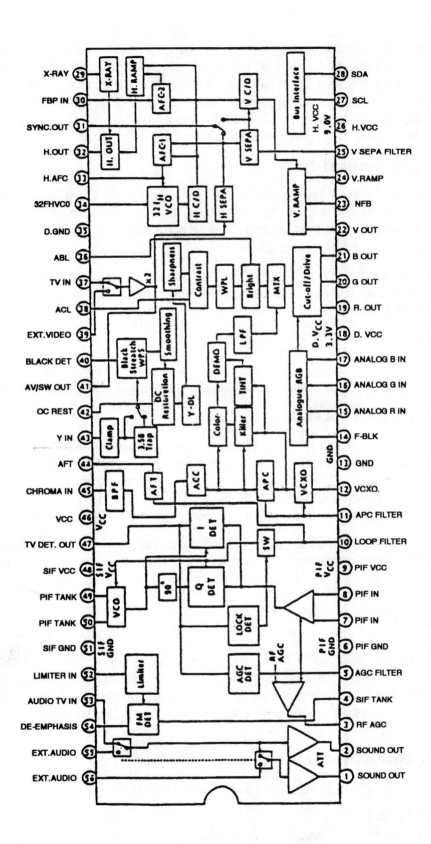

Figure 10-10 ICX2200 Block Diagram

Video Circuit

ICX2200, the workhorse of the video system, is similar to processors Zenith has used in the past because it processes picture and sound IF, video and chroma, generates horizontal-vertical drive, serves as an RGB generator and driver, and OSD mixer. It also includes features like black level expansion, white level compression, an internal 3.58 MHz trap and bandpass filter. User controls are adjusted over the bus lines that go to pins 27 and 28. Refer to figure 10-5 for a layout of the circuit in its awesome details. Figure 10-10 gives you a peek inside the chip itself.

Vertical and Horizontal Drive

The vertical section employs a narrow range countdown circuit to generate vertical drive. It operates within +/- 10% of 60 hertz but defaults to 60 hertz in the absence of a signal. Zenith engineers designed the circuit like this so that the vertical size and position of the on-screen menu would change very little from a signal to a no-signal condition.

The horizontal section utilizes a dual PLL loop with a horizontal coincidence detector to control the gain of the phase detector. The coincidence detector ensures fast signal acquisition during channel change or in loss-of-signal situations.

Video Muting

The microprocessor controls a video-muting feature to keep the on-screen menu stabilized and the raster blanked when the customer selects the auxiliary video input and no signal is present.

Luma and Chroma Processing

If you read about luma processing in the C-11 and C-12 chassis, you pretty much know how the GX chassis processes luminance. The IF signal from the tuner enters at pins 7 and 8 of ICX2200 where it is processed and exits at pin 47 to reenter at pin 37. Auxiliary video appears at pin 39. The switched video out, which is either tuner or external, goes out on pin 41 where it is processed by Q2202, Q2201, the delay line (DL2400) and associated components. IXC2200 does the switching internally.

Luminance alone reenters ICX2200 at pin 43 while the chroma puts in its appearance at pin 45. On-screen information from the microprocessor in the form of R, G, and B data enter at pins 15 through 17, respectively. R, G, and B drive for the picture tube make their way out on pins 19 through 21 respectively.

Audio Signal

U1200 separates the 4.5 MHz audio IF signal from the video signal. The resultant signal enters ICX2200 at pin 52 at which point the signal undergoes some additional processing and exits at pin 2 as composite audio. Auxiliary audio appears at pin 55. ICX2200 has a set of internal switches that permits it to select between tuner or composite audio just as it switched between external and internal video.

Key Voltages and Waveforms

Check key operating signals and voltages at the following points:

composite video out at pin 41
composite audio out at pin 2
horizontal drive out at pin 32

vertical drive out at pin 24
blue video out at pin 3 of 2C5
green video out at pin 2 of 2C5
red video out at pin 1 of 2C5
+9 volts B+ at pins 9, 23, 33, 46, and 48
serial data at pin 30
serial clock at pin 31

Video Output Module

The video output module is a breakaway portion of the main module just like other video modules we have studied. Get a close look at what's on it by studying figure 10-11. Two things immediately strike you: (1) there is no service switch, and (2) there are no black and white tracking controls. The servicer must set the tracking via the service menu. The EAROM stores those adjustments and permits the microprocessor to retrieve them and send them over the I2C bus to the video processor.

Now let me call your attention to Q5104.

This transistor receives a signal from the microprocessor when the TV has been turned off that causes it to turn on and temporarily conduct to discharge the picture tube. The circuit works nicely, but the transistor is prone to failure. When it fails, it produces a set of symptoms that might lead you to think

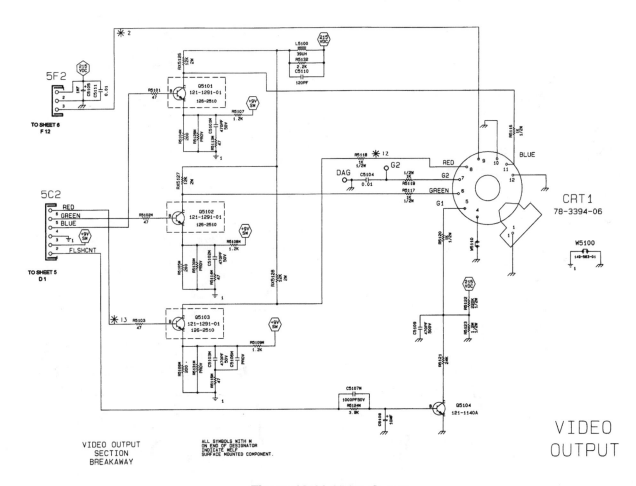

Figure 10-11 Video Output

Servicing Zenith Televisions

the picture tube is bad. Before you condemn the picture tube, check Q5104 for emitter-to-collector short. If it is defective, replace it with Zenith part 121-1140A or ECG399 or NTE399.

AFC Search Problems

I want to deal with one more problem that relates to ICX2200. I have debated about where to put it and decided this is as good a place as any. The complaint may be something like this, "When in the cable mode, the set searches for the channel several times and then locks off the channel." If you change the BAND/AFC to CATV AFC OFF, you solve the problem until the customer runs auto search again because the set will revert to CATV AFC ON. According to a service bulletin, PIFVCO in these

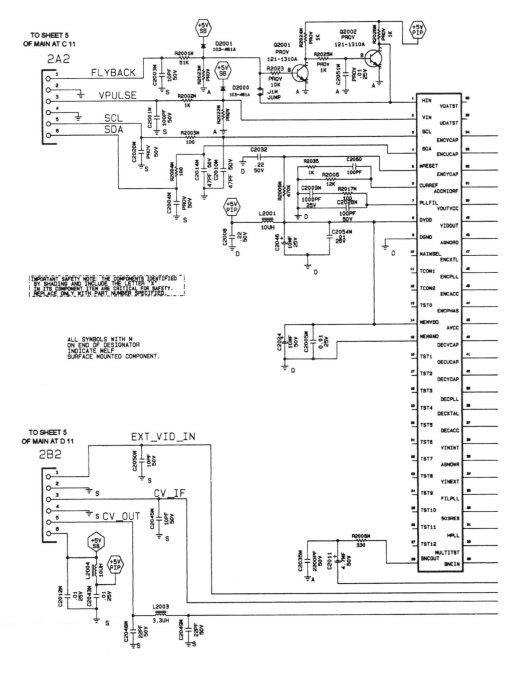

Figure 10-12 PIP Module

The Y-Line Chassis

TV's was set to 63 and AGC to 40. Readjusting the PIFVCO slightly fixed the problem. If you run into such a complaint, make the adjustment one increment (either up or down) at a time and check your results. You shouldn't have to make a radical adjustment. Be sure to record the original reading as a reference. Don't trust your memory!

Picture-In-Picture Module

The PIP function operates by remote control. The user may select any of the sources available, but it initially fires up with the same source as the main picture. However, pressing the source key on the remote control lets you select any one of the other signal sources for the PIP.

Zenith sets from 27" through 35" chassis use the 9-1586 (figure 10-12) PIP module. It is a vertically oriented plug-in module located on the left side of the chassis (side opposite the flyback), plugging in at the 2A2 and 2B2 connectors.

Composite video from ICX2200 comes into IC2000 on pin 36 while auxiliary video comes in on pin 34. The main microprocessor exercises control over this IC via the clock (pin 4) and data (pin 3) lines. Horizontal and vertical pulses, signals necessary for PIP operation, make their appearance on pins 1 and 2 respectively. Video exits ICX2000 at pin 49 where it goes to the base of Q2000. Its emitter sends the signal on to pin 41 of ICX2000.

The PIP module is relatively trouble-free, but like other Zenith PIPs, it has occasionally experienced problems. For example, there have been reports of a TV coming on with vertical bounce or reduced vertical deflection accompanied by weak video and retrace lines in the picture. All standby voltages checked good. If you made a few other checks, you found the high voltage slightly low as well as the derived voltages. The set even had audio. Moreover, it turned itself off – it did NOT shut down – in something like 20 seconds. The problem was the PIP module pulling the +5-volt line low and corrupting the data line.

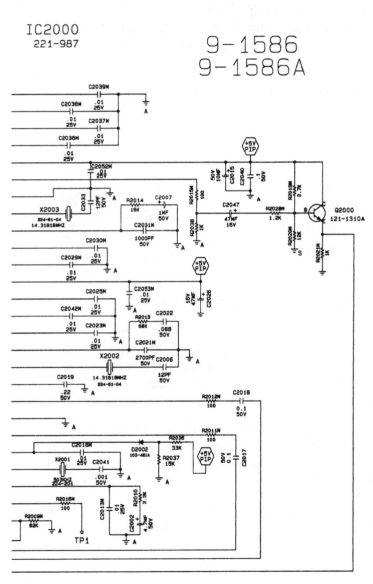

Figure 10-12 PIP Module (Continued)

I suppose all things are possible. It is, therefore, possible to fix it. However, the little thing doesn't cost much. It is only about the size of your hand! Since you are likely to waste more time trying to fix it than it's worth, just replace it as a module and be done with it.

If you replace the PIP, be sure to unplug the set before you remove it and install the new one. Don't merely turn the TV off. The microprocessor has the ability to detect hardware that has been hooked up to the bus line. This feature is called an "auto detect" feature, and it runs every time the microprocessor resets. If the TV has just been turned off, the microprocessor may not recognize the new PIP, and it may not work correctly when the set is turned back on.

Audio Circuit

The GX chassis comes equipped with one of three audio systems: monophonic, low-cost stereo, and MTS stereo. Regardless of the audio feature, the audio module plugs into connectors 4B9 and 4A9 on the rear of the chassis right beside the tuner. The little module is about the size of a man's hand and contains all the circuitry necessary to process the audio and drive the set's speakers.

If you change the module or if you change the main module, you MUST enter the service menu and check the settings of menu items 32 through 38 to make sure they match the bar code on the jack pack label (figure 10-13). If you don't make the check and the items don't match, one of two conditions will exist: the customer might not have audio or the audio won't match the set's features.

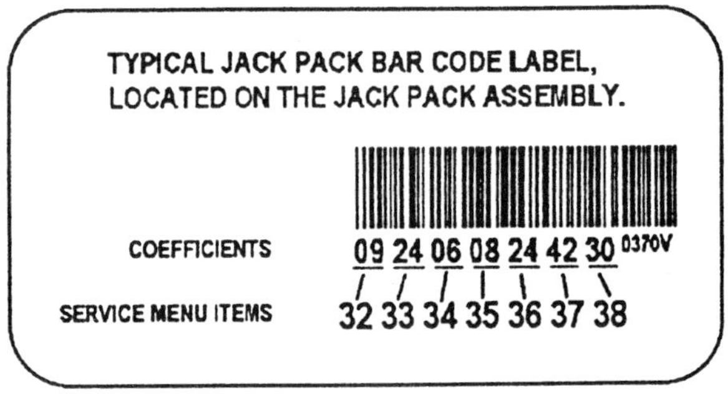

You should also be aware that a defective jack pack will cause the TV to be dead. If, for example, the audio output shorts, the fuse (FX3402) in the +15-volt line of the power supply opens. Do you remember our discussion of the power supply? If you do, you recall that the +5 volts for the microprocessor is derived from the +15-volt supply. No 15 volts, no 5 volts, and a dead set. If you replace the module without doing a bit of troubleshooting first and put the original jack pack onto the new module, the new module will obviously "fail" at plug-in. Then, when you work on a set that has a blown FX3402, remove the jack pack and check pin 9 of IC804 (the audio output IC) to its heat sink or to wire W1605. If you get a reading of just a few ohms, you may be reasonably certain that IC804 has shorted and caused the fuse to fail.

Figure 10-13 Jack Pack Label

Be aware that when you replace the audio output module you must permit the microprocessor to reset in order to recognize the replacement, just as you do when you replace the PIP module. It is best, then, to unplug the TV before you attempt to change the audio module. The microprocessor resets when AC is reapplied, causing it to run its routines, one of which is the auto detect routine.

As I indicated earlier, the GX chassis comes equipped with one of three audio jack packs, monophonic (figure 10-14), non-MTS stereo (figure 10-15), and MTS stereo (figure 10-16).

Monophonic Audio

If the set has mono audio then audio from pin 2 of ICX2200 is coupled to pin 4 of IC801, which is the audio output amplifier. It amplifies and sends the signal on its way to the speaker by way of connector 9R4. Pins 5, 6, 7, and 8 of IC801 are tied to ground for heat sink purposes.

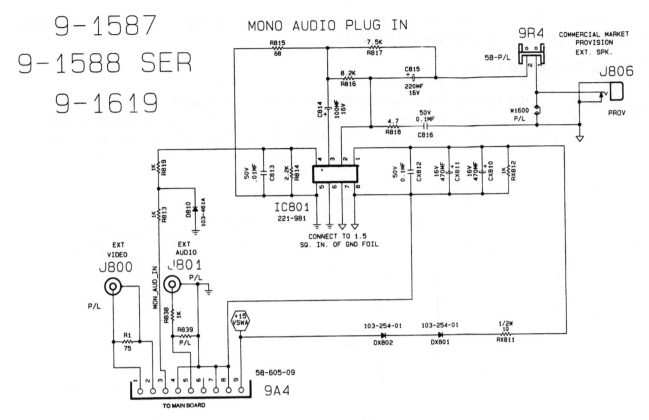

Figure 10-14 Monophonic Audio Jack Pack

MTS Stereo

In the TV mode, composite audio from ICX2200 enters the jack pack on pin 4 of connector 9A4 and goes to pin 17 of IC1400. IC1400 processes the baseband audio signal to produce left channel (on pin 35) and right channel (on pin 35) signals. These signals proceed to pins 8 and 4 of the Aux/TV switch, IC1401.

Auxiliary audio is also present at pins 1 and 11. Transistors Q1401 and Q1400 control the switching arrangement. The switching signal originates at the microprocessor and enters the jack pack at pin 3 of 9B4 to the base of Q1400. When the consumer selects TV audio, the microprocessor sends a high to the base of Q1400 to turn it on. This action places a low on pins 12 and 13 of IC1401 which is the proper voltage level for processing audio from the TV's tuner. When the consumer selects auxiliary audio, the microprocessor pulls the base of Q1400 low, placing a low on pins 12 and 13 which is the proper voltage level for aux audio processing.

The selected left and right audio exit IC1401 and go to pins 1 and 2 of IC1402 via 1K resistors and 10mfd capacitors. IC1402 is a volume limiter. It limits the level of the input audio signal to prevent excessive and above normal audio programming.

Resistors R1440M and R1441M are connected to pin 10 to control the peak threshold level of the limiting exercise. Pin 11 is the "enable line." If it is pulled lower than 1.5 volts, the circuit becomes active. If it is higher than 2 volts, the circuit is in the off mode. The signals leave the sound limiter at pins 4 and 5 and go back into IC1400 at pins 37 and 38. Right channel audio exits on pin 5 and left channel audio on pin 6 and proceed to IC804, a dual-channel audio amplifier, that amplifies the signals and sends them on to the speakers. The I2C bus controls volume and mute functions. Q800 is used to mute the audio during turn on and turn off to keep the speakers from making a popping sound that can be annoying at high volume levels.

Non-MTS Stereo Audio

Composite audio from ICX2200 enters pin 5 of IC1601, the decoder amplifier. Unlike the MTS stereo decoder, there are no operating parameters to set via the service menu because a series of external components set the operating parameters. Transistor Q1601 and its associated components comprise

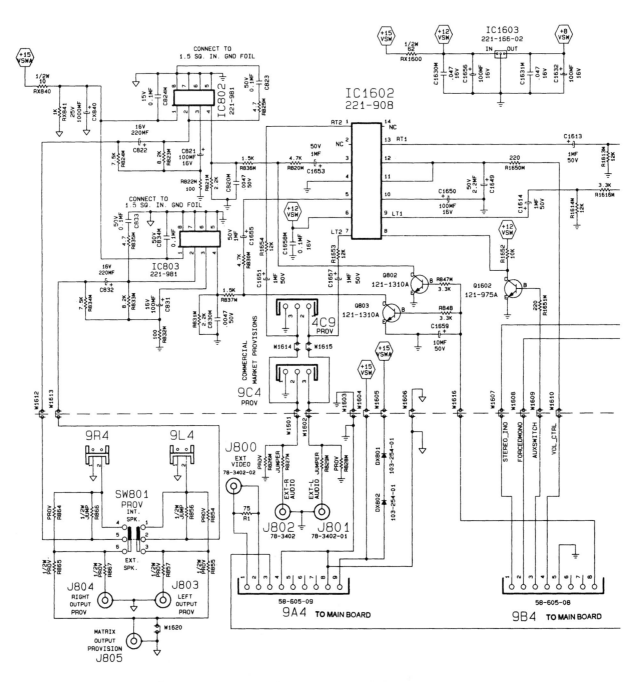

Figure 10-15 Audio Module / Non-MTS Stereo Jack Pack

a de-emphasis filter for the stereo decoder because the best separation for the non-MTS system takes place between 3 kHz and 5 kHz.

IC1601 processes the audio signal, producing left (pin 14) and right audio (pin 13) audio signals. When they leave IC1601, the audio signals go directly to IC1602, a voltage-controlled amplifier. External audio puts in an appearance at pins 1 and 7 with the aid of Q1602. IC1602 does the job of switching between these two sources. If you look at the schematic (figure 10-15), you see that this chip also handles the volume (pin 11) and mute (pin 12) functions when it receives the proper instructions from the microprocessor.

There is one more stopover on the way to the speakers. Left and right audio exit IC1602 at pins 4 and 3, respectively, and are coupled to pins 5 and 4 of the audio output amplifier, IC803. Note that pins 5

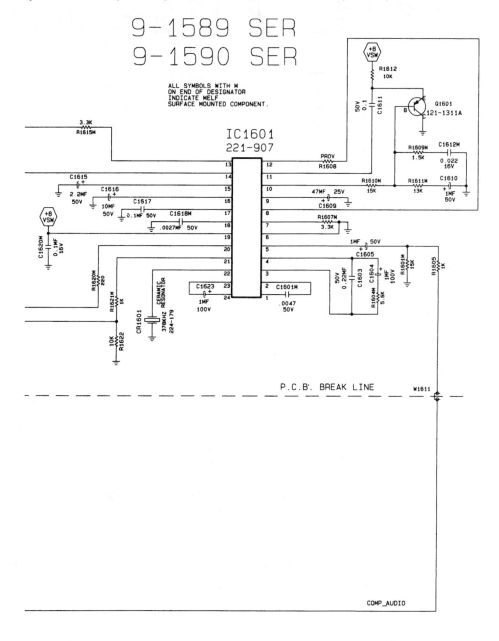

Figure 10-15 Audio Module / Non-MTS Stereo Jack Pack (Continued)

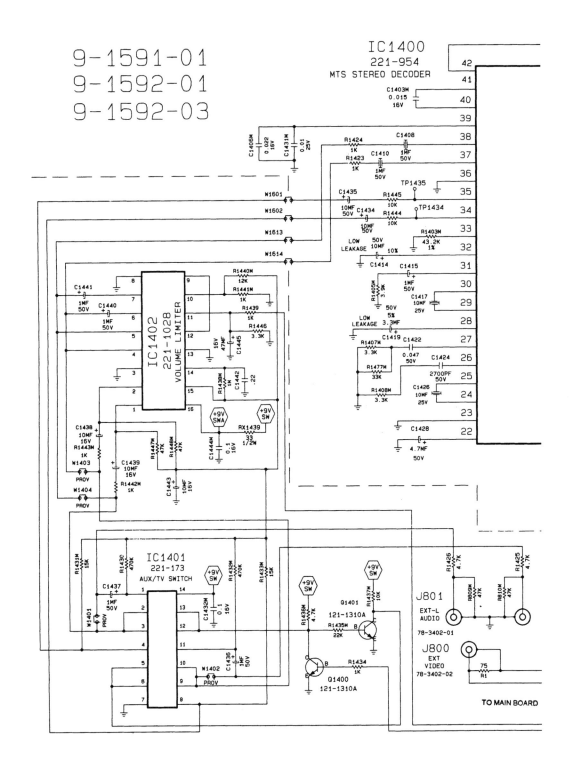

Figure 10-16 Audio Module / MTS Stereo Jack Pack

The Y-Line Chassis

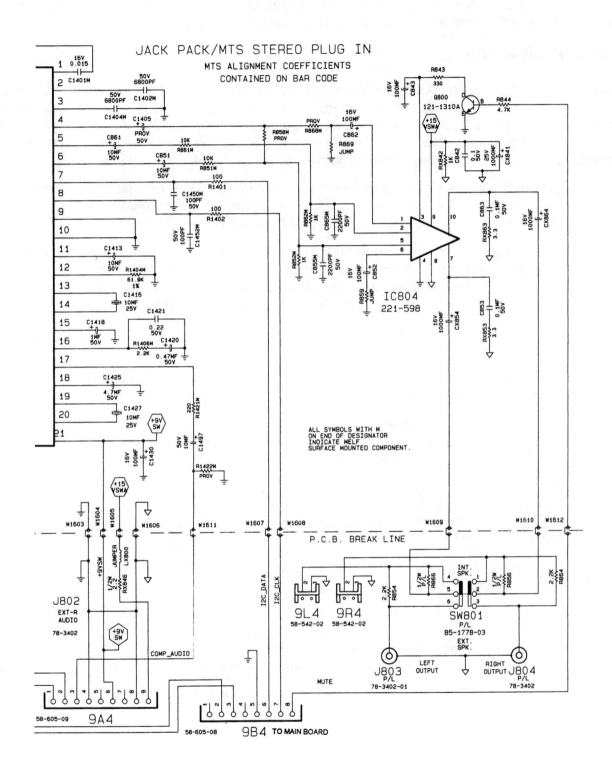

Figure 10-16 Audio Module / MTS Stereo Jack Pack (Continued)

through 8 are tied to ground for heat sinking purposes. Transistors Q802 and Q803 mute the audio at turn on and turn off to keep the speakers from popping due to transient switching noise. These transistors receive their instructions from the microprocessor.

Repair History

It's time to take a look at the GX's repair history. I have notes on most of the modules in this family. Since the repair for one is applicable to the others, I won't distinguish among the modules. You should also be aware that I have discussed some of these repairs in the body of this chapter. I thought about including some information from *ZTips*, Zenith's technical tips program, but decided to recommend that you buy the program. It is inexpensive and has several useful service programs in it.

(1) Channels will periodically tune off channel and back on. TV will not play back some copy guarded tapes. Also poor closed caption performance. Zenith issued a service bulletin telling us servicers to replace the EEPROM (part number A-18382-07).

If you have to replace it, have enough sense to replace the shield that covers it and the microprocessor. I have seen servicers leave out screws, fail to install shields, and fail to properly route the wires that go to and from the main chassis. Shields, screws, and wires are placed as they are for a reason. Be professional; replace them.

(2) Bright (white) screen with retrace lines. Check Q5106 before you do anything else. See the discussion in the text for details.

(3) No audio. Check the audio going into the jack pack module. If it is good, you most likely have a defective audio output IC, but don't forget to check the components tied to it.

(4) Snowy picture on all channels. You would be correct to suspect the tuner. But before you replace it with a new one, open the shields and peek inside. One I serviced had a crack in the printed circuit board just where the RF connector solders into it. There is no need to use a new tuner to replace one you can fix yourself, is there?

(5) Distorted picture. In this instance a horizontal bar 4 to 6 inches wide made a random appearance on the screen. You guessed it, a defective picture tube. Make a note that this phenomenon can occur in just about any picture tube Zenith uses.

(6) Dead set. The start-up resistor (RX3404) for the power supply had opened. I have seen the problem twice.

(7) Scrambled picture that gets worse on some channels. Looks very much like what we used to call "an AGC overload." ICX2200 caused this problem. Pick up on it by applying a little heat from a soldering pencil or hot air gun directly to the IC. If the picture clears up, you know you have found the problem. Alternately, you may have to apply a little freeze spray.

Other than the usual damage caused by power surges and customer abuse, that's all I have on the GX chassis. It seems to be holding up quite well.

CHAPTER 11

THE Z-LINE

The Z-Line made its appearance in 1997and included four chassis, the GA, the GH, GX, and the GZ. The GA and GX chassis have been designed for use in 19" through 27" sets. The GH chassis appears in 27" through 36" sets.

The GZ chassis has been designed for use exclusively in Zenith's high-end direct-view and projection sets and belongs to Zenith's "advanced video imaging" line. Its features include a comb filter, auto skin tone circuitry, black level enhancement, white level and peak white level compression, and a video menu preset mode to allow the viewer to optimize the picture for various programs and conditions. Audio features may include SoundRite, Dolby ProLogic, and a three-channel 20-watt audio amplifier. Additional features include a two-tuner PIP, channel labels, menu icons, Gemstar, a very sophisticated jack pack, and a "Z-Trak" remote control. I could spend several paragraphs just talking about the GZ chassis features, but you get the point. It's very similar to the GZ chassis that belongs to the Y-Line. After a bit of arguing with myself, I decided not to discuss it because it really lies outside the scope of this book. If you are interested in the GZ chassis or expect to service them, a bibliography and information about the service manuals and training material is provided.

Someone asked, "How can you tell which chassis is which for these new Zenith televisions?" That's a good question, one that figure 11-1 should help you answer. The "P" under chassis stands for "purchased" and refers to those televisions Zenith purchases from other manufacturers and sells under their brand name. The other designations are self-evident.

The GA family of direct-view television receivers uses these modules:

9-1789 for 25" mono sets
9-1790 for 25" mono sets
9-1791 for 27" stereo sets
9-1869 for 19" mono sets
9-1870 for 19" mono sets
9-1871 for 20" stereo sets
9-1831 for 25" stereo sets

The GH chassis includes these modules:

9-1816 for 27" sets
9-1911 for 27" sets
9-1817 for 32" sets
9-1818 for 36" sets

The GH modules are complimented by a 9-1819 MTS stereo or a 9-1820 MTS stereo jack pack and/or a 9-1822 single-tuner PIP or a 9-1823 two-tuner PIP.

Servicing Zenith Televisions

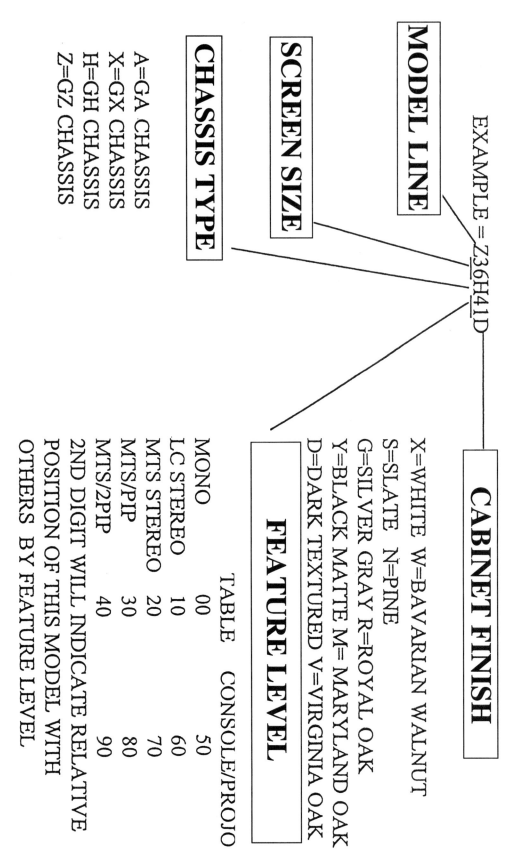

Figure 11-1 Model Number System for Regular Z-Line Color Televisions

The GX chassis includes these modules:

9-1722 for the 19/20" sets,
9-1807 for the 20" sets,
9-1808 for the 25" sets without a comb filter,
9-1809 for 25" sets with a comb filter,
9-1810 for 27" sets without a comb filter,
9-1811 for 27" sets with a comb filter,
9-1812 for 32" sets with a comb filter,
9-1813 for 35" sets with a comb filter.

The GX modules come equipped with a 9-1814 picture-in-picture module and a 9-1733 MTS stereo jack pack.

Since there are differences – and sometimes major differences – among these chassis, I feel the need to treat them as individual entities, making note of what they have in common to minimize discussion space while I highlight their differences. I hope this departure from the procedure followed in the preceding chapters won't be confusing.

Available Literature

As I indicated in the last chapter, Zenith's technical training material and service manuals have made giant leaps forward with the advent of "repair to the component level" modules. I have absolutely no reservations about recommending both to you. In fact, I strongly suggest that you secure them when you can. I operate an authorized service center for the repair of over two dozen brands, and in my opinion Zenith's new literature ranks near the top of the list with respect to readability and usability.

Since I will be skipping the GZ chassis, I'll list its literature first. If you see very many of these complicated TVs, you certainly ought to order the training manual which is *Technical Training Program: GZ Directview and Projection TV* (manual number 923-3314TRM). Order the service manual by requesting CM-149/GZ using part number 933-3321. It contains operating instructions, troubleshooting aids, a principal component location guide, interconnect diagrams, schematics, block diagrams, and that all-important parts list.

Order the remaining training manuals by requesting CT149TRG for the GA and GH chassis, the latter doing double duty for the GX chassis. The service manuals are CM-149/GA, (part number 923-3322 for the GA chassis), CM-149/GH, (part number 923-3323 for the GH chassis), and CM-149/GX (part number 923-3315 for the GX chassis). One or more supplements should come with each manual because Zenith added a model or two (or more) to each chassis after the main manual was printed. The supplements make note of parts and circuit changes and/or additional parts used in the model(s) in question.

GA Chassis

The place to begin is at the beginning. Let us then turn our attention to the GA chassis. Would you believe I don't have a source for reproducing a picture of it? But I am able to give you a rough sketch that also serves as a component location guide and an interconnect diagram (figure 11-2). However, it looks like the chassis in the previous Y-Line in that all components have the same physical location on the circuit board. If I were able to produce a picture of it, I believe you would find that is quite similar to the GH chassis.

Servicing Zenith Televisions

The GA series was developed for use in TVs using 19" through 27" picture tubes. Naturally, these chassis have to offer a variety of features. The question is, "How is it possible to offer a choice of features without making lots of physical changes to the circuit board?" Zenith engineers solved the problem by programming the EAROM to accommodate a remarkable variety of feature levels and supported those features by employing plug-in modules, wired-in components, and input-output jacks, circumventing the need to alter significantly the physical layout of the module.

The GA chassis makes use of three ICs to handle the major functions of the chassis. ICX2200 handles the audio/video, snyc, and sweep drive processing as it has in previous Zenith chassis. IC6000 is the system control microprocessor, and IC6001 is the EAROM. The keyboard and IR receiver are

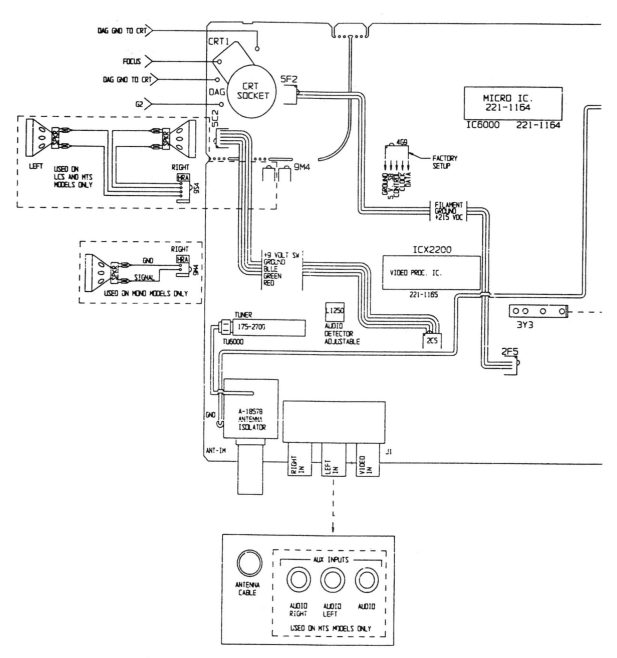

Figure 11-2 GA Chassis Location Guide and Interconnect Diagram

tied directly to IC6000. IC2100/2101 handles vertical sweep. There are, of course, other ICs in the chassis, but these three handle the signal, sync, and sweep processing.

The customer menus are displayed over video and appear on the left side of the screen. Unlike the classic Zenith menus, these scroll up vertically from the bottom of the screen or appear instantly. How the customer menus appear depends on the setting of item 26 ("Scroll") in the service menu. The color of the menu depends on the "level" of features, the level having been set in the manufacturing process. Level 0, private label sets, have a dark red background; Zenith levels 1 and 2 have a blue background.

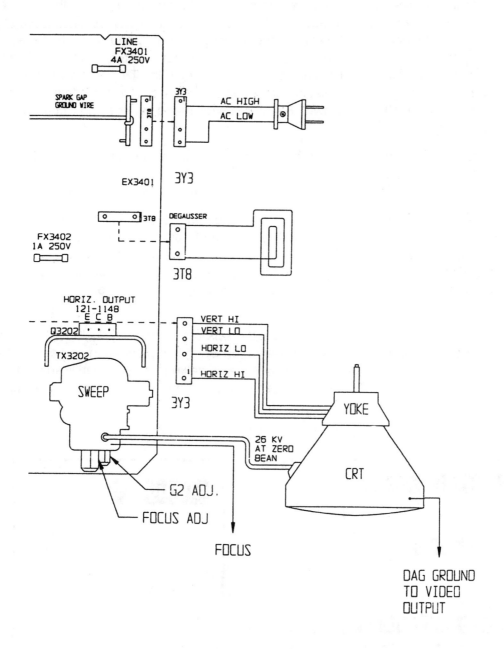

Figure 11-2 GA Chassis Location Guide and Interconnect Diagram (Continued)

Servicing Zenith Televisions

There are two sections to the menu. "Video Menu" includes contrast, brightness, color, tint, sharpness, and picture preference. "Set Up Menu" includes auto program, add/del/surf, channel labels, clock set, timer setup, parental control, captions/text, audio mode, language, and auto demo.

Since these features have already been discussed, please excuse me when I limit my comments just to the "auto demo" feature. It is not just a sales floor gimmick because it is used to demonstrate the menu structure to the customer in his or her home. Place it in the "off" mode for normal television viewing.

Service Menu

Gaining access to the service menu is simple. Press and hold the menu key on the remote control until the user menu disappears. Then press keys 9-8-7-6-Enter, and a display similar to figure 11-3a pops up. You can also gain access to the service menu using the front panel keys. If the set has a 10-button keyboard, press and hold the Menu key until the customer menu disappears. Then without delay, press – at the same time – Adjust Right and Channel Up. If the TV has a six-button keyboard, press and hold the Menu key until the display disappears. Then, and without delay, press – at the same time – Volume Up and Channel Down. Exit the service menu by pressing Enter on the remote control or the front panel.

Now take another look at figure 11-3a. The bar at the top indicates the part number of the software that drives the set. The bar at the bottom contains a date and a number. The date tells you when the module went through the factory, and the number tells you that the module has been tested.

The service menu always comes up on "Hpos" the third item in the menu. To gain access to all of the items in the menu, you must use the Select Up key to select F Mode, item number 00 (figure 11-3b), and then use the adjust key to turn it on. If you don't turn Factory Mode on, you will be able to access only the first seven menu items. Once Factory Mode has been turned on, you have access to any of the 31 items in the service menu. By the way, the 31 items appear at the top left of the screen one item at a time (figure 11-3).

If you forget to turn factory mode off before you exit the service menu, your customer might complain that the set is doing some strange things, like coming on when it is plugged in and cannot be turned off. Don't panic! Just check the setting of Factory Mode by calling up the Setup Menu. If you see a pair of dashes at the top of the list (figure 11-4), then you know you have accidentally left it on. Turning

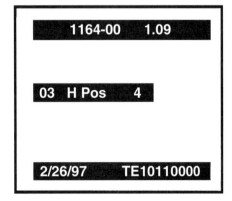

Figure 11-3a Accessing the Service Menu

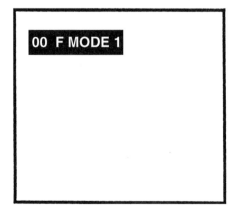

Figure 11-3b Gaining Access to all Service Menu Items

it off is as simple as setting the on-screen clock. You don't even have to make a service call to do it. As a matter of fact, you and your customer should be able to perform "the procedure" right over the phone.

Having gone through the preliminaries, let's now take an in-depth look at each item in the service menu. Table 11-1 lists the 31 items, gives you the range of settings, and the preferred setting for the screen size on which you are working. I don't see the need to comment on each and every item because several are simply set to the "on" or "off" position. Others, however, need a bit of explanation.

01 (Pre Px) stores the customer menu adjustments in the nonvolatile memory of the EAROM.

02 (V Pos) and 03 (H Hos) positions the on-screen display vertically and horizontally.

Setup Menu

Auto Program
Add/Del/Surf
Ch Labels
Clock Set
Timer Setup
Parental Ctl
Captions/Text
Audio Mode
Language
Auto Demo

Figure 11-4 Factory Mode On

ITEM	RANGE	19/20"	25'	27"
\multicolumn{5}{c}{FactoryMode 0 (Blue)}				
00 F Mode	0-1	0	0	0
01 Pre Px	0-1	1	1	1
02 V Pos	0-24	10	10	10
03 H Pos	0-13	10	10	10
04 Level	0-2	1*	1*	1*
05 Band	0-7	1**	1**	1**
06 AC On	01	0	0	0
\multicolumn{5}{c}{Factory Mode 1 (Black)}				
07 RF Bpf	0-1	0	0	0
08 3.58t	0-1	0	0	0
09 RF Brt	0-63	36	26	28
10 Aux Brt	0-63	36	26	28
11 V. Size	0-63	18	18	18
12 V. Phase	0-7	0	0	0
13 H. Phase	0-31	16	16	16
14 Aud Lvl	0-63	46	46	46
15 RF Agc	0-63	40	40	40
16 H Afc	0-1	1	1	1
17 WhComp	0-1	1	1	1
18 60 HzSw	0-3	2	2	2
19 PifVco	0-127	28	28	28
20 R Cut	0-254	0	10	5
21 G Cut	0-254	0	10	5
22 B Cut	0-254	0	10	5
23 G Gain	0-254	90	90	90
24 B Gain	0-254	90	90	90
25 C Type	0-5	2	2	2
26 Scroll	0-1	1	1	1
27 6 Keys	0-1	1	1	1
28 SpkrSw	0-1	0	0	0
29 5 Jacks	0-1	0	0	0
30 St&Sap	0-1	0	0	0

Table 11-1 31 Items in the Service Menu

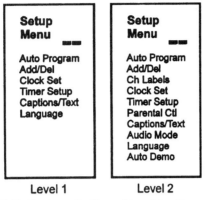
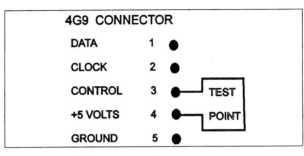

Figure 11-6 Setting Levels

Figure 11-5 Factory Settings Used for Programming

04 (Level) is an item to which you MUST pay attention. It has three positions, 0, 1, and 2. The factory uses these settings to program the module for its correct features as figure 11-5 illustrates. Level 0 is used for private label sets. If you set the level to 0 and exit the service mode, you must use remote control codes 21 or 121 to get the remote control to operate the TV. Zenith reserves levels 1 and 2 for its own use.

Setting the level is simple. After you have entered the service menu and selected item 04, place a jumper between pins 3 and 4 of connector 4G9 (figure 11-6). Then use the Adjust button to change the level. Connector 4G9 is located on the left side of the module just to the left of the microprocessor on 19" and 20" sets and is on the left edge of the module on the 25" and 27" ones. Be careful when you attach the jumper because you might damage the module if you short the wrong pins together, and be certain to remove the jumper after you have set the level. If you are not certain which level is correct, look on the back of the set and find the sticker that has the model and serial number on it, and you will find the correct level for that particular set.

05 (Band) has eight settings: 0 is broadcast fixed; 1 is CATV afc; 2 is HRC afc; 3 is ICC afc; 4 is broadcast afc; 5 is CATV fixed; 6 is HRC fixed; 7 is ICC fixed.

09 (RF Brt) and 10 (Aux Brt) set the range of customer control for brightness in the RF mode (signal from the antenna) or the auxiliary (signal from the jack pack) mode.

14 (AudLvl or Audio Level) sets the gain for the MTS Stereo.

20 through 24 are used to set the black and white tracking (gray scale).

25 (C Type or "chassis" type) has six possible settings: 0 for mono, 1 for mono with auxiliary inputs, 2 for non-MTS stereo, 3 for non-MTS stereo with auxiliary inputs, 4 for MTS stereo, and 5 for MTS stereo with auxiliary inputs.

26 (Scroll) selects the way the customer menu appears on the screen. It can be set to appear all at once or one item at a time.

27 (6 Keys) adapts the chassis to use either a six-key or a ten-key control panel.

28 and 29 are not being used. Set both to 0.

30 (St&Sap) selects the Stereo icon or Sap icon. However, since it is not being used, set it to 0.

Power Supply

The 19" and 20" sets using modules 9-1869/70/71 are line connected, meaning they are hot chassis (figure 11-7). You MUST use an isolation transformer when servicing them. The 25" and 27" chassis use an isolated power supply (figure 11-8), but you still must use an isolation transformer to service it.

19/20" Power Supply

The standby circuit in figure 11-7, a half-wave configuration very much like the one used in the C-6 chassis that I discussed in Chapter IV, produces +13 volts, +12 volts, and +5 volts. The +12 volts switched supply developed by the scan circuit takes over its current demands when the set comes on. Diode DX3409 isolates the standby and switched voltage sources. Make a note that lightning might damage DX3409 resulting in lowered or missing standby voltages.

The full power supply consists of DX3401 through DX3404, capacitor CX3406, ICX3401, and associated components. ICX3401, Zenith part number 126-3643, is also known as an STR30130. It comes on line when the microprocessor sends a high to the base of QX3402, the relay driver transistor, causing it to saturate and provide a path to ground for the coil of the relay (KX3401). The relay "makes" and provides an AC path to the bridge diodes. The bridge diodes and CX3405 supply about +150 volts to pin 3 of ICX3401 which is biased to output +130 volts at pin 4 for the horizontal output transistor and two other circuits.

Zenith taps the +130 volts with RX3414 to develop +33 that the tuner uses to develop its tuning voltage. Four resistors and a zener diode also tie into the +130-volt source to develop +9 volts for the horizontal oscillator. This circuit is located just below ICX3401 in figure 11-7. Be sure to check the +9 volts when the horizontal oscillator doesn't start.

Let me point out one more thing about the circuit before I leave it. Because of the circuit configuration, a power surge caused by lightning, for instance, might be coupled across its components into the horizontal section of the TV signal processor. When it enters via the horizontal vcc input, the surge will more than likely destroy the circuit, meaning you have to replace the chip. For example, I just fixed one that came in with the "dead set" complaint. The relay clicked when I tried to turn it on, indicating that system control was working. I found that FX3402 had opened for no apparent reason.

Replacing it restored the +130 volts to the horizontal section but didn't fix the TV because horizontal drive still wasn't coming up. Notice from the schematic that +9 volts from the +130 volts enters ICX2200 at pin 26 (figure 11-14). Well, that voltage seemed to be missing. After a few checks to rule out open or shorted components, I wicked out pin 26 and the +9 volts returned. The power surge had wiped out ICX2200.

All in all, I believe Zenith has designed a simple circuit that ought to provide a lot of trouble-free use for the people who buy it. When it does give trouble, we servicers are faced with a circuit that is relatively easy to fix.

Troubleshooting the 19/20" Power Supply

Confirm the operation of the power supply by checking the following voltages at the prescribed points:

+5 volts standby at pin 3 of ICX3402
+12 volts standby at pin 1 of ICX3402
+13 volts standby at the anode of CX3411

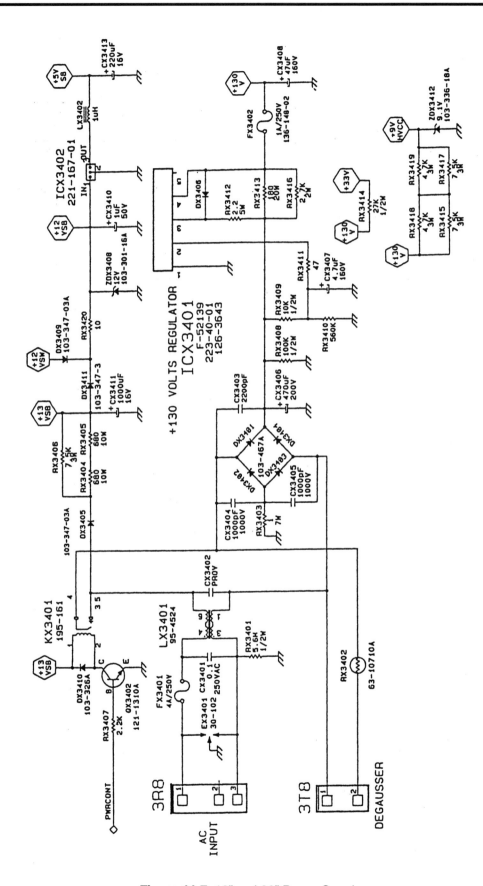

Figure 11-7 19" and 20" Power Supply

Confirm the power on command via the IR receiver or keypad by checking for +5 volts pin 52 of IC6000 that should be coupled to the base of QX3402. In other words, +5 volts at pin 52 of IC6000 should become 0.7 volts at the base of the relay driver transistor.

When the set turns on, check for the switched voltages at these points:

+5 volts at the cathode of ZD3206
+9 volts at pin 3 of IC3201
+12 volts at pin 1 of IC3201
+14 volts at the + side of C3222
+25 volts at the cathode of D3202
+130 volts at FX3402 and the collector of the horizontal output transistor
+150 volts at CX3406
+180 volts at + terminal of C3207

Horizontal sweep should develop +12 volts, +14 volts, +25 volts, and the +180 volts for the video output transistors.

25/27" Power Supply

The larger chassis in the GA line naturally require a heftier power supply. With this in mind, turn your attention to figure 11-8. When you do, you find a variation of the power supply we have been discussing since Chapter 1. Because I devoted a lot of time to the "how it works" agenda in the last chapter, I'll keep my comments here to a minimum.

Like other switch-mode power supplies, it starts up as soon as AC is applied. Since each secondary winding of TX3401 has its own rectifier and filter, the output voltages are always available. Zenith uses PNP and NPN transistors to switch the voltages on to turn the TV on or switch the voltages off to turn the TV off.

When it is off, the TV needs the standby voltages only. The PNP transistors are therefore held off by the NPN transistor. When it receives a command to turn the TV on, the microprocessor sends a high to the base of NPN transistor Q3402. It turns on and pulls the bases of PNP transistors Q3403 and Q3404 low, turning them on. These transistors then permit the voltages they control to become available to the rest of the chassis.

Use these points to confirm the operation of the power supply in standby:

+5 volts at the cathode of ZD3401
+15 volts at the emitter of Q3403
+123 volts at the anode of CX3420

Check for the correct power on sequence by seeing if pin 52 of IC6000 goes high when you attempt to turn the TV on. The power on high signal should be coupled to the base of QX3402 where the voltage ought go from 0.0 volts to 0.7 volts.

Turn the set on and make the following checks to confirm the presence of the necessary switched voltages:

+15 volts at the collector of Q3404
+9 volts at the anode + terminal of C3428
+ 5 volts at the cathode of ZD3402
+15 volts at the collectors of Q3403 and Q3404

When it comes on line, horizontal sweep should bring up these sweep-derived DC voltages:

+23 volts at the junction of CX3268 and RX3242
+35 volts at CX3272
+215 volts at the junction of CX3296 and RX3277

System Control

The microprocessor avails itself of CMOS technology and comes in a 52-pin DIP package. It runs at 8 megahertz and has an 8-bit core. It has a built-in on-screen generator, infrared input hardware,

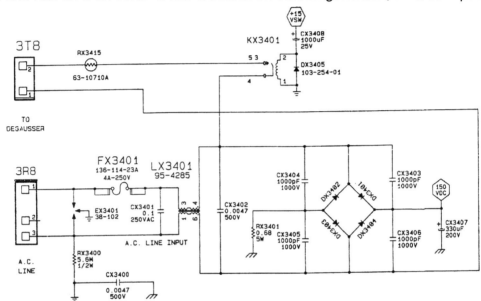

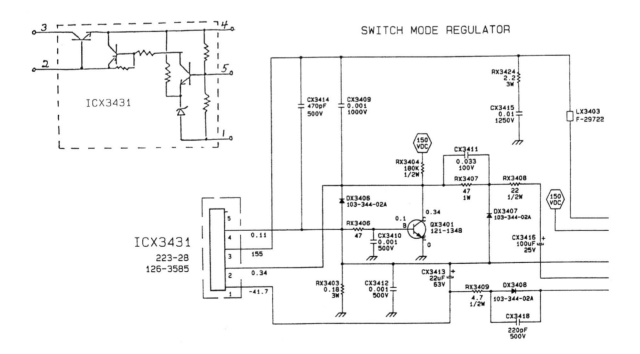

Figure 11-8 25" and 27" Power Supply

closed caption decoder, multiple 8-bit A/D converters, multiple D/A converters (pulse-width modulators), several general-purpose input/output ports, and two I2C communication buses.

I show the schematic just for the 25/27" circuit (figure 11-9) in order to conserve space. It comes from the training manual as opposed to the service manual just because it is easier to copy. There are differences between the 25/27" system control circuit and the 19/20" sets, but there are enough similarities to permit you to use one schematic to troubleshoot both. Just be aware of the fact that there are some differences. Consult a PHOTOFACT or a Zenith service manual if you need to know precisely what those differences are.

IC6000 communicates with the various ICs in the TV by means of the two communication buses. Bus one (pins 37 and 39) supports just the EAROM (IC6001) as it did in the Y-Line products. Bus two (pins 38 and 36) supports the tuner, the video processor, and MTS audio module. The clock crystal connects to pins 24 and 25.

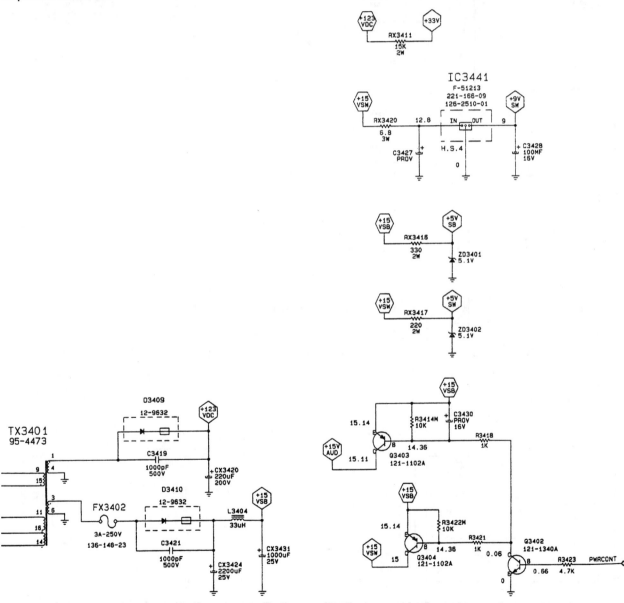

Figure 11-8 25" and 27" Power Supply (Continued)

Other than the ground and vcc pins, you need to know that the power control pin is pin number 32, and you have almost all of the information you need to run a pretty thorough check on the microcontroller. You might recall from our discussion of the service menu that these chassis come equipped with either a six-key (shown in figure 11-9) or a 10-key customer control panel (figure 11-10). The keypads are analog-to-digital devices making use of a resistor ladder circuit to produce a distinct DC voltage level for each key depressed. The microprocessor senses the voltage level input to it and responds

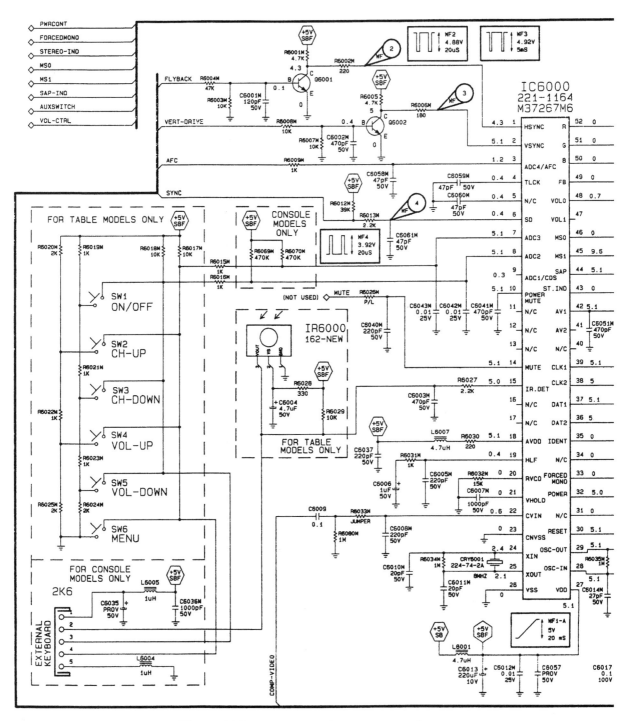

Figure 11-9 25" and 27" Microprocessor Circuit

accordingly. Like the keypad used in the Y-Line, the keypads make use of two, separate resistor-ladder circuits and are referenced to +5 volts.

Troubleshooting System Control

Let's talk a little bit about troubleshooting the system control circuit. You should have enough information now to do accurate troubleshooting if you use the schematic in figure 11-9, the pin out chart in figure 11-11, and the explanation of the pinout functions in figure 11-12.

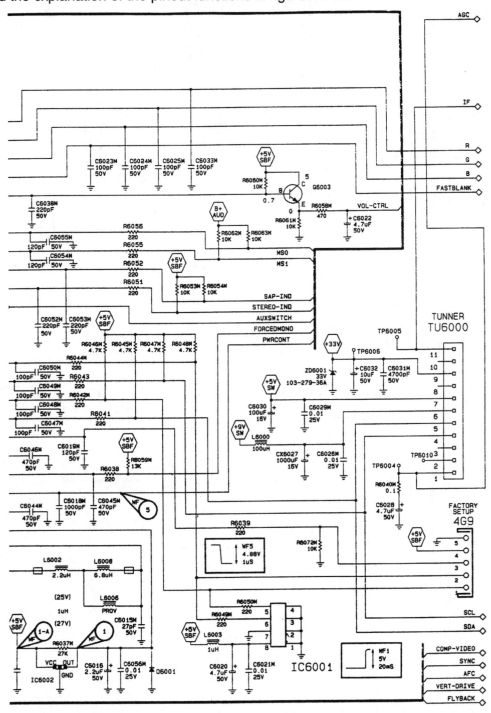

Figure 11-9 25" and 27" Microprocessor Circuit (Continued)

Servicing Zenith Televisions

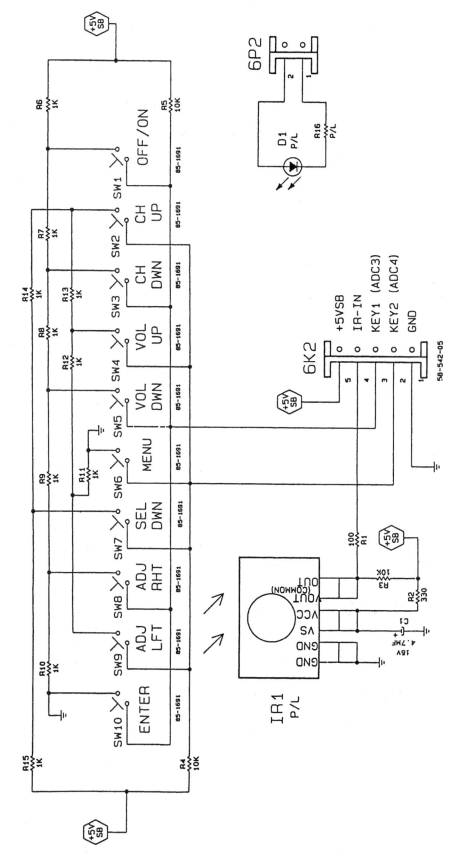

Figure 11-10 10-Button Keyboard with IR

The Z-Line Chassis

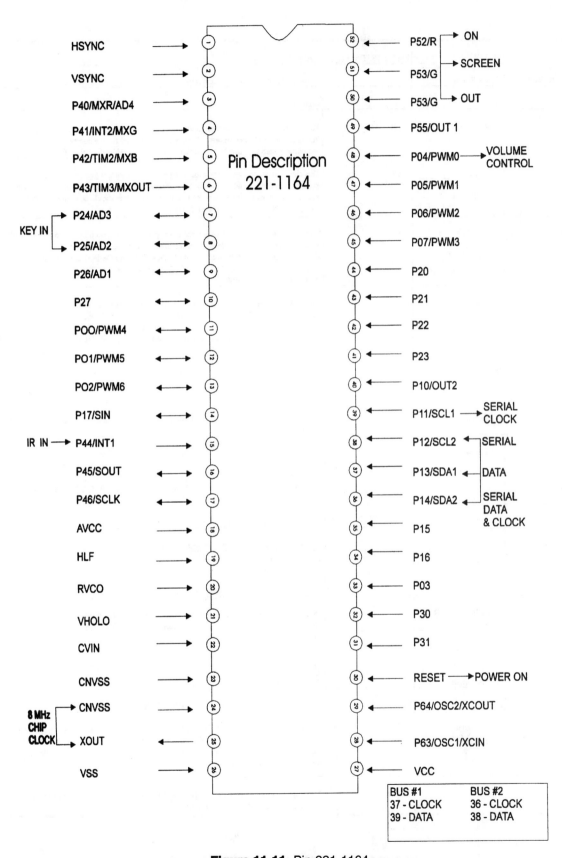

Figure 11-11 Pin 221-1164

Pin	NAME	INPUT/OUTPUT	NAME
VCC, AVCC VSS	Power Source		Voltage of 5 V ± 10% VCC, and 0 to VSS
CNVSS	CNVSS		Connected to VSS
Reset	Reset Input	Input	To enter the reset state, the reset input pin must be KEDT at a "L" for 2µ S or more (under normal VCC conditions). If more time is needed for the quartz-crystal osystal oscillator to stabilize, this "L" condition should be maintained for the required time.
XIN	Clock Input	Input	This chip has an internal clock generating circuit. To control generating frequency, and external ceramic resonator or a quartz-crystal oscillator I connected between pins XIN and XOUT. If an external clock is used, the clock source should be connected to XIN pin another XOUT pin should be left open.
XOUT	Clock Output	Output	
POO-P07	I/O Port PO	I/O	Port PO 8-pin is an I/O port with direction register allowing each I/O bit to be indivually programmed as input or output. At reset, this port is set to input mode. The output structure of PO3 is CMOS output, that of POO-PO2 and PO4-PO7 are N-Channel Open-Drain Output. POO-PO2 and PO4-PO7 are also used as PWM output pins PWM4-PWM6 and PWMO-PWM3 respectively.
P10-P17	I/O Port P1	I/O	Port P1 IA an 8-Bit I/O port has basically the same functions as port PO. The output structure of P15-P17 is CMOS output, that of P11-P14 is N-Channel open-drain output, FOOT P10, Refer to CRT output. P11-P214 are used as SCL1, SCL2, SOA1 and SOA2 respectively, when multi-master I2 C-Bus interface is used. P17 is also used as serial input pin SIN.
P20-P27	I/O Port P1	I/O	Port P2 is an 8-Bit I/O port and has basically the same functions as port PO. The output structure is CMOS output P24-P26 are al so used as analog input pins AO3-AO1 respectively.
P30-P31	I/O Port 3	I/O	Port P3 is an 8-Pin I/O port and has basically the same functions as port PO. The output structure is CMOS output.
P40-P46	I/O Port 4	Input	P40-P46 is a 7-Bit port and has basically the same functions as PO. Only when serial I/O is used, however, the output structure of P45 and P46 is N-Channel open drain output. For P40-P43, refer to video signal input for maxing. P44 is also used as external output interrupt input pin 1. P45 and P46 are used as pins SOUT, SCLK respectively, when serial I/O is used at this time, SCLK is an I/O, SOUT is an input pin.
CVIN		Input	Input composite video signal through a capacitor.
VHOLO			Connect a capacitor of 100pF between VHOLD and VSS.
RVCO			Connect a resistor between RVCO and VSS.
HLF			Connect a filter using of a capacitor and resistor between HLF and VSS.
OSC2	Clock Input for CRT Display	Input	There are I/O pins of the clock generating circuit for the CRT display function. OSC1 is used as sub-clock input XCIN and input port P63. OSC2 is also used as sub-clock output XCOUT and input port P64.
OSC2	Clock Input for CRT Display	Output	
HSYNC	HSYNC Input	Input	Horizontal synchronizing signal input for CRT display.
VSYNC	HSYNC Input	Input	Vertical synchronizing signal input for CRT display.
MXR MXG MXB MXOUT	Video Signal Input for Mixing	Input	Video signal input pins. MXR, MXG, MXB, MOUT are also used as input ports P40, P41, P42, P43, respectively. MXR is also used as analog input pin AD4. MXG is also used as external interrupt pin INT2, MXB and MXOUT are used as external clock input pin TIM2 and TIM3 respectively.
R, G, B OUT 1	CRT Output	Output	Color signal output pin for CRT display, R,G,B, OUT1 and OUT2 are also used as output ports P52, P54, and P10 respectively.

Figure 11-12 Pin Out Functions of 221-1164

Begin by checking for the usual "must haves:" vcc (pin 27), vss (pin 26), clock (pins 24 and 35), and reset (pin 30). Don't forget to check both communication buses (pins 36-39) for activity. And, remember to scope pin 1 for horizontal sync and pin 2 for vertical sync. Remember that IC6000 turns the set on by outputting a high on pin 52. These several checks should let you know if the microcomputer is healthy or stands in need of a "transplant."

You may be wondering why I said you should check pins 1 and 2 for horizontal and vertical sync. I really cannot overstress the importance of checking pin 2 for vertical sync. If this signal is missing because vertical deflection has failed or because of a problem in the circuit that routes vertical sync to pin 2, the TV will shut down. The newer TVs have special software built into the microprocessor that initiates a shutdown procedure when it does not detect vertical sync input. If you keep up with modern electronics, you know Zenith is not the only manufacturer to use such software. There is more about how to deal with this problem when we discuss the vertical deflection circuit. By the way, the circuit and software have been added to protect the picture tube from damage caused by a failure of the vertical deflection circuit.

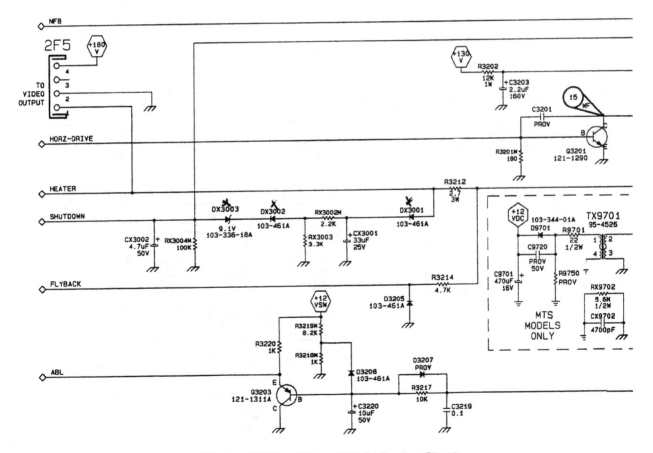

Figure 11-13a 19" and 20" Deflection Circuits

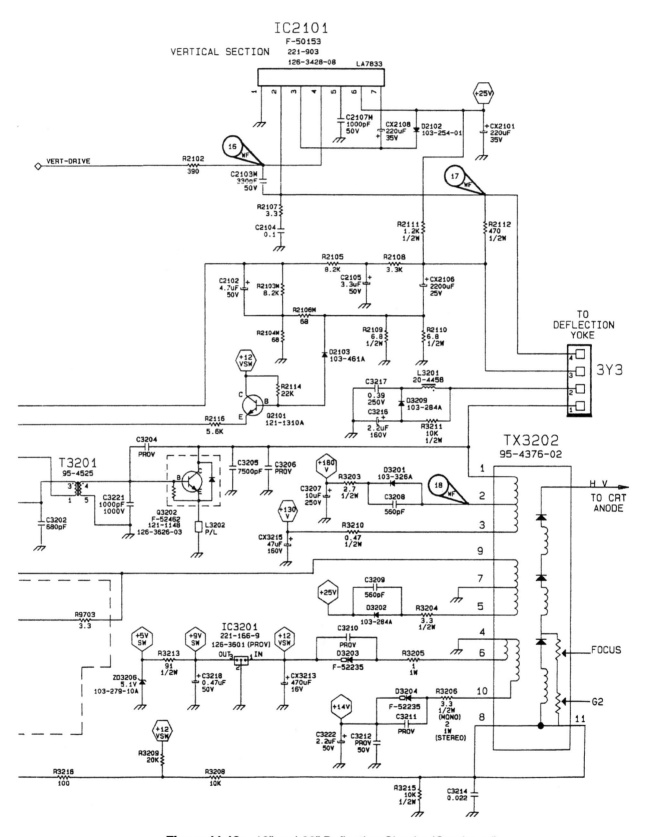

Figure 11-13a 19" and 20" Deflection Circuits (Continued)

Tuner

Now let me say a few words about the tuner by giving you a signal/voltage reference for each of its 11 pins:

pin 1 is the AGC input (AGC is developed by the signal processor.)
pin 2 is labeled TP6010 (I'm not real sure what the reference is!)
pin 3 is not connected
pin 4 is serial clock (The signal should be about 5 volts P.P.)
pin 5 is serial data (The signal should be about 5 volts P.P.)
pin 6 is connected to the switched +9 volts
pin 7 is connected to the switched +5 volts
pin 8 is not connected
pin 9 is connected to the +33-volt line (the tuning voltage)
pin 10 is not connected
pin 11 is IF out

Deflection Circuits

I don't believe we're going to encounter anything radically new in the deflection circuits of the GA chassis. Since there are differences between the 19/20" sets and the 25/27" ones, I'm going to give you the schematics for both. Figure 11-13a treats the former while figure 11-13b treats the latter.

Vertical Deflection

The vertical deflection circuits in the GA chassis are very similar to the circuit Zenith used in the C-6 chassis (Chapter IV). Vertical drive exits pins 22 of ICX2200 and goes to pin 4 of IC2101 (19/20" sets) or IC2100 (25/27" sets). Vertical drive exits pin 2 of these vertical output chips and goes to the vertical winding of the yoke. B+ for the output chips is developed from the +25-volt scan-derived voltage source. Both vertical output ICs belong to the LA78xx family.

There is one feature about the Z-Line televisions of which you should be aware. I talked about it when we discussed system control and will now elaborate. If the set turns on and almost immediately shuts down and cannot be turned on again unless the microprocessor has been reset by disconnecting and reconnecting AC and if the power supply voltages are good, immediately suspect lack of vertical deflection. The training literature is pretty specific about this phenomena. If the vertical pulse is missing at pin 2 of IC6000, a special software package activates to protect the CRT from potential damage, resulting in a TV that shuts down and cannot be turned back on until the microprocessor resets. Make a quick check by using the scope at pin 2 of IC6000 where you ought to find a DC voltage of about +5 volts. If the vertical pulse is absent, evidenced by a very low DC voltage, you must troubleshoot the vertical circuit and/or the circuit that supplies the vertical pulse to pin 2 of the microprocessor. If, on the other hand, you find no abnormality in either circuit, shift your troubleshooting expertise to X-Ray protection circuit.

You may be wondering how to troubleshoot a no-vertical deflection problem when the set won't stay on. I don't believe the literature tells you this, but you can supply an external +5 volts to pin 2 of IC6000 and force the TV to stay on, but you must turn the G-2 control down to keep lack of vertical deflection from damaging the CRT.

Servicing Zenith Televisions

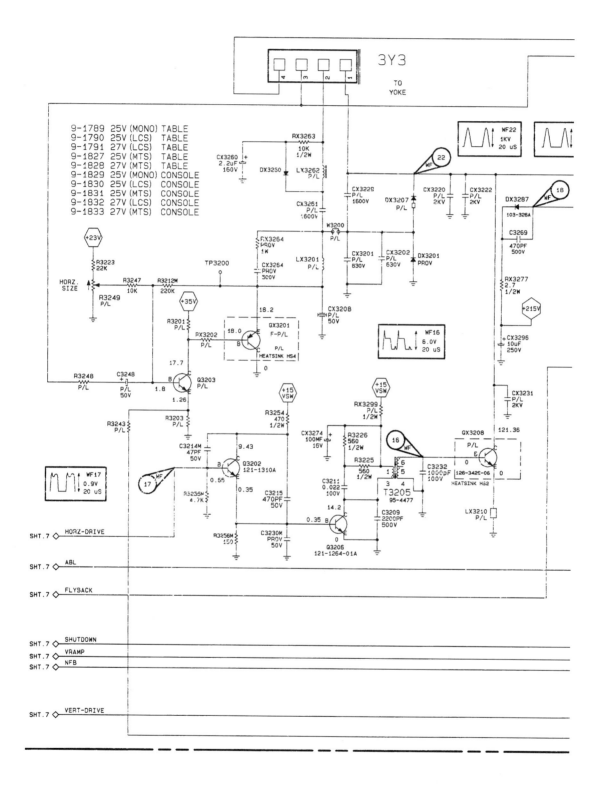

Figure 11-13b 25" and 27" Deflection Circuits

The Z-Line Chassis

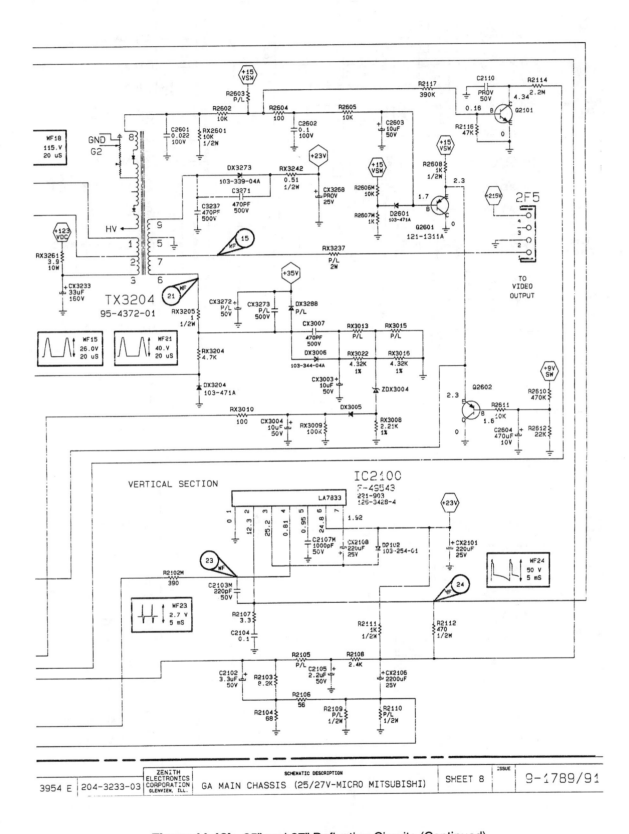

Figure 11-13b 25" and 27" Deflection Circuits (Continued)

Servicing Zenith Televisions

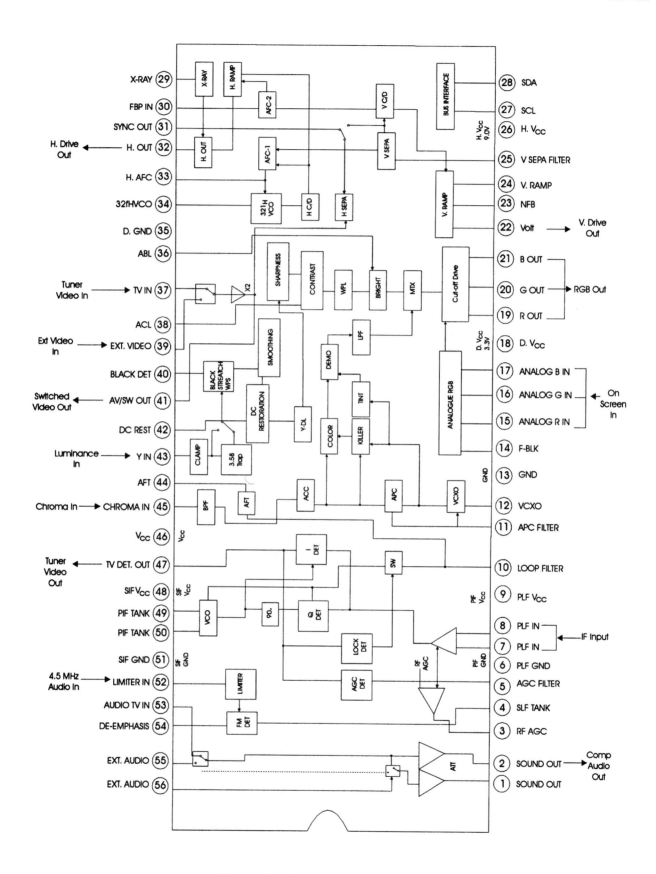

Figure 11-14 Block Diagram 221-1165

The Z-Line Chassis

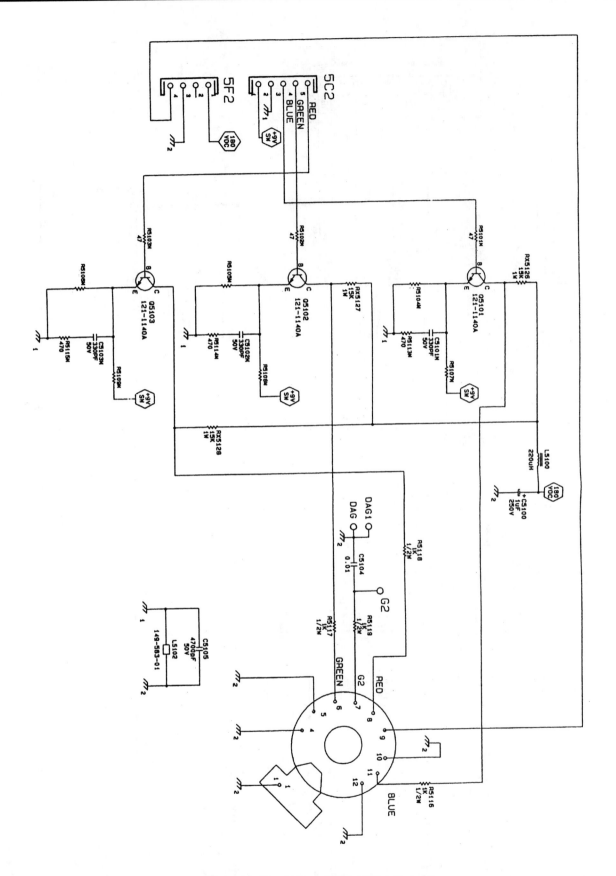

Figure 11-16a 19" and 20" Video Output

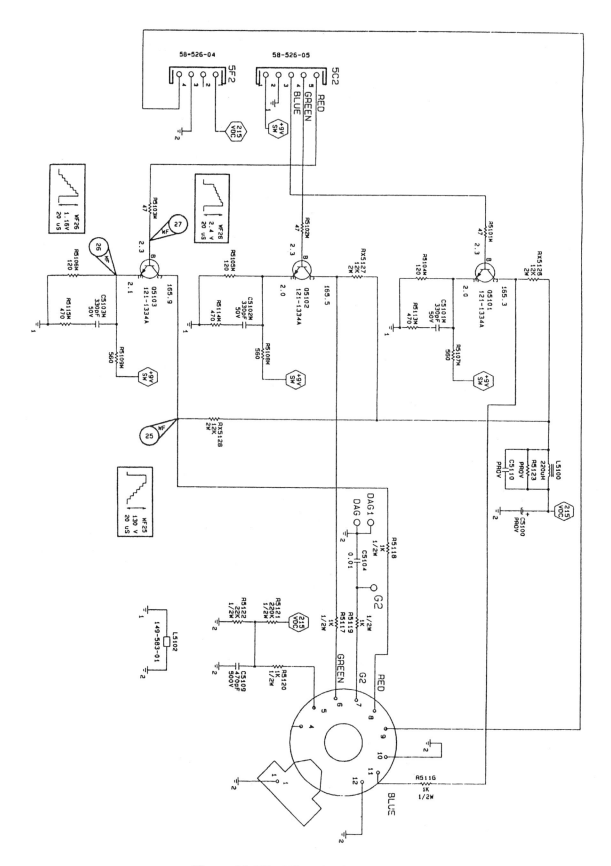

Figure 11-16b 25" and 27" Video Output

Horizontal Deflection

Let's deal with the 25/27" sets first. Horizontal drive exits pin 32 ICX2200 and goes right to Q3202 in the horizontal driver circuit. It is configured as an emitter follower to match circuit requirements to the horizontal driver transistor and to provide current gain. Resistor RX3254 in the collector circuit plays an important part in the circuit because the voltage levels available from the video processor don't allow for what Zenith engineers call "a more convenient current limiting at the base of Q3206" (923-3313TRM, p.15). RX3299 in the collector circuit of Q3206 is another resistor to which I want to call your attention. It has been chosen to limit the transistor's base current so that it is neither underdriven or overdriven. The "P/L" beside the part number in the schematic alerts you to the fact that you need to check the parts list for the correct value of the resistor for the chassis on which you are working.

The 19/20" horizontal deflection circuit pretty much follows suit, but there are a few differences of which you need to be aware. For example, horizontal drive exits ICX2200 and is routed directly to the horizontal driver transistor, eliminating the pre-driver circuit altogether.

Pincushion Correction

Now shift your attention to transistors Q3201 and Q3203. These two components and a spate of others form a diode modulator circuit to correct horizontal pincushion. Zenith follows its usual procedure by winding pincushion correction into the yoke of the smaller sets.

XRP Shutdown

You may follow this discussion by consulting figure 11-13b. The high-voltage shutdown network consists of diodes DX3006, DX3004, DX3005, capacitor CX3003, and an associated resistor network. Its output feeds the shutdown input of IXC2200 (pin 29). The anode of ZFX3004 is a good place to check the shutdown voltage should you need to check or monitor it. Change the component designations, and you have the XRP shutdown circuit for the 19/20" sets.

Video Processor and Associated Circuits

The GA chassis use a TA1268N video processor. Figure 11-14 gives you a peek inside this 56-pin wonder. I have read and reread the training material about the TA1268N and have come to the conclusion that there is nothing new here – nothing I haven't said several times over. Therefore, my discussion is limited by pointing out its key operating signals:

composite video out at pin 47
video in at pin 37
luminance in at pin 43
chroma in at pin 45
composite audio out at pin 2
horizontal drive out at pin 32
vertical drive out at pin 24
blue drive out at pin 3 of 2C5
green drive out at pin 2 of 2C5
red drive out at pin 1 of 2C5
serial and clock signals in at pins 27 and 28
+9 volts B+ in at pins 9, 46, and 48

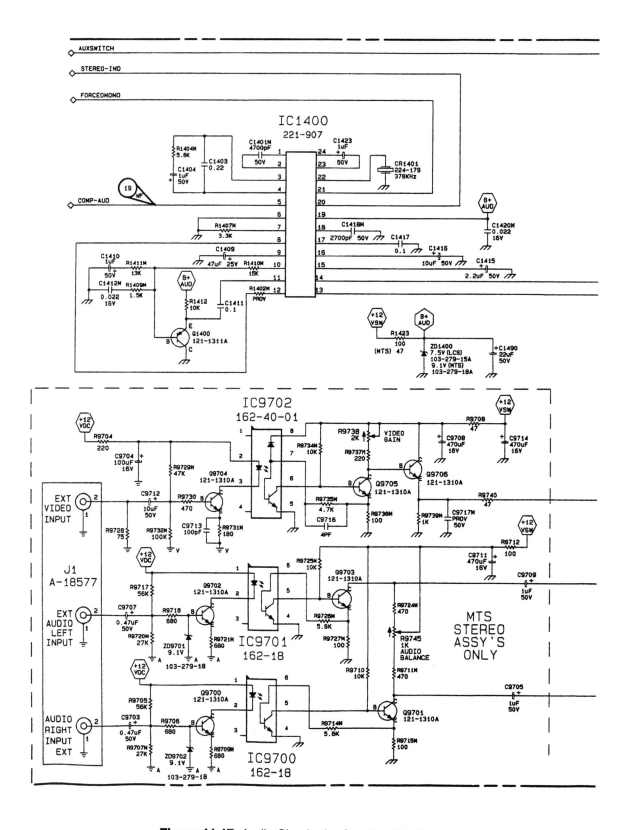

Figure 11-17 Audio Circuits for Smaller GA Chassis

The Z-Line Chassis

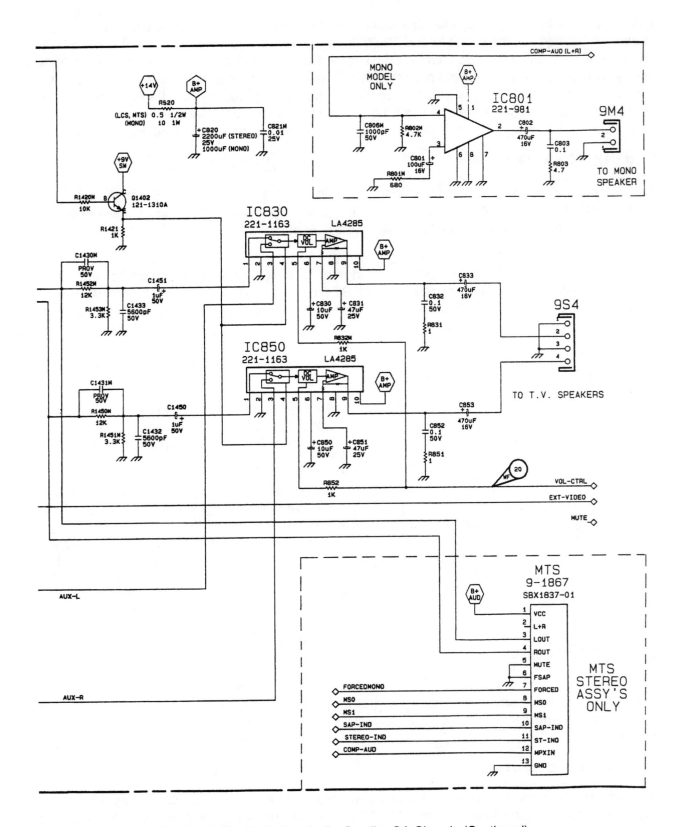

Figure 11-17 Audio Circuits for Smaller GA Chassis (Continued)

275

Video Output Module

The video output circuit board assembly rounds out my discussion of the video processing circuitry. Figure 11-16a lays out the circuit for the 19/20" televisions, and figure 11-16b does the same for the 25/27" sets. The circuit has been simplified compared to some of the other video output modules we have looked at, hasn't it? For example, there is no transistor that turns on to bleed the charge off the picture tube when the TV turns off as there was in the Y-Line products, and there is no luminance driver transistor. It appears that the mixing of the luminance and color information is accomplished inside ICX2200, lowering the parts count and eliminating certain potentially troublesome parts.

Setting The Gray Scale

Following the trend in consumer electronics, Zenith has also done away with most mechanical adjustments. In the event that you need to tweak or set the gray scale, follow the procedure outlined in the service literature as opposed to the "fly by the seat of your pants" method most of us have used for years.

Begin by setting G-2 to the proper level by turning the control fully counterclockwise (to zero). Apply a color bar signal from an NTSC generator to the RF connector and turn the color off via the customer menu. Adjust the G-2 control so that the bar pattern range from completely black to a "not overdriven" (or not saturated) white. There should be a distinct difference between the black and white bars. When you have correctly set the voltage, the fifth bar will be light, the sixth bar will be medium, and the seventh and eighth bars will be dark. Return the color control to its normal setting, but leave the generator hooked up and check your results.

Set the color level to minimum, the tint to midrange, the signal generator to a pure white signal. Then enter the service menu and set items 9,10, 20, 21, 22, 23, and 24 to the settings recommended in table 11-1 for the screen size of the TV on which you are working. These are the default settings.

Now go to the cutoff controls in the service menu. But before you start tweaking, observe which color predominates on the screen, and DO NOT adjust its cutoff control. Do, however, adjust the other two controls for the best white screen display. Confirm your work by setting the generator to produce a color bar output. With the color control still turned to minimum, check that the set displays a good gray scale from black to white according to the G-2 alignment. If the black level is too high, decrease by ONE step at a time the RF brightness setting in the service menu until the black level has been correctly set.

Now exit the service menu and return the color level control to its normal position because you have completed the task.

Audio Circuits

The GA chassis come equipped with three audio systems: mono, non-MTS stereo, and MTS stereo. Figure 11-17, taken from Zenith training material, illustrates all three for the small screen size GA chassis while figure 11-18 deals with non-MTS stereo and MTS stereo audio for the larger screen GA chassis.

Monophonic Audio

The sets that have mono audio only employ one basic IC in the audio circuit, designated IC801. It receives baseband audio from pin 2 of ICX2200, processes, amplifies, and sends it on to the speaker connector 9M4. Be aware that pins 5-8 are tied to ground for heat sinking purposes. Of course, the microprocessor handles volume and mute functions via the clock and serial lines to ICX2200.

Non-MTS Stereo Audio

The circuit gets a little more complicated when the non-MTS stereo has been incorporated into the module. Baseband audio exits ICX2200 and goes to pin 5 of IC1400, the decoder IC. There are no adjustments for this IC because peripheral components set the operating parameters. Q1400 and its associated components form a de-emphasis filter for the stereo decoder because the best stereo separation takes place between 3 kHz and 5Khz.

Right channel audio exits at pin 13 while left channel audio exits at 14. These signals go to IC830 and IC850, respectively, where they are amplified and ushered on to the speakers via connector 9S4. Volume and mute control signals come from the microprocessor on pin 5 of the audio output amplifiers.

MTS Stereo

MTS stereo-equipped models add the 9-1867 MTS stereo decoder to the circuit. If you have seen one of these chassis, you know that it is hardwired directly onto the circuit board. It receives baseband audio from ICX2200 at pin 12 and processes the signal to produce left and right output signals at pins 3 and 4, respectively. Volume and mute control come from the microprocessor, making their appearance at pin 5.

Troubleshooting the Audio Circuits

If you come across a no-audio or garbled audio problem and find that you have the necessary voltages and signals to the audio circuit, immediately check item 25 in the service menu to make sure it has been set to match the audio features for the set on which you are working (Check table 11-1 and associated comments.) If the setting is correct, proceed to item14. Set it to a value of 46, tune in a good strong off-the-air signal, and place a high impedance meter on pin 54 of ICX2200 or the +side of capacitor C1211. Then adjust L1205 for as close to +4 volts DC as you can.

Jack Pack

Figures 11-17 and 11-18 also include the jack pack schematics which I pass along without comment because we have seen configurations like them before.

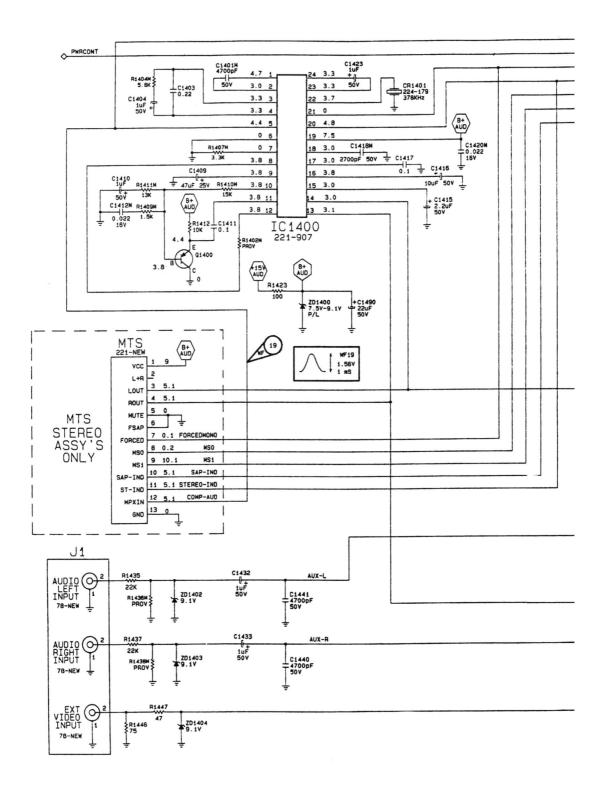

Figure 11-18 Audio Circuits for Larger GA Chassis

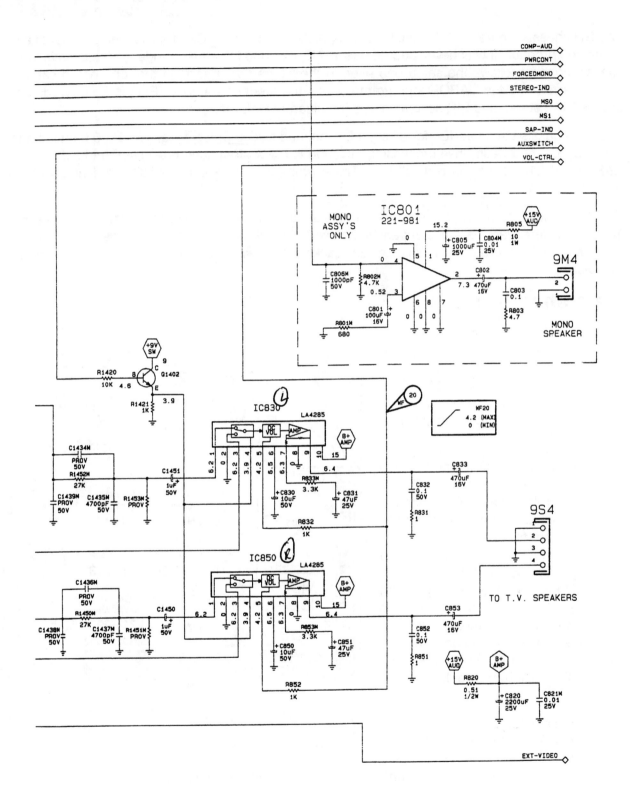

Figure 11-18 Audio Circuits for Larger GA Chassis (Continued)

GH Chassis

The GH chassis (figure 11-19) is a bit more sophisticated than the GA chassis, being closer to a top-of-the-line product. It was developed for use in 27", 32", and 36" stereo models. Zenith says the GH chassis concept follows the same one used in the Y-Line GX chassis in that one board serves all chassis sizes and the EAROM's program reflects the model's features. Plug-in boards, "daughter boards," like stereo and PIP, accommodate the different features.

The GH chassis uses four ICs for all signal, sync, and sweep processing. ICX2200 handles audio/video, sync, and sweep drive processing. IC6000 is the system microprocessor, and IC6001 is the EAROM. IC2100 develops vertical drive. The power supply makes use of ICX3431 as a switching regulator.

The 9-1819 MTS stereo audio jack pack module accompanies the one-tuner PIP models while the 9-1820 MTS stereo jack pack accompanies the two-tuner PIP models. Zenith follows what it has been doing by putting the audio circuitry and the jacks with their associated components on the same modules.

Customer Menu

Since it reflects added features, the customer menu is a bit more complicated than the one found in the preceding chassis (figure 11-20). Even though the entries are self-explanatory, I suspect I still need to comment on a few.

"Surround" in the audio menu has two selections, "Front" and "Off." In the "Front" mode, according to the sales pitch, the stereo signal is enhanced to create "a rich, full, room-filling sound." The "Off" mode receives and reproduces normal stereo sound. "SoundRite" keeps the audio at the same level as the program switches from program mode to commercial mode. "Speakers" turns the internal speakers off and on. Should I mention that you ought to check this menu if the customer complains of "no sound"?

"Color Temp" in the video menu lets the customer choose between "Warm" or "Cool." The cool setting adds a bit of a blue to the picture and is the setting I prefer.

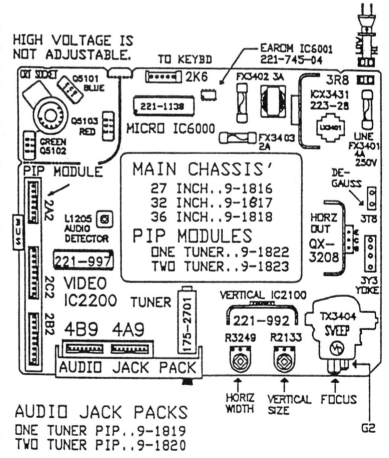

Figure 11-19 GH Chassis

"Picture Pref" lets the customer choose the settings of the video menu. The "Custom" setting are those selected by the customer while the "Preset" setting are those set by the factory.

SUMMARY OF MENU ITEMS

```
Setup Menu              Audio Menu      Video Menu      Pip Menu
                                                         Color
Auto Program            Bass            Contrast         Tint
Add/Del/Surf            Treble          Brightness       Size
Ch Labels               Balance         Color
Clock Set               Audio Mode      Tint
Timer Setup             Surround        Sharpness
Parental Ctl            SoundRite       Color Temp       Source Menu
Captions                Speakers        Picture Pref
Captions/Text                                            Main Source
Language                                                 PIP Source
Auto Demo
```

Figure 11-20 Customer Menu

Service Menu

If you have forgotten how to get into the service menu, I suggest you read (or reread) the discussion of the GA service menu. Once you get into it, the display looks like the screen depicted in figure 11-21. The black bar at the top contains the part number for the software. Put a "221" in front of it, and you have the part number of the microprocessor (221-1138-01). Remember that the service menu comes up on 'HPOS" which is the third item in the menu. You must turn "F Mode" (factory mode) on to gain access to all of the items in the menu. Simply use the select key to highlight it and the adjust key to turn it on (from 00 to 01), and proceed to the item you want to work with. But don't forget to turn it off before you exit the service menu. You can quickly tell if factory mode has been left on by looking at the customer setup menu. If you see a pair of dashes at the top of the menu, then Factory Mode is on (figure 11-22). Turn it off by setting the clock. Figure 11-23 lists each item in the service menu, its typical setting, and description of its function.

Alignment Procedures

Things can really get out of hand if the items in the service menu aren't correctly set. If you work on Thomson products, you know what I'm talking about. Because there may come a time when you need to do an alignment or two, I'm going to take the time to go through the alignment procedures with you. Manual CM-149/GH (pages 5 and 6) contains the same information in a more detailed format.

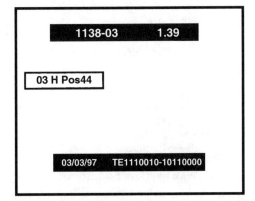

Figure 11-21 Accessing the Service Menu   **Figure 11-22** Factory Mode On

GENERAL SETTINGS (Green)

ITEM	RANGE	TYPICAL SETTING	DESCRIPTION
00 F Mode	0 - 1	0	Factory Mode 0 is Off
01 Pre Px	0 - 1	1	0 is Custom, 1 is preset
02 V Pos	0 - 31	10	Moves the On-Screen Displays Vertically
03 H Pos	0 - 75	44	Moves the On-Screen Displays Horizontally
04 Level	0 - 2	1	1 is Zen, 0 is P Lbl
05 Band	0 - 7	1	CATV
06 AC On	0 - 1	0	1 is On, 0 is Off

TECHNICAL SETTINGS (Black)

ITEM	RANGE	TYPICAL SETTING	DESCRIPTION
07 C.Phas	0 - 254	80	Caption Phase
08 C.Srch	0 - 1	0	0 is Off, 1 is On
09 C.Line	0 - 32	17	Caption Line
10 Rf Bpf	0 - 1	1	RF Bandpass Filter 1 is On, 0 is Off
11 3.58T	0 - 1	1	3.58 Mhz Trap
12 RF Brt	0 - 63	27	RF Sub brightness
13 AxBrt	0 - 63	24	AUX Sub brightness
14 MaxCon	0 - 63	63	Maximum Contrast
15 VPhase	0 - 7	0	Vertical Phase
16 HPhase	0 - 31	20	Horizontal Phase
17 AudLvl	0 - 63	46	Sound Attenuation (Gain for LCS, MTS audio)
18 Aud Adj	0 - 63	63	Sound Balance
19 RF Agc	0 - 63	40	RF AGC
20 H Afc	0 - 1	1	H Afc 0 is Off, 1 is On
21 Wh Comp	0 - 1	1	White Compression 0 is Off, 1 is On
22 60HzSw	0 - 3	2	60 Hz Switch
23 PifVco	0 - 127	40	PIF Voltage Controlled Oscillator

COLOR TEMP COOL STARTING VALUES

ITEM	RANGE	TYPICAL SETTING	DESCRIPTION
24 R Cut	0 - 254	0	Red Cutoff
25 G Cut	0 - 254	0	Green Cutoff
26 B Cut	0 - 254	0	Blue Cutoff
27 G Dvr	0 - 254	75	Green Gain Cool
28 B Dvr	0 - 254	65	Blue Gain Cool

COLOR TEMP WARM STARTING VALUES

ITEM	RANGE	TYPICAL SETTING	DESCRIPTION
24 R Cut	0 - 254	0	Red Cutoff
25 G Cut	0 - 254	0	Green Cutoff
26 B Cut	0 - 254	0	Blue Cutoff
27 G Dvr	0 - 254	75	Green Gain Cool
28 B Dvr	0 - 254	40	Blue Gain Cool

PIP SETTINGS (White)

ITEM	RANGE	TYPICAL SETTING	DESCRIPTION
29 PipRas	0 - 254	69	PIP Raster
30 Pip Sw	0 - 15	9	PIP Switch Delay
31 PipLuD	0 - 7	1	PIP Luma Delay

AUDIO SETTINGS (Yellow)

ITEM	RANGE	TYPICAL SETTING	DESCRIPTION
32 InLev	0 - 15	9	Input Level
33 StVco	0 - 63	31	Stereo VCO
34 SapVco	0 - 15	0	SAP VCO
35 SapLpf	0 - 15	0	SAP LP Filter
36 StLpf	0 - 63	26	Stereo LP Filter
37 Spectr	0 - 63	30	Spectral High Freq. Separation
38 WideBa	0 - 63	30	Wide band Low Freq. Separation

Figure 11-23 Service Menu Items, Settings, and Descriptions

Video Detector

If the TV doesn't have a picture, enter the service menu and check the default setting of these items: 05 should be set to 0; 19 should be set to 40; and 23 should be set to 40. When you need to realign the detector, begin by connecting a high-impedance DC voltmeter to pin 44 of ICX2200 or R1219 and applying a good, standard signal to the RF connector. Then adjust item 23 (PIFvco) for a reading of 2.5 volts on the voltmeter. This is also the AFC crossover point.

AGC Delay

Apply a channel 6 signal of 750 microvolts to the RF connector. Then adjust item 19 (RF AGC) until the tuner AGC drops 1 volt DC from its typical reading. Use the + side of C6028 to monitor the voltage. Note: If you adjust the setting beyond 40, the tuner will most certainly overload under certain conditions causing beats to appear in the picture.

Audio Detector

If the TV doesn't have audio, enter the service menu and confirm that item 17 has been set to 46 and item 18 to 63. If you need to align the audio detector, hook a DC meter to pin 54 of ICX2200 or the + side of C1211 and apply a good signal to the RF connector. Then adjust L1205 for a reading of 4.0 volts DC on your meter. Don't forget to confirm that the settings in the service menu correspond to the bar code on the back of the television (figure 11-24).

Stereo Level Adjust

Menu item 17 should not be changed unless ICX2200 has been replaced. If you need to align it, begin by attaching a high impedance AC meter with a 47k-ohm load resistor to pin 3 of connector 4A9 and the ground lead to pin 4. Connect a 4700pf capacitor from the + side of capacitor C1211 to the ground lead jumper W49 on the main circuit board. Apply AC power and an RF signal with good video and audio at 400 Hz and 100% modulation. Then go to the service menu and adjust item 17 for a reading of 490 to 500 millivolts AC on the meter.

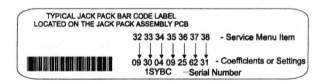

You may perform stereo level adjustment with the jack pack removed. Begin by attaching the positive probe of the meter to jumper W1611 and the negative to jumper W1603. Connect the 4700 capacitor from the + side of C1211 to ground lead jumper W49. Apply AC power and the above-mentioned signal to the RF port and make the necessary adjustment to item 17.

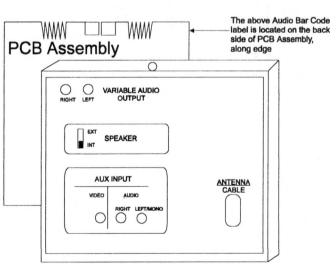

Figure 11-24 Confirming the Bar Code

Gray Scale Adjustment

Read the procedure dealing with gray scale adjustment in the previous section (GA chassis).

Power Supply

The power supply (figure 11-25) should evoke the feeling, "I've seen it before." Since I have addressed it extensively in previous sections, I've provided a schematic taken from Zenith's training material and a few checks to confirm its operation and move on. If you just started reading at this point and want to know more about the power supply, please consult the power supply section of Chapter 10 for the Y-Line series of televisions.

Check the standby voltages first:

+150 to +155 volts DC at CX3404
+130 to +138 volts DC (depending on screen size) at CX3420
+18 volts DC at C3422
+15 volts DC at the emitter of Q3404
+5 volts DC at pin 3 of IC3442

Check the power on sequence by monitoring the voltage at pin 52 of the IC6000. It should go high when the set receives an on command via the remote receiver or the keyboard. Then confirm that the high has been coupled to the base of Q3402, the transistor that turns on the switched voltages. Check them by looking for +9 volts DC at pin 3 of IC3441 and +15 volts DC at the emitter of Q3404.

Finally, confirm the presence of the scan-derived voltages: +23 volts DC at the junction of CX3268 and RX3244; +35 volts DC at CX3272; and +200 volts DC at the junction of CX3296 and RX3277. The CRT filament voltage appears at connector 2F5 and should read in the neighborhood of 6 volts AC on a true RMS meter. If you use a scope, expect to find a pulse of about 22 volts peak-to-peak.

System Control

The microprocessor and system control (figure 11-26) follow the same general lines Zenith has been using for some years. IC6000 appears in the familiar 52-pin DIP configuration. It makes use of CMOS design, has a 12-megahertz clock, and an 8-bit core. It generates on-screen information that is sent out at pins 24, 25, and 26 to ICX2200 and has two communication buses: bus one at pins 33 and 34 and bus two at pin 30 and 31. And, it follows previous microprocessor design by having multiple A/D converters, multiple general purpose I/O ports, and a built-in closed-caption decoder.

Troubleshoot system control problems by making the following checks:

+5 volts at pins 22 and 39
ground at pin 21 (Use this connection to check voltage at pins 22 and 39)
reset voltage at pin 35
oscillator at pins 36 and 37 (Check the peak-to-peak level and frequency.)
IR input at pin 2
keyboard input at connector 2K6 (Keyboard inputs go to pins 19 and 20.)
power control at pin 52 (It should go high to turn the set on.)
horizontal pulse at 28
vertical pulse at pin 29 (The TV will shut down if this pulse is absent!)
bus activity at pins 30 and 31 and at pins 33 and 34 (bus I and bus II)
R, G, and B output at pins 24, 25, and 26

The Z-Line Chassis

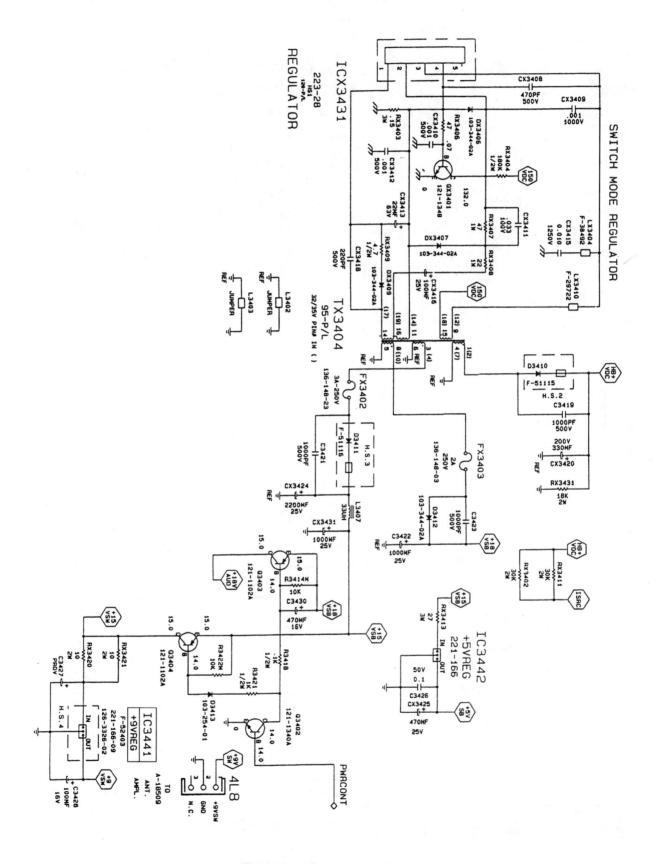

Figure 11-25 Power Supply

Servicing Zenith Televisions

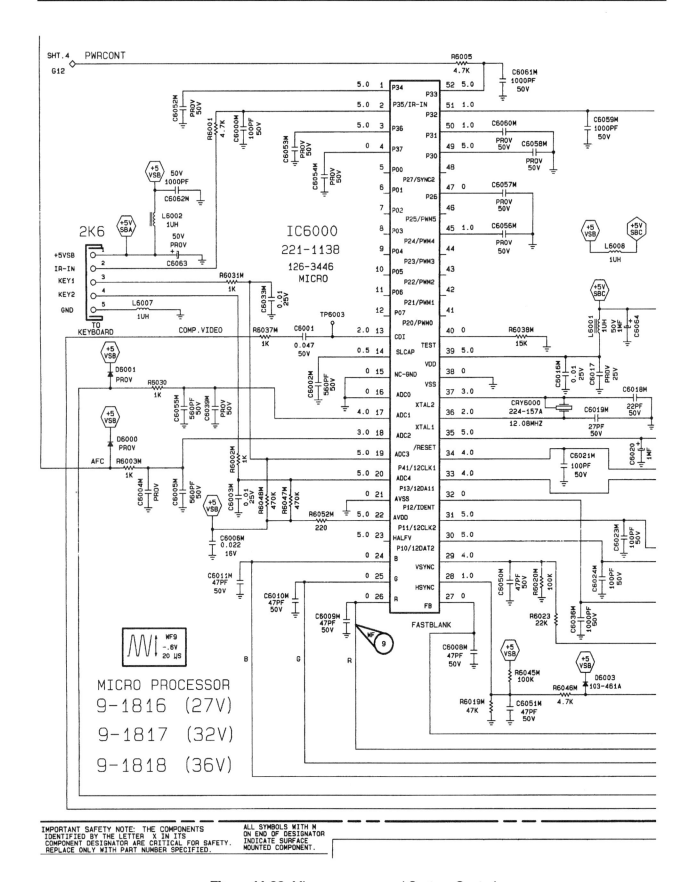

Figure 11-26 Microprocessor and System Control

The Z-Line Chassis

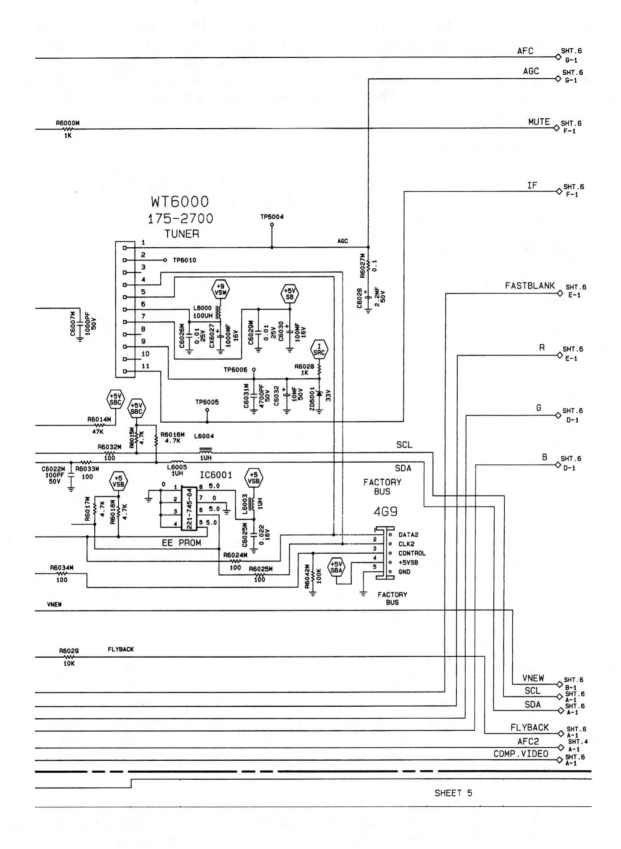

Figure 11-26 Microprocessor and System Control (Continued)

I have never had a keyboard go bad on a Zenith direct-view set, but I have replaced them on projection units. If you should need to troubleshoot a keyboard, use figure 11-27 as a guide. I have tried to repair those few that have caused trouble on the projection sets and had limited success. Well, to be honest, I usually just gave up and ordered new ones. The resistors in the ladder network are precision devices, to say the least, and require replacement with exact equivalents. I am bald enough without adding hair loss over something like this!

Deflection Circuits

I am repeating myself when I say there is nothing substantially new about either of the deflection circuits (figure 11-29), but I will make a few comments about both because we are now dealing with the larger direct-view televisions.

Horizontal Deflection

Horizontal drive exits pin 32 of ICX2200 and goes almost directly to the base of Q3202 which is configured as an emitter follow to boost the current gain of the signal before it is applied to the base of Q3206 (the horizontal driver). As we have seen in previous chassis, R3254 in the collector circuit helps to limit base current. Therefore, replace it with the correct value for the screen size on which you are working. Resistor RX3299 in the horizontal driver circuit is also critical. Its value ensures that Q3208 is neither under driven nor over driven.

The size of the CRT determines the EHT output of the flyback. Expect about 30 KV for 27" screens and about 32 KV for the 32" and 36" screens. As you should expect, B+ to the horizontal output transistor also depends on the size of the CRT. Look for about +130 volts for the smaller screen size and about +137 volts for the larger size screens.

Now shift your attention to horizontal width control (R3249) and the pincushion circuit. Q3201 and Q3203 are a part of a diode modulator circuit used to correct horizontal pincushion. Adjust horizontal width by varying the setting of R3249.

DX3006, ZDX3004, CX3003, and the associated resistor network form a circuit that monitors the high voltage and provides input to the shutdown section (pin 29) of the video processor IC. Shutdown occurs when this network develops 3.5 volts DC and inputs it to pin 29. Expect a voltage of between 1.5 volts to 2.5 volts DC when the set is working normally. The literature says the voltage is 3.0 volts at zero beam current.

Vertical Deflection

The vertical deflection circuit is configured around IC2100, an integrated circuit that permits a common design for all screen sizes with a change in just three resistors, RX2126, RX2115, and R2122. The first two resistors are in parallel and are used to sample vertical yoke current. Their value trims the total resistance so that the proper yoke current produces a signal at RX2126 that is about 1.5 volts peak to peak. If this value is held constant, the vertical circuit works with the same voltages no matter what the CRT size is. The value of R2122 varies to maintain vertical linearity for the screen size of the chassis in which it is installed.

Vertical drive exits ICX2200 at pin 24 and is routed through Q2200, a transistor that inverts the drive pulses and sends them on their way to pin 3 of IC2100. Now shift your attention to capacitors C2114 and C2115. These capacitors develop the sawtooth waveform for vertical deflection and provide a point at their junction where linearity correction can be applied. Diode DX2101 and capacitor CX2100 form the usual voltage-doubler circuit to provide the voltage necessary to ensure vertical retrace.

The Z-Line Chassis

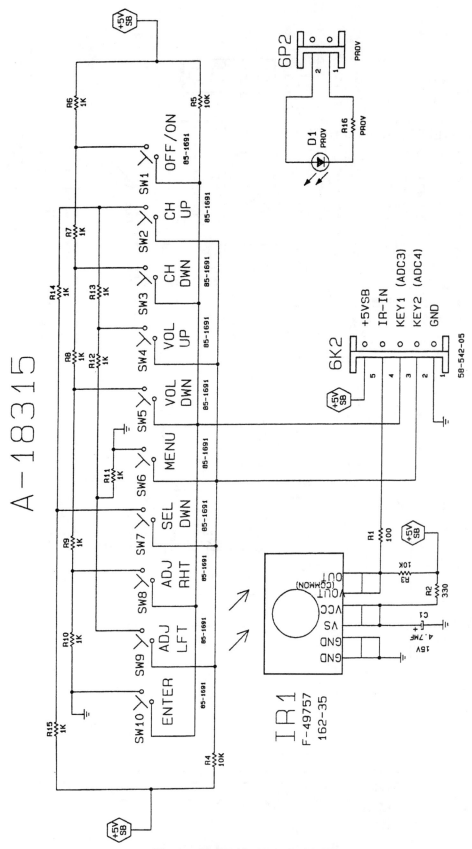

Figure 11-27 Keyboard with IR

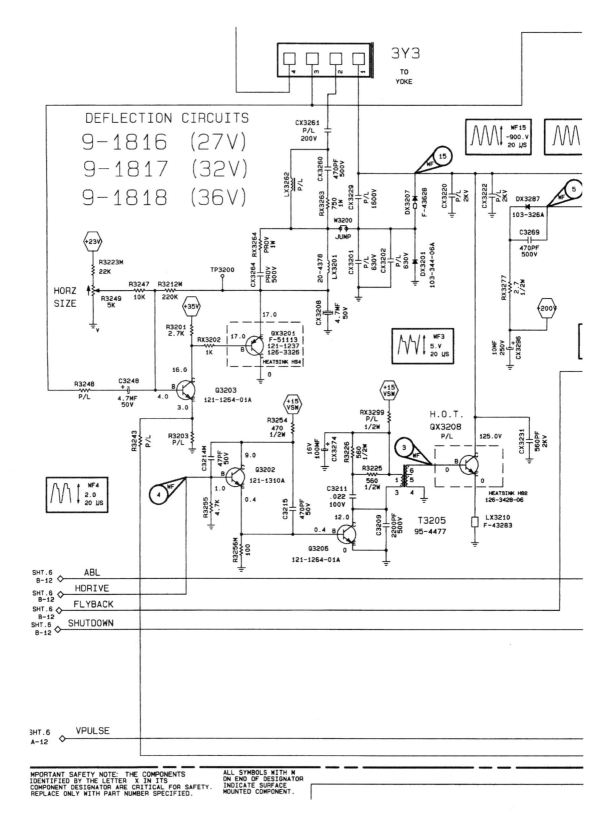

Figure 11-29 Deflection Circuits

The Z-Line Chassis

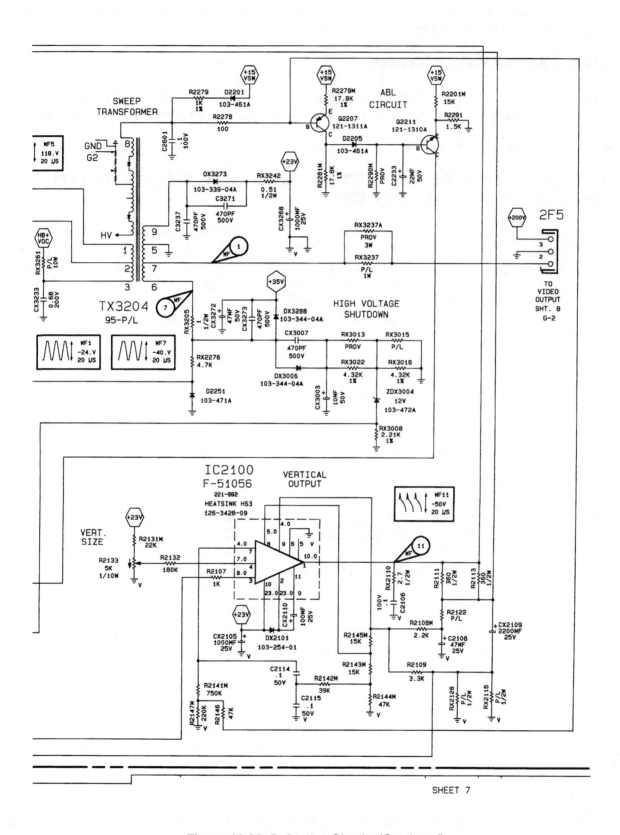

Figure 11-29 Deflection Circuits (Continued)

Before I leave this section, let me underscore the fact that lack of vertical deflection will most definitely cause the set to shut down. If you haven't read it or don't recall it, let me refer you to discussion of the GA chassis in the previous section of this chapter.

Video Processing Circuit

Let's begin the discussion of the video processing circuits by "getting the cart before the horse" because I want to call your attention to the video output module (figure 11-30) before I discuss anything else.

The Video Output Module

YOU MUST NEVER OPERATE THE TELEVISION WITH THE VIDEO OUTPUT MODULE REMOVED FROM THE CRT.

If you do, you will more than likely damage the picture tube and set yourself up for possible electrical shock! You see, certain elements inside the CRT are tied to the high voltage through a divider network inside the neck of the tube and terminate at pin 4 of the socket. If the socket is removed and the TV turned on, the tube will arc internally and more than likely be damaged. If you look at figure 11-30, you see that resistor R5110 connects pin 4 to ground and is the component that ties the divider network to ground. The "P/L" beside its designation means its value depends on the screen size. According to the training literature, R5110 is just a piece of wire on the 32" sets and is a 15-meg resistor on the 36" sets. This warning applies to the 32" and 36" sets only and not to the 27" sets.

Let me call your attention to a couple of circuit components, the functions of which may not be clear. First, each video output transistor has a 120k resistor that connects between collector and base. I'll use the red drive circuit to illustrate the point I'm trying to make. R5143 is the resistor in question and has been placed in the circuit to provide feedback for the red video output transistor. If it opens or increases in value, it might cause you to think that the tube has lost its red gun because the picture takes on a washed-out look. (If you work on Zenith projection televisions, you know what I'm talking about!) Before you condemn the CRT, check R5143 (or R5142 or R5141).

Second, take a look at the components tied to pin 5 of the CRT socket, the external connection to the control grid (the G1 connection). You should read about 0 volts DC when you check it. The components tied to it form a series LCR circuit that consists of the capacitive coupling between each of the cathodes and the G1 element, L5101, and R5120. The LCR circuit provides high-frequency drive for the picture tube's control grid.

I'm sure you have noticed the lack of controls on the video output module, meaning you cannot set the gray scale by tweaking pots. It has to be done by adjusting the appropriate registers in the service menu. Incidentally, the G2 voltage should be set so that it falls between 300 and 400 volts.

The Video Processing IC

Figure 11-31 lays out the complete video processing circuit in all of its gory details. However, expect slight variations as you move from model to model because of PIP and tuner options. I can't see going into an involved discussion of ICX2200 because, as far as I can tell, there is nothing really new here. But, I feel I must comment on the details that are unique to the GH chassis.

Let's begin with the luminance signal path.

The video IF signal from the tuner enters the chip at pins 7 and 8, and the detected signal exits at pin 47. Q1203 and Q1204 process the composite signal and send it to connector 2B2 on its way to the

PIP module. The comb filter located on the PIP module separates the luma and chroma signals and nudges both signals along to IC2001, the video switch which is also located on the PIP module. The switched signals go from IC2001 to pins 2 and 4 of connector 2C2. The luminance portion then loops back from 2C2 through Q2201 into pin 43 of ICX2200 while the chroma goes to pin 45.

Auxiliary video from the jack pack comes in on pin 1 of 2B2. The switched video out – either from the tuner or external video – is available at pin 41 of ICX2200 at about a 2-volt peak-to-peak level. Q2203 drops the voltage to a 1.5-volt level before it is sent on its way to the microprocessor.

The following table should assist you in the event that you need to troubleshoot either ICX2200 or its associated circuits:

composite video out on pin 47
switched video out to the microprocessor on pin 41
tuner composite video in on pin 37
luminance in from the PIP at pin 43
chroma in from the PIP on pin 45
composite audio out at pin 2
horizontal drive at pin 32
vertical drive at pin 24
blue drive at pin 3 of 2C5
green drive at pin 2 of 2C5
red drive at pin 1 of 2C5

Continue by checking for B+ of about 9 volts DC at pins 9, 23, 46, and 48, and don't forget to check for serial and clock data at pins 27 and 28 respectively. You must remember that the microprocessor maintains control of ICX2200 via the I2C bus. Let me put it differently. The IC simply won't work if clock and data signals are missing.

PIP Circuits

Module 9-1822 is the one tuner PIP (figure 11-32). Integrated circuits IC2000 and IC2001 do the bulk of signal processing. The former handles the analog signal processing, logic, and memory functions. The latter is the video switch. Module 9-1823 is the two tuner PIP (figure 11-33) and has all the circuits the one tuner PIP has plus a second tuner and IC1201 that processes the video from the second tuner.

Things get kind of complicated here, but I hope to keep the discussion as simple as possible.

All video passes through the PIP module. Main tuner video for ICX2200 appears at pin 3 of connector 2B2 and goes from there to pin 3 of the video switch (IC2001). Composite video from the second tuner goes to pin 11 of the video switch. Auxiliary video from the jack pack comes in pin 1 of connector 2B2 and goes to pin 1 of the video switch. External luminance and chrominance come in on pins 5 and 6 and go out on pins 15 and 16 of the video switch respectively. As you see, the video switch permits the user to select among the signals from tuner one, tuner two, or external Y/C signals.

If you have a problem with the video path, it might be a good idea to begin your troubleshooting right here since all of the set's video passes through the PIP and the video switch. At least, it provides a sort of "halfway" point that lets you divide the circuit into half as you attempt to find the reason for the "no video" symptom.

Servicing Zenith Televisions

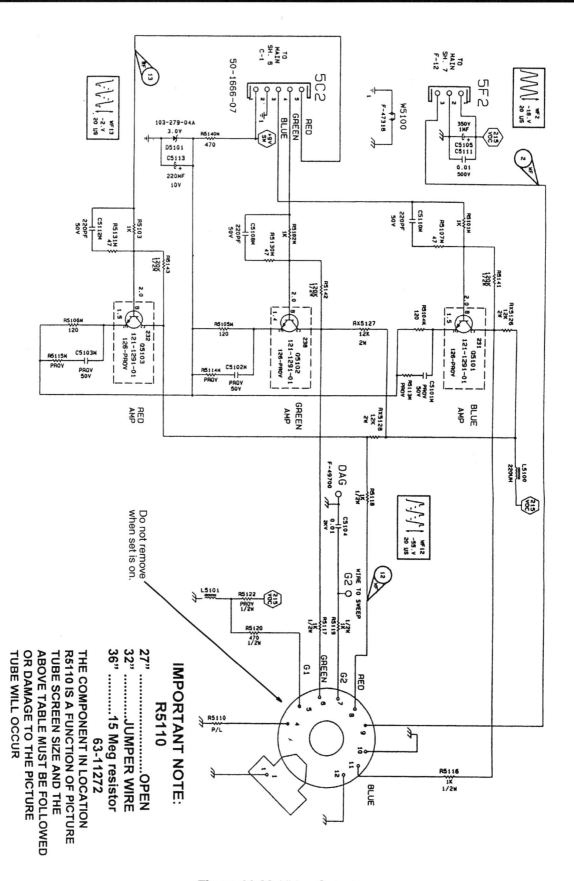

Figure 11-30 Video Output

The PIP has to have other signals in order to operate properly. For example, it receives control signals from the microprocessor over bus one at pins 2 and 4 of IC2001. A pulse from the flyback arrives at pin 2 of IC2000 through transistors Q2001 and Q2008, and a vertical sync pulse comes to pin 2 through transistor Q2002. And, of course, it has to have the necessary B+ voltages.

Let's trace the composite video through this maze first. It exits pin 14 of the video switch and is processed by Q2003 before it gets to the comb filter (DL100). The luminance portion of the signal comes out pin 5 of DL100 and loops back into the video switch at pin 8 while the chrominance portion comes out pin three and makes its loop back into the video switch at pin 10. These two signals then pass through the switch and exit at pins 18 and 17 respectively where they receive processing by emitter followers Q2004 and Q2005 before they enter the PIP IC (IC2000).

PIP composite video input 1 exits the switch at pin 15 while the composite video input signal number two exits on pin 16. These signals pass through Q2006 and Q2007 on their way to pins 34 and 36 of IC2000.

Luminance that consists of the main video plus the PIP video appears at pin 40 of IC2000. Chroma out that consists of the main chroma information plus the PIP chroma appears at pin 51. These two signals go to pins 2 and 4 of connector 2C2 on their way back to the main circuit board.

Audio Circuit

The audio circuit (figure 11-33) consists of a single module that plugs into the main circuit board at connectors 9A4 and 9B4 on the back edge of the board. The module contains three ICs: (1) IC1400 is the stereo decoder, audio processor, Aux/TV switch, and SoundRite volume limiter; (2) IC803 is the variable audio signal amplifier; (3) and IC804 is the dual audio amplifier that drives the speakers IC803 is really the only new feature in the circuit. This eight-pin chip supplies left and right channel variable audio signals to the jack pack for connection to an external amplifier and is necessary because IC804 is turned off via the customer menu. If the audio were switched by the customary "external/internal" switch on the jack pack, IC803 wouldn't be necessary because IC804 could supply audio to the external jack pack.

The circuit employs two circuits to eliminate speaker "pop" at turn on and turn off. Q801 mutes the variable audio signal at turn on while diodes D820 and D830 connect to a voltage divider network to reduce the pop at turnoff. Incidentally, Q800 performs the same function for the main audio amplifier. You might make a note to check it first when you service a set for a "no audio" complaint. I serviced two Zeniths just this week for a "no sound" problem caused by collector-emitter leakage of the mute transistor.

Finally, don't forget that there are five adjustments to the stereo module via the service menu, eliminating all mechanical adjustments on the module itself. If you have to replace either the audio module or the main module, make absolutely certain that these five adjustments agree with the bar code on the audio module. Expect to have audio problems if you don't.

Well, that's it for the GH chassis. It's time to turn our attention to the last of the Z-Line televisions.

Servicing Zenith Televisions

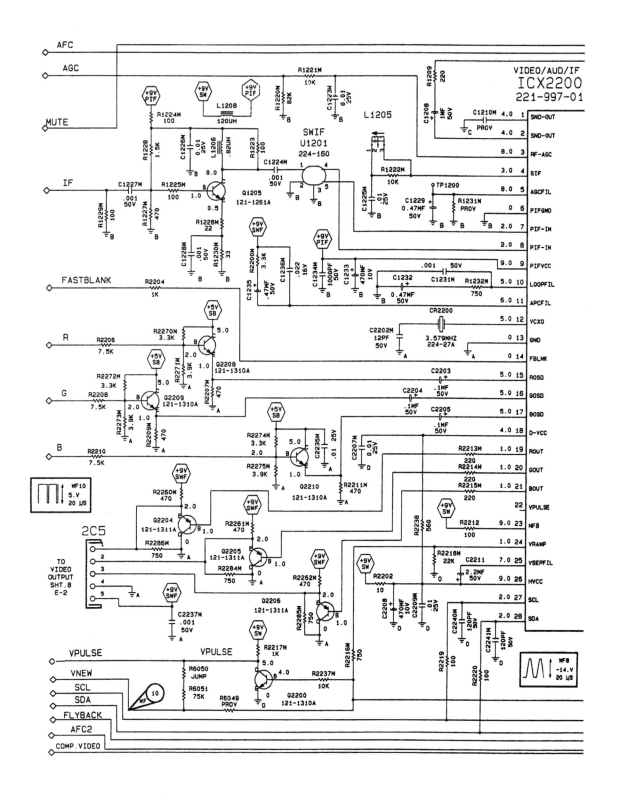

Figure 11-31 Video Processing Circuit

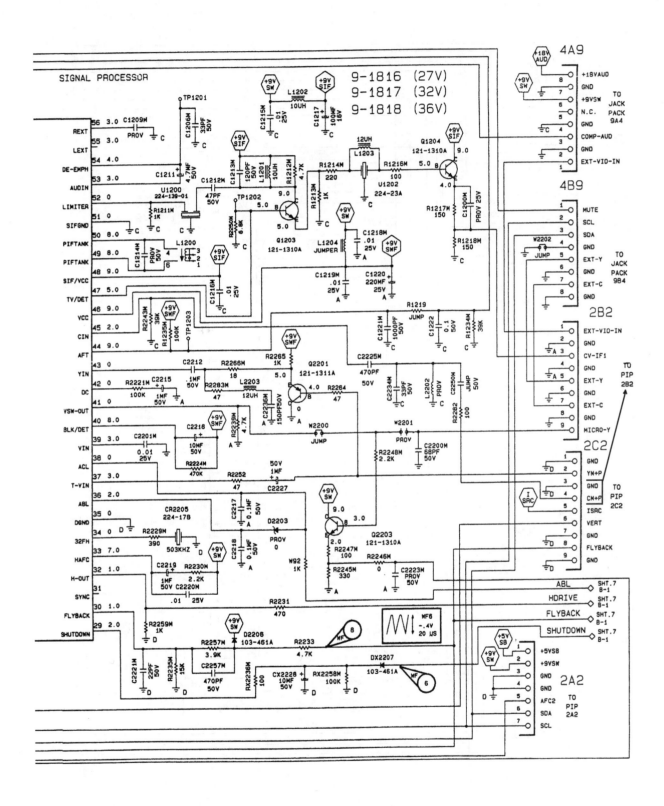

Figure 11-31 Video Processing Circuit (Continued)

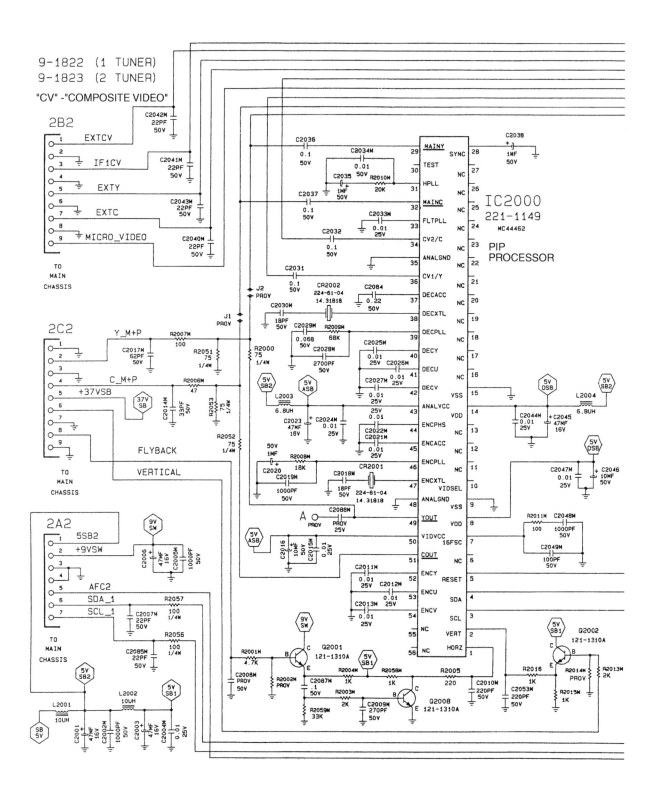

Figure 11-32 PIP Circuits

The Z-Line Chassis

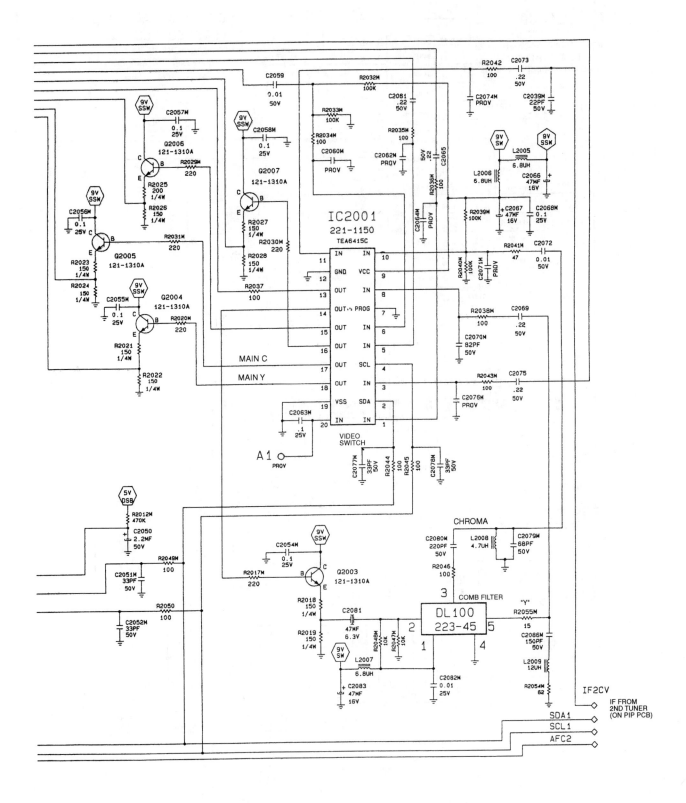

Figure 11-32 PIP Circuits (Continued)

Servicing Zenith Televisions

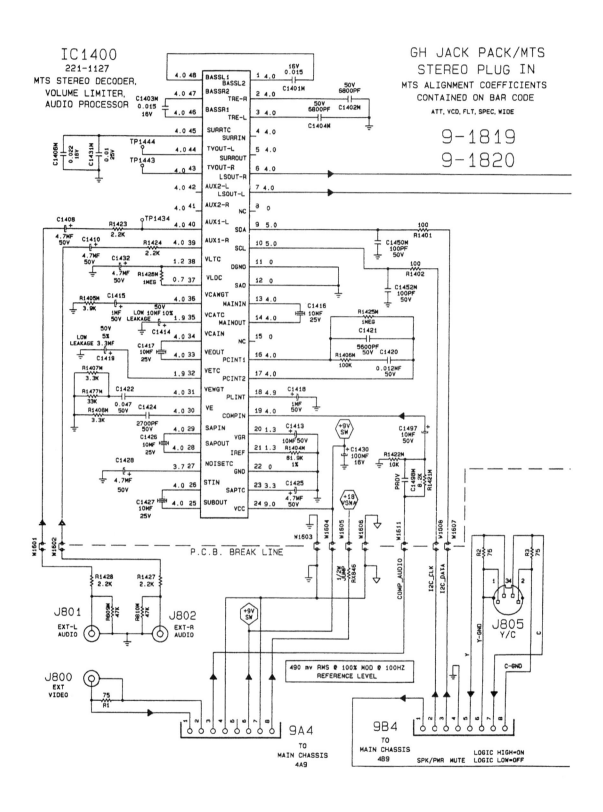

Figure 11-33 Audio Circuit

The Z-Line Chassis

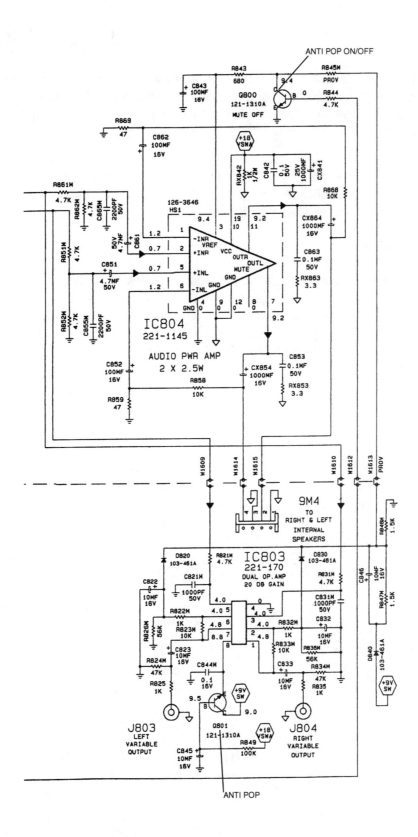

Figure 11-33 Audio Circuit (Continued)

GX Chassis

The GX chassis looks very much like the chassis we have been discussing (figure 11-34). It was designed for use in 19" through 35" mono and stereo televisions and even appears in a few commercial models. Like previous models, one board fits all. System control has been programmed during the manufacturing process to allow for various features, aided, of course, by plug-in modules like the PIP and stereo pack.

The module features four major integrated circuits: the now familiar ICX2200 for signal, sync, and deflection drive processing; IC6000 for system control; IC6001 for memory (EAROM); IC2100 for vertical sweep processing; and ICX3431 for the power supply.

Service Menu

When it comes up, the service menu looks like the screen depicted in figure 11-35. The black bar at the top contains the part number for the software that drives the chassis's hardware. The black bar at the bottom contains two numbers. The number on the left is the date the module went through the factory and the number on the right indicates the module's test status. As usual, the service menu comes up on the third item in the menu (HPos). Follow the usual procedure to gain access to all of the 39 items in the menu.

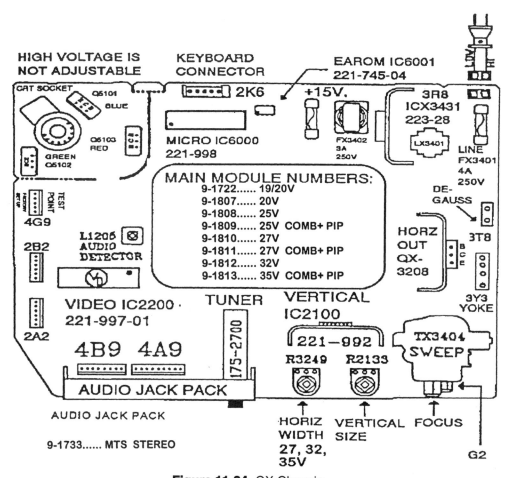

Figure 11-34 GX Chassis

A few of the service menu items need comment (table 11-3). If you change modules or if the customer complains of the loss of some menu features, check to see that item 04 (Level) agrees with the level specified on the bar code. There are three levels.

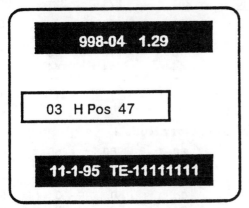

Figure 11-35 Accessing the Service Menu

Level 0 is the private label setting and is characterized by red menus with borders and italics. It has the following options: auto channel set, set channels, clock set, caption/text, three languages, and sleep timer.

Level 1 is the low-end Zenith setting and has blue menus with borders and italics. Its features include auto program, channel add/delete, clock set, on/off timer, caption/text, three languages, and sleep timer.

Level 2 is the high-end Zenith setting and is characterized by gray menus with borders and italics. Expect to find these menu features: icons, auto program, channel add/delete, channel labels, clock set, on/off timer, parental control, caption/text, three languages, auto demo, channel surf, and sleep timer.

If you need to change the level, access the service menu, select item 04, short pins 3 and 4 of connector 4G9 (figure 11-34) together, and use the adjust keys on the remote control to make the change. Then remove the shorting cable and exit the service mode.

I probably should remind you that there are eight settings for item 05 (Band). If you need to know what they are, consult the discussion of the service menu for the GH chassis.

If you change the PIP module, be sure that items 29 through 31 are correctly set. If you have to change the MTS module or the main module, confirm that items 32 through 38 correspond to the bar code on the jack pack.

If the module has a comb filter, confirm that item 11 (3.58 Trap) has been set to 1.

Service Procedures

The chassis has just two mechanical controls on it, one to adjust vertical size (R2133) and one to adjust horizontal width (R3249). Adjust the former for about ½ inch overscan at the top and bottom of the raster, and adjust the latter for about ½ inch overscan on each side.

As far as I can tell, there are no differences between alignment procedures of the GH and GX chassis for the video detector, AGC delay, and audio detector. But, there are differences for the stereo level adjust. Remember, you must make the adjustment if you change ICX2200, and you can make it with the jack pack on or off the module.

If you elect to remove the jack pack, attach a high impedance AC meter with a 47k ohm load to pin 3 of connector 4A9, and use pin 4 for ground. Ground pin 7 of 4A9, and ground pin 2 of 4B9 through a 10k resistor. Turn AC power off and back on again to reset the microprocessor. Apply a RF signal with video and audio at 400 hertz and 100% modulation to the antenna input. Then enter the service menu and adjust item 17 (AudLvl) for a reading of 490 to 500 millivolts AC after you have had the set on for three to five minutes.

	ITEM	RANGE	19 - 20"	25"	27"	32"	35"
	GENERAL SETTINGS (Blue) Factory Mode 0						
00	Fact Mode	0-1	0	0	0	0	0
01	Pre Px	0-1	1	1	1	1	1
02	V Pos	0-31	10	10	10	10	10
03	H Pos	0-75	44	44	44	44	44
04	Level	0-2	1*	1*	1*	1*	1*
05	Band	0-7	1	1	1	1	1
06	AC On	0-1	0	0	0	0	0
	TECHNICAL SETTINGS (Black) Factory Mode 1						
07	C.Phas	0-254	80	80	80	80	80
08	C.Srch	0-1	0	0	0	0	0
09	C.Line	0-32	17	17	17	17	17
10	Rf Bpf	0-1	0	0	0	0	0
11	3.58T	0-1	0	0	0	0	0
12	RF Brt	0-63	30	30	30	30	30
13	Ax Brt	0-63	30	30	30	30	30
14	MaxCon	0-63	63	63	63	63	63
15	VPhase	0-7	0	0	0	0	0
16	HPhase	0-31	16	16	16	16	16
17	Aud Lvl	0-63	46	46	46	46	46
18	AudAdj	0-63	63	63	63	63	63
19	RF Agc	0-63	40	40	40	40	40
20	H Afc	0-1	1	1	1	1	1
21	WhComp	0-1	0	0	0	0	0
22	60hzSw	0-3	2	2	2	2	2
23	PifVco	0-127	63	63	63	63	63
24	R Cut	0-254	19	19	19	19	19
25	G Cut	0-254	0	0	0	0	0
26	B Cut	0-254	39	39	39	39	39
27	G Gain	0-254	90	90	80	93	89
28	B Gain	0-254	70	65	60	98	80
	PIP SETTINGS (White)						
29	PipRas	0-254	1	1	1	1	1
30	PipSw	0-15	11	11	11	11	11
31	PipLuD	0-7	3	3	3	3	3
	**AUDIO SETTINGS (Yellow)						
32	In Lev	0-15	9	9	9	9	9
33	St Vco	0-63	24	24	24	24	24
34	SapVco	0-15	7	7	7	7	7
35	SapLpf	0-15	8	8	8	8	8
36	St Lpf	0-63	24	24	24	24	24
37	Spectr	0-63	50	50	50	50	50
38	WideBa	0-63	30	30	30	30	30

* Level 0 is used for Private Label sets.
 Level 1 and 2 is used for Zenith.
**These are default settings. See Bar Code on jack pack for correct Factory Alignment
For sets with the comb filter, Item 11, The 3.58 Trap should be set to 1.

Table 11-3 Service Menu Recommended Settings

If you want to leave the jack pack on the module, attach the positive probe of the meter to W1611 and the ground lead to W1603. Both of these connections are on top of the jack pack. Ground W53 on the main chassis to reduce high-frequency noise. Interrupt AC to reset the microprocessor. Apply an RF signal to adjust item 17 for the proper reading on the meter.

If you replace the module or the picture tube, be prepared to do the G2 and gray scale set-ups. The literature recommends that you follow a definite procedure to do both. The quality of the picture will more than likely suffer if you don't.

Let's begin with the G2 adjustment. Set brightness and contrast via the customer menu to midrange and color level to minimum. Connect a NTSC signal generator set to display a color bar signal to the antenna, and turn the chroma off (or use a pattern like Sencore's "Ten Bar Staircase" from a signal generator). Then adjust the G2 control so that the bar patterns range from completely black to a "not overdriven" (or saturated) white. You should see a distinct difference between the black and white bars and be able to distinguish different shades of gray among the bars. Finally, return the controls in the customer menu to their normal settings.

Now let's do the gray scale, but only after we have set G2. Set the color level to minimum and tint to midrange via the customer menu. Set items 26 and 27 in the service menu to the default settings for the screen size of the TV on which you are working (See table 11-3). Then apply a signal from a NTSC generator to the RF connector with the generator set to a pure white signal (i.e. chroma off). If you are not in the service menu, access it at set items 24, 25, and 26 to 0 (their cutoff values).

Take a look at the screen and note which color predominates, but do not adjust the value of its register. Do, however, adjust the other two for best white screen display. Then set the generator to produce a color bar signal and turn the chroma off via the customer menu. Check that the set displays a good gray scale from black to white. If the black level is too high, slightly readjust the G2 control to correct the picture. Finally, return all controls to their normal setting.

Power Supply

Since the GX power supply (figure 11-36) is like several we have already looked at, I won't comment on it. Merely consult the previous discussions if you need additional information or tips on how to repair it. But if you need to troubleshoot it, make these checks.

Check the standby voltages first:

+150 volts DC at RX3404 (or at C3405 or pin 3 of ICX3431)
+130 volts DC at CX3420 or the collector of the horizontal output transistor
+15 volts DC at the emitter of Q3403
+5 volts DC at pin 3 of IC3442

Then check for a high at pin 52 of the microprocessor after you issue an on command. The high should be coupled to the base of Q3402 causing it to turn on. When it turns on, expect to find +9 volts DC at pin 3 of IC3441 and +15 volts DC at the collector of Q3404.

Finally, check for the derived, or sweep, voltages: +23 volts DC at the junction of CX3268 and RX3242; +35 volts DC at CX3272; +245 volts DC at the junction of CX3296 and RX3217; and filament voltage at pins 1 and 2 of connector 2F5.

Servicing Zenith Televisions

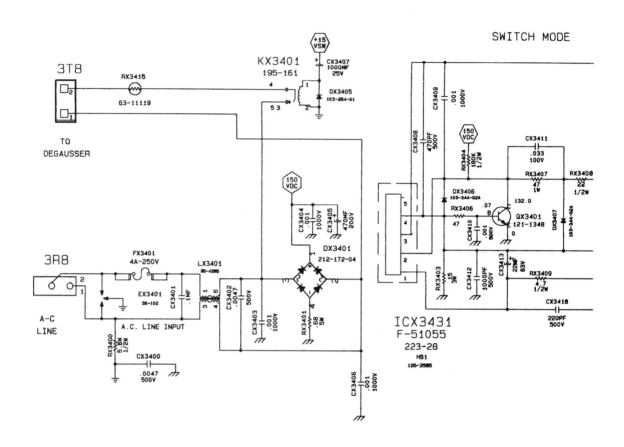

9-1722...... 19/20V
9-1807...... 20V
9-1808...... 25V
9-1809...... 25V COMB+ PIP
9-1810...... 27V
9-1811...... 27V COMB+ PIP
9-1812...... 32V
9-1813...... 35V COMB+ PIP

Figure 11-36 Power Supply Schematic

The Z-Line Chassis

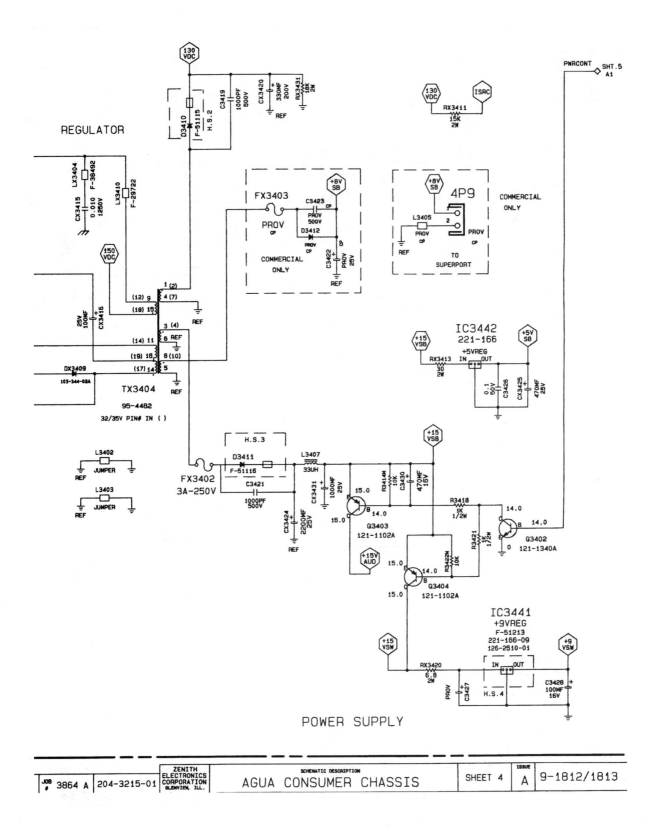

Figure 11-36 Power Supply Schematic (Continued)

Servicing Zenith Televisions

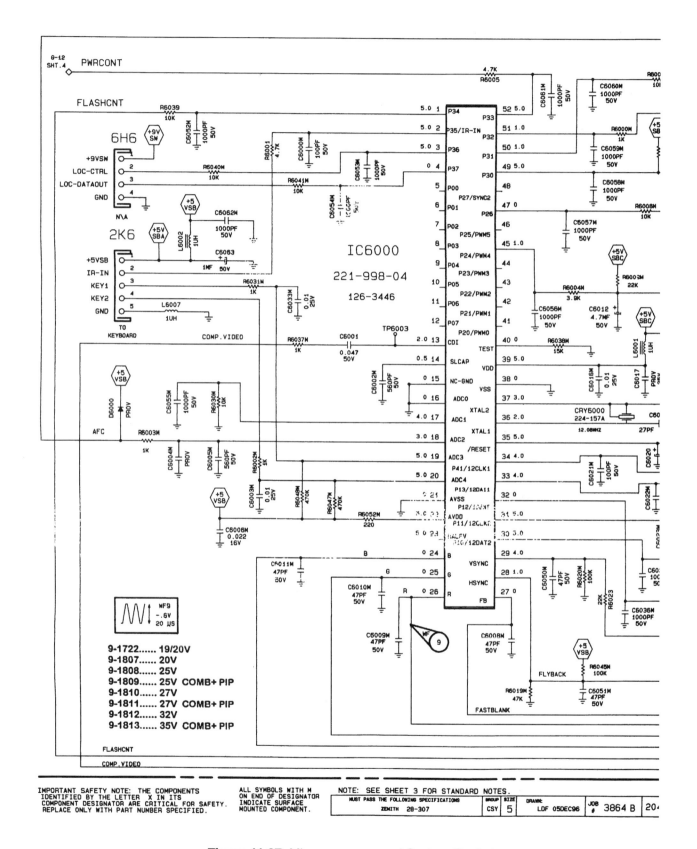

Figure 11-37 Microprocessor and System Control

The Z-Line Chassis

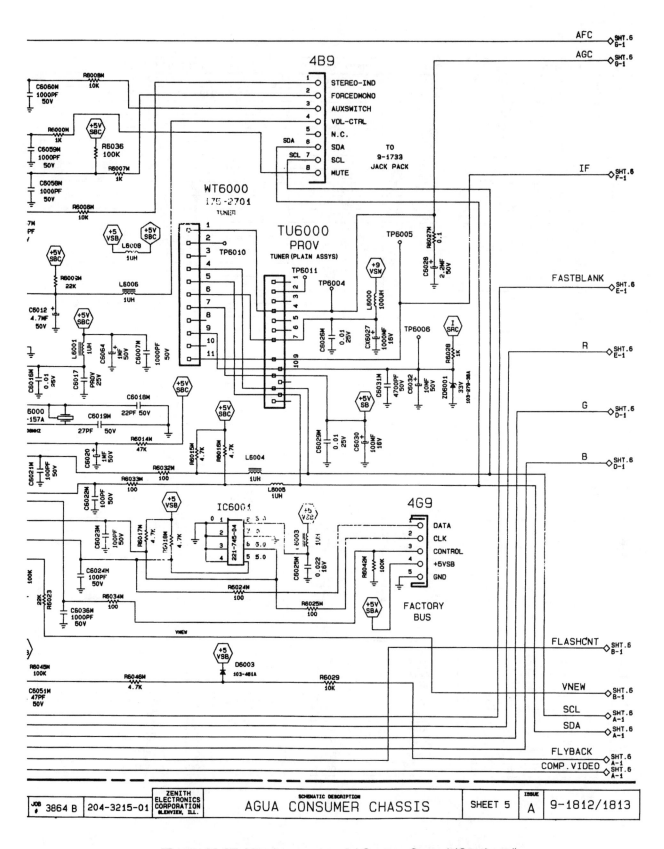

Figure 11-37 Microprocessor and System Control (Continued)

PIN #	SPECIFIER	NAME	PORT MODE	INPUT / OUTPUT	VOLTAGE LEVELS
1	P34	FLASHCNT	PORT	OUTPUT	TV OFF = 0V, TV ON = 5V
2	P35 / IR-IN	IR-IN	SPECIAL	INPUT	NO IR = 5V, IR PULSES = 0V
3	P36	LOC-CTRL	PORT	OUTPUT	0V
4	P37	LOC-DATOUT	PORT	OUTPUT	0V
5	P00		PORT	OUTPUT	0V
6	P01		PORT	OUTPUT	0V
7	P02		PORT	OUTPUT	0V
8	P03		PORT	OUTPUT	0V
9	P04		PORT	OUTPUT	0V
10	P05		PORT	OUTPUT	0V
11	P06		PORT	OUTPUT	0V
12	P07		PORT	OUTPUT	0V
13	CDI	COMP VIDEO		INPUT	1.85V DC OFFSET, 1.5V P-P
14	SLCAP			INPUT	0V
15	NC-GND			INPUT	0V
16	ADC0			INPUT	0V
17	ADC1			INPUT	0V
18	ADC2	AFC		INPUT	2.5V WHEN OPTIMALLY TUNED
19	ADC3			INPUT	0V TO 5V
20	ADC4			INPUT	0V TO 5V
21	AVSS			INPUT	0V
22	AVDD			INPUT	5V
23	HALFV			OUTPUT	0V
24	B			OUTPUT	0V TO 5V
25	G			OUTPUT	0V TO 5V
26	R			OUTPUT	0V TO 5V
27	FB			OUTPUT	0V TO 5V
28	HSYNC			INPUT	0V TO 5V, POSITIVE PULSE
29	VSYNC			INPUT	0V TO 4.3V NEGATIVE PULSE
30	P10 / I²C-SDA₂		SPECIAL	INPUT / OUTPUT	0.2V TO 5V
31	P11 / I²C-SCL₂		SPECIAL	INPUT / OUTPUT	0.2V TO 5V
32	P12 / IDENT		PORT	INPUT	BUS2 UNFROZEN = 0V, BUS2 FROZEN BY TE = 5V
33	P13 / I²C-SDA₁		SPECIAL	INPUT / OUTPUT	0.2V TO 5V
34	P14 / I²C-SCL₁		SPECIAL	INPUT / OUTPUT	0.2V TO 5V
35	/ RESET			INPUT	5V
36	XTAL1			INPUT	0V TO 5V
37	XTAL2			INPUT	0V TO 5V
38	VSS			INPUT	0V
39	VDD			INPUT	5V
40	TEST			INPUT	0V
41	P20 / PWM0		PORT	OUTPUT	0V
42	P21 / PWM1		PORT	OUTPUT	0V
43	P22 / PWM2		PORT	OUTPUT	0V
44	P23 / PWM3		PORT	OUTPUT	0V
45	P24 / PWM4	VOL-CTRL	PWM	OUTPUT	0V TO 5V AT PIN, 0.8V TO 5V AT C6012
46	P25 / PWM5		PORT	OUTPUT	0V
47	P26	STEREO-IND	PORT	OUTPUT	STEREO PRESENT = 0V, STEREO NOT PRESENT = 4.3V
48	P27 / SYNC2		PORT	OUTPUT	0V
49	P30	FORCED MONO	PORT	INPUT / OUTPUT	STEREO = 0V, MONO = 5V
50	P31	AUX SWITCH	PORT	OUTPUT	AUX = 0V, RF = 5V
51	P32	MUTE	PORT	OUTPUT	MUTE OFF = 0V, MUTE ON = 5V
52	P33	PWR CONT	PORT	OUTPUT	TV OFF = 0V, TV ON = 5V

Table 11-4 Microprocessor Port Modes, I/O, Voltage Levels

System Control

If I were to describe the microprocessor, 52-pin DIP, CMOS design, 12 MHz clock, etc., you might think it is identical to other microprocessors we have discussed, particularly the one for the GH chassis. But it is not. The GH chassis uses part number 221-1138 while the GX chassis uses part number 221-998. If you study figure 11-37 and compare it to figure 11-26, you'll see that there are obvious differences, but you troubleshoot it the same way. Table 11-4 contains about all you need to know to do a pretty thorough job of checking it out.

Zenith follows its tradition by establishing communications with other parts of the chassis over two communication buses. Bus one (pins 33 and 34) communicates with the tuner, video processor, PIP and MTS module. Bus two (pins 30 and 31) supports the EAROM (IC6001). The microprocessor controls brightness, color, contrast, tint, and sharpness by sending commands to the video processor over bus one. It receives input over the A/D keyboard Zenith has been using for the last couple of years or the IR receiver.

If you think you have a system control problem, use the following chart to help you locate it:

+5 volts at pins 22 and 39
ground (vss) at pin 21
oscillator at pins 36 and 37 (check peak-to-peak level and frequency)
reset at pin 35

These, as you know, are what engineers call "the must haves." If you still can't locate the cause of your particular problem, continue by checking the power control at pin 52, which should go high to turn the set on. Check both communication buses at pins 30 and 31 and pins 33 and 34. Check horizontal sync at pin 28 for a 5-volt peak-to-peak positive-going pulse and vertical sync at pin 29 for a negative-going pulse of about 4.3 volts peak-to-peak. Finally, check pins 24 through 26 for the R, G, and B OSD output to the ICX2200.

Deflection Circuits

According to the training material, the vertical deflection circuit is the same as the one used in the GH chassis (figure 11-38), and as far as I can tell, that is the way it is. Should you need the information, please refer to the technical data I gave you in the previous section. If you are curious about the vertical output chip Zenith has been using for the last few years, take a look at figure 11-39. The training material doesn't make the same observations about the horizontal deflection circuit, but

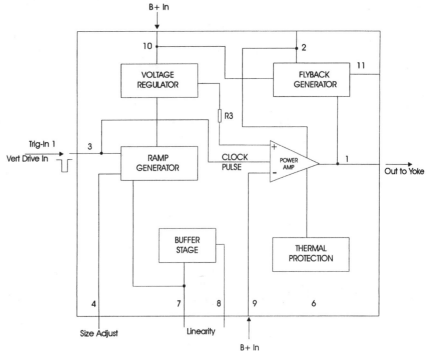

Figure 11-39 Vertical Output Chip

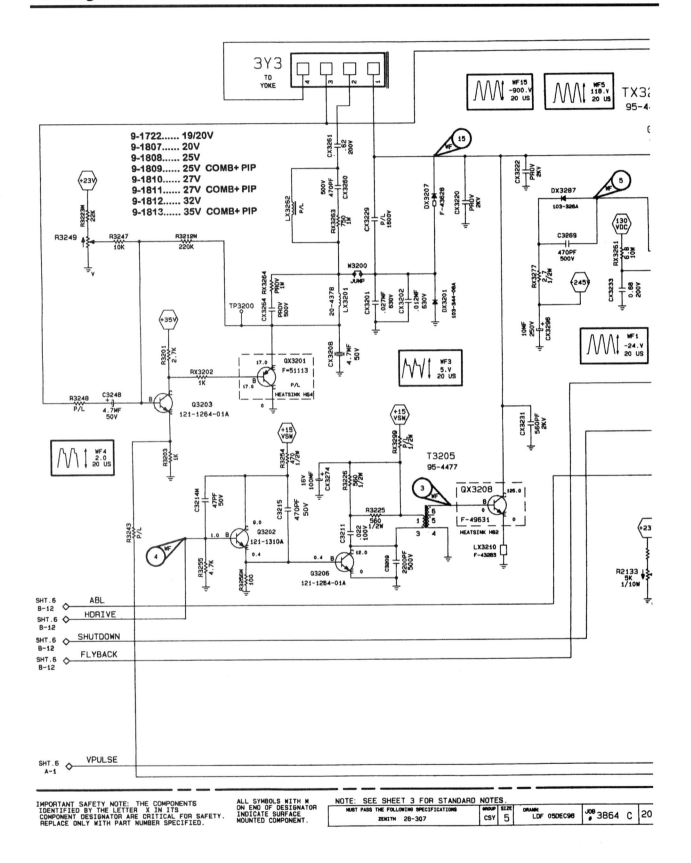

Figure 11-38 Deflection Circuits

The Z-Line Chassis

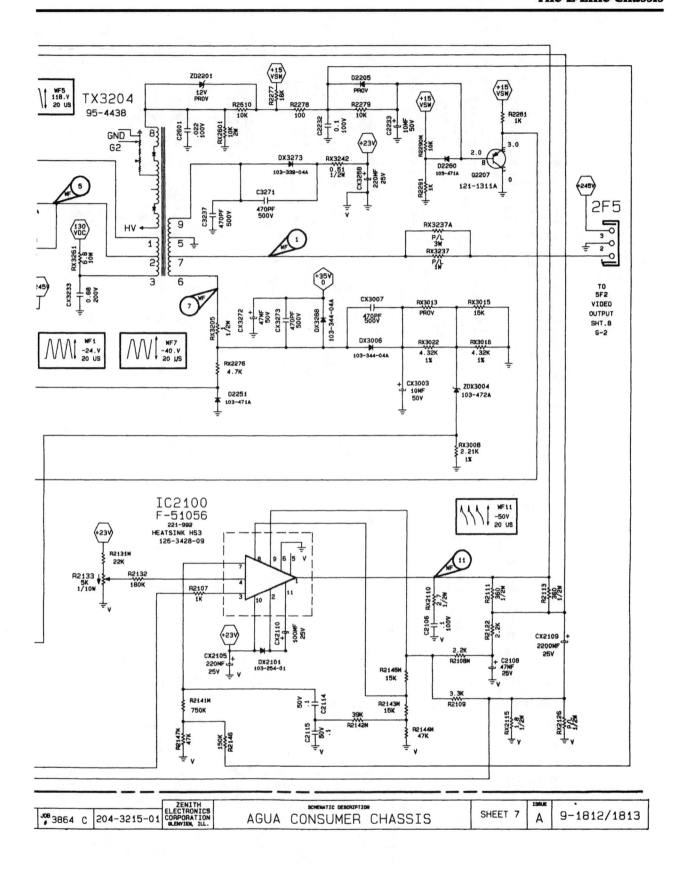

Figure 11-38 Deflection Circuits (Continued)

Servicing Zenith Televisions

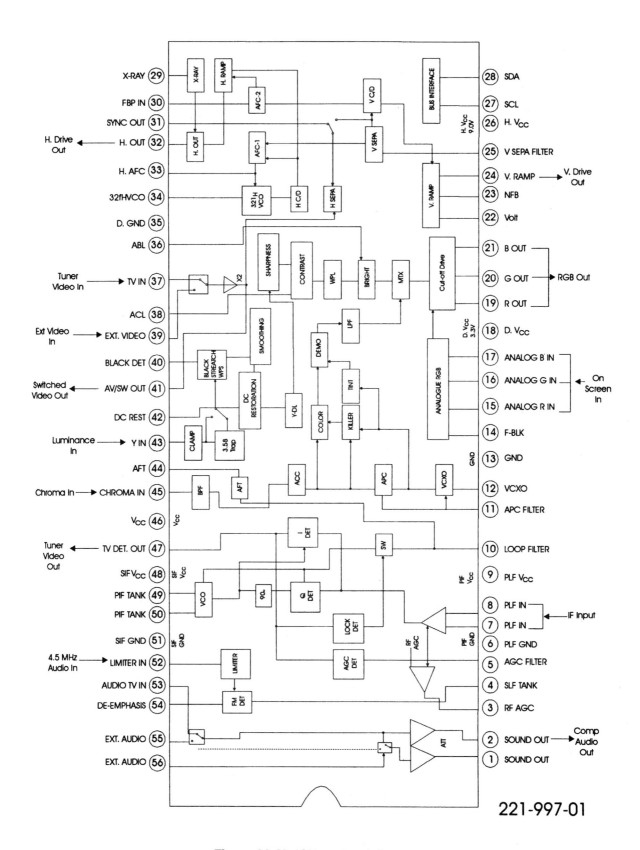

Figure 11-40 ICX220 Block Diagram

if you read the circuit descriptions in the service manual, you get the impression that the horizontal deflection circuits are "identical," including component designations and signal path. Nevertheless, there are a few things to which I call your attention even though I know I am repeating myself.

Please pay attention to resistor RX3299 in the horizontal driver circuit. Zenith uses it to control the base current of Q3208 (the horizontal output transistor). Its value depends on the screen size because the best drive for the horizontal output transistor is a function of its collector current, and that is chassis dependent. I make this point here even though I have made it before because overdriving Q3208 is one sure way to cause it to fail.

The horizontal deflection circuit develops about 28 KV for the 19" through 27" sets and about 30 KV for the 32" and 35" ones. R3249, the horizontal width control, is found only on the 27" through 35" sets, the chassis that have pincushion correction circuits. The smaller televisions have horizontal pincushion correction wound into the design of the yoke.

Video Processing Circuits

The GH and GX chassis use the same video processing IC (part number 221-997-01). Figure 11-40 is an internal block diagram of the IC and should give you a feel for how complicated the chip really is. Figure 11-41 is a schematic representation of the entire video processing circuit for the chassis we are discussing.

Servicing the Video Processor

Use this table to check for key operating signals in and around ICX2200:

composite video out at pin 41
composite audio out at pin 2
horizontal drive out at pin 32
vertical drive out at pin 24
blue drive out at pin 3 of 2C5
green drive out at pin 2 of 2C5
red drive out at pin 1 of 2C5
+9 volts B+ at pins 9, 23, 33, 46, and 48
serial data at pin 30
serial clock at pin 31

As I told you when we discussed the GH chassis, if you don't see a viewable picture, enter the service menu and check the settings of items 05, 19, and 23 which should be set to 0, 40, and 63 respectively. If you need to check the alignment of the video detector (or align it), connect a high impedance meter to pin 44 of ICX2200 or R1219, apply a good signal to the TV, and adjust item 23 (PIFvco) for 2.5 volts DC.

Video Output Circuit Board

The video output module (figure 11-42) comes as a breakaway portion of the main circuit board. Following previous manufacturing techniques, Zenith has put no mechanical controls in the circuit meaning you must make gray scale adjustments through the service menu. Have you spotted Q5104, our old friend from the Y-Line? Don't forget it and the problems it causes when it fails. Refer to the preceding chapter if you need the information.

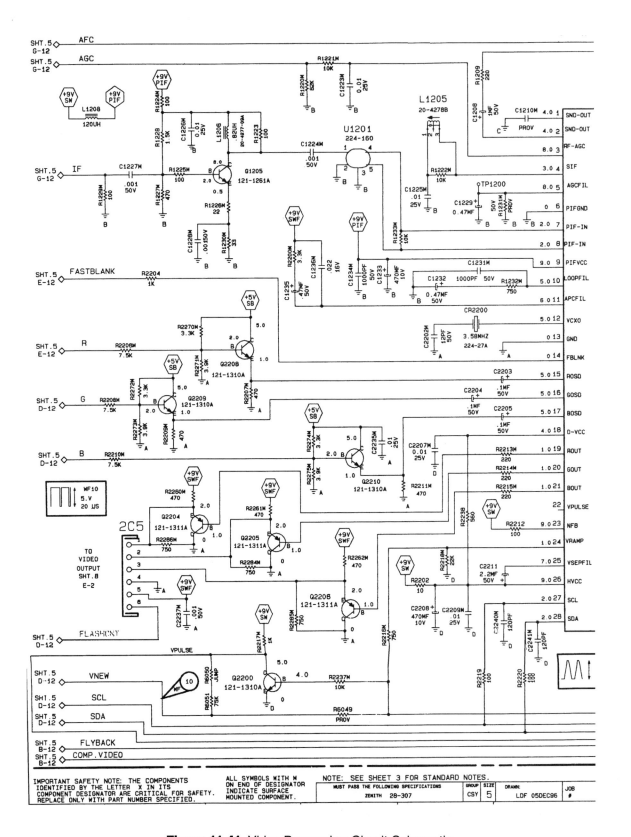

Figure 11-41 Video Processing Circuit Schematic

The Z-Line Chassis

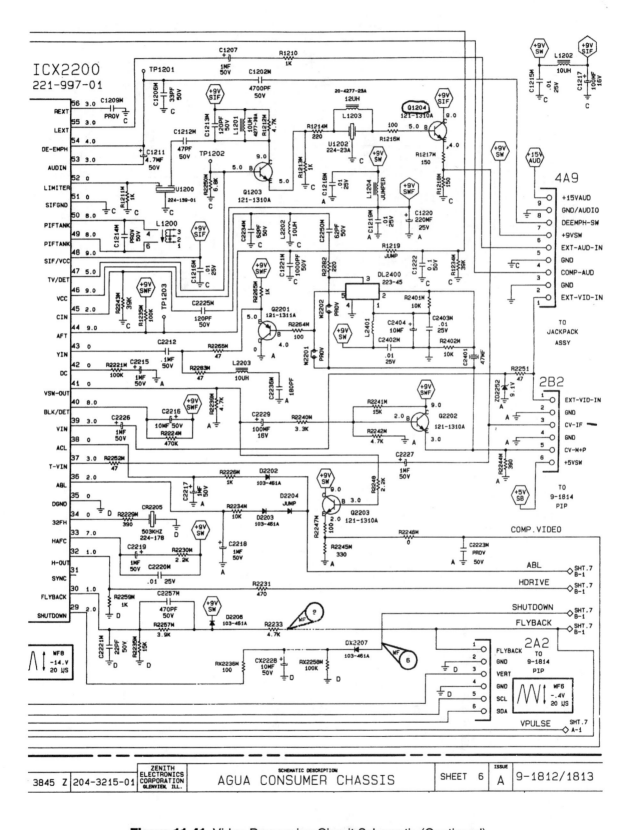

Figure 11-41 Video Processing Circuit Schematic (Continued)

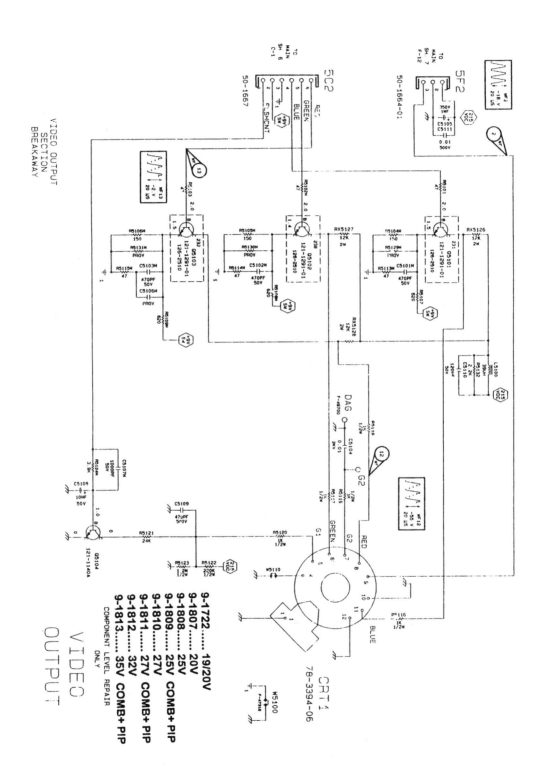

Figure 11-42 Video Output Circuit Board

Notice that there are no warnings about keeping the video output module on the CRT base while the TV is on. If my information is correct, there should be! In other words:

DON'T REMOVE THIS CIRCUIT BOARD WHILE THE TV IS PLAYING OR TURN THE TV ON WITH THE BOARD OFF THE PICTURE TUBE.

This warning applies to the larger tubes, not the smaller ones. If you haven't read the information I presented when I dealt with the GH chassis, I suggest you do it now because it is applicable here, especially the discussion about the external connection at pin 4 of the CRT socket.

PIP Circuit

The PIP circuit (figure 11-43) is a bit different from the last one we looked at, but it works about the same way. It initializes with the same source as the main picture, but the user has the choice of selecting any one of the other signal sources via the PIP source menu.

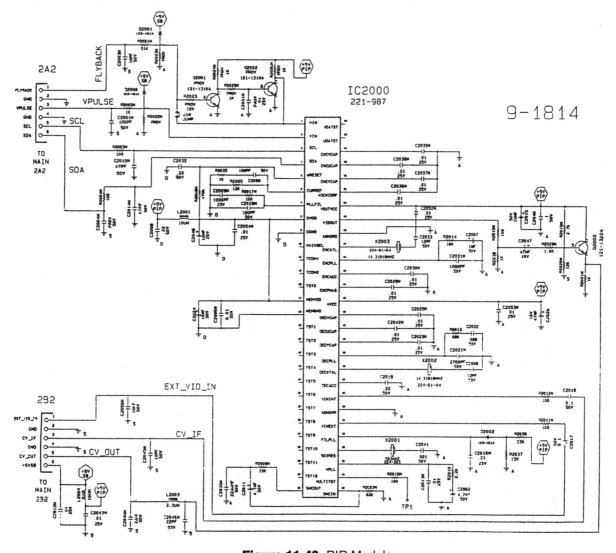

Figure 11-43 PIP Module

The GX chassis uses a microprocessor that has an auto detect feature. If the set has a PIP, the microprocessor detects it and routes all video signals to it. It is therefore important to unplug the TV before you install a new PIP module or reinstall the old one. If you don't let the microprocessor reset, it won't detect the PIP module and will behave like a non-PIP set. Turning the set off with the front control power button or the remote control won't work. You must unplug it and plug it back in.

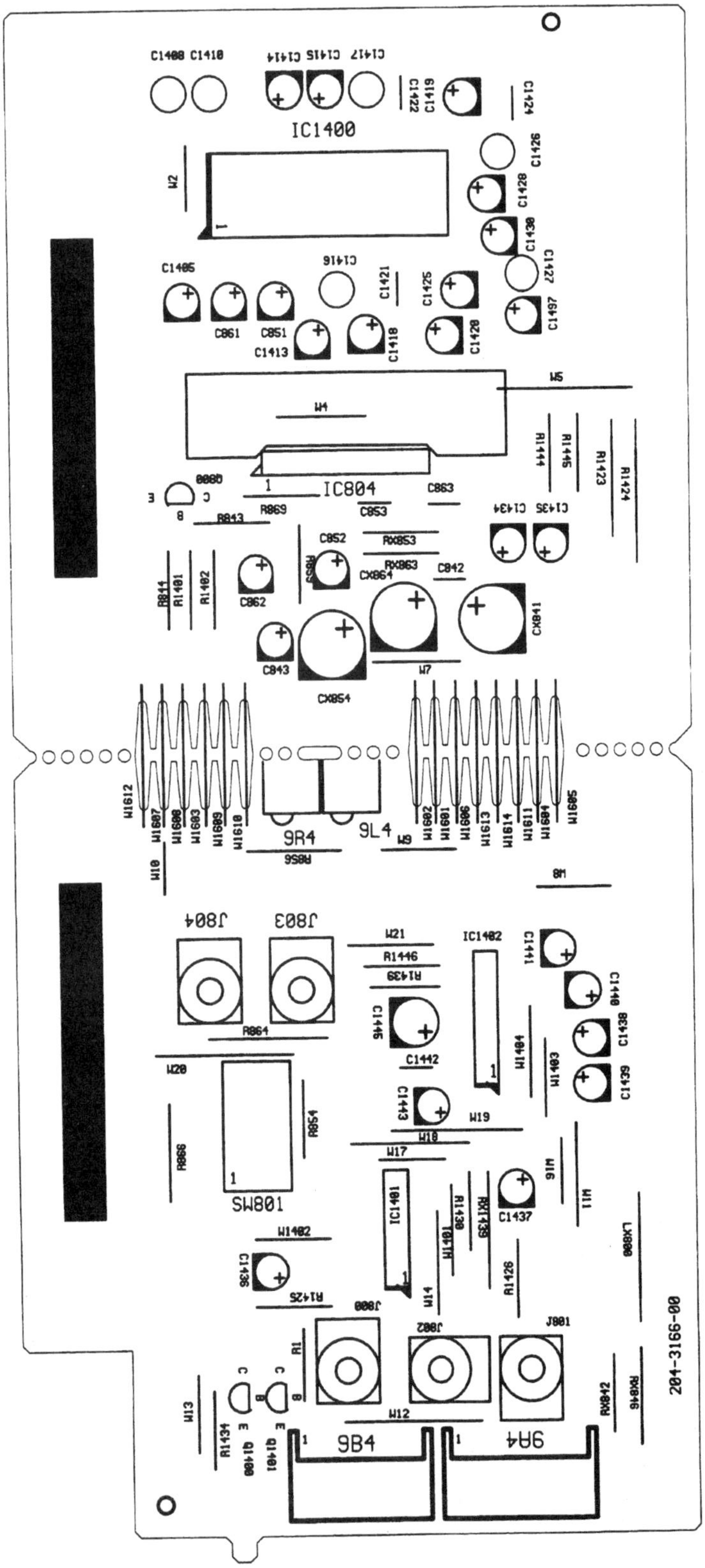

Now just a few words about the circuit. Composite video from ICX2200 enters IC2000 at pin 36. Auxiliary video comes in at pin 34. The bus from system control appears at pins 3 and 4. The horizontal pulse is routed to pin 1 while the vertical pulse is routed to pin 2. Video out, which consists of the main video plus PIP video, exits at pin 49. These few facts should give you "a leg up" if you have to troubleshoot the circuit.

Audio Circuit

Since I have mentioned the jack pack-audio module several times, I thought you might like to take a look at how it is laid out (figure 11-44) and get an idea about the electronics it contains (figure 11-45). It folds in half at the center of the two boards where you see the perforations and wire jumpers and fastens to the jack pack by means of small philips screws.

The MTS stereo circuit makes use of several integrated circuits: IC1400 (the stereo decoder), IC1401 (the aux/TV switch), IC1402 (the volume limiter), and IC804 (the dual power amplifier).

Let's assume that you have the set in the TV mode as opposed to the aux mode. Composite audio exits pin 2 of ICX2200 and comes into the audio module on pin 4 of connector 9A4 and is routed to pin 17 of IC1400 which is the "heart" of the audio circuit (figure 11-46). Do you remember that all adjustments for it are accomplished via the service menu? After it receives the audio

Figure 11-44 Audio Module/Jack Pack

The Z-Line Chassis

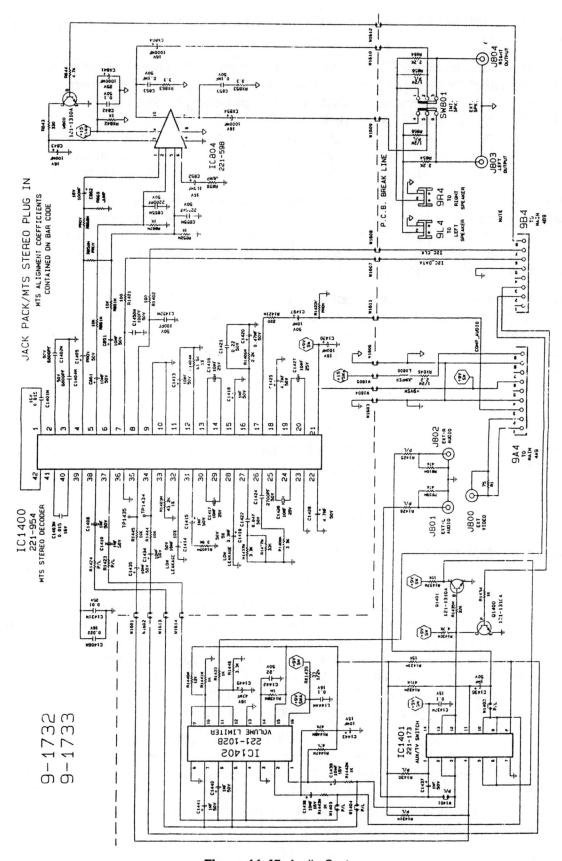

Figure 11-45 Audio System

signal, IC1400 processes the baseband audio and outputs left channel audio at pin 35 and right channel audio at pin 34.

Left and right channel audio exit IC1400 and go to pins 8 and 4 the switcher IC (IC1401). Auxiliary audio from the left channel jack, if it were available, would be at pin 11. Right channel audio takes another path. The microprocessor controls the switcher by issuing commands that are coupled from its pin 50 to pin 3 of connector 9B4 and on to transistors Q1400 and Q1401. When the microprocessor issues a high signal to Q1400, the transistor conducts pulling pins 12 and 13 of the switcher low, putting it into the TV position. If the microprocessor issues a low signal, Q1400 turns off while Q1401 turns on placing a low on pins 5 and 6, putting the switcher into the auxiliary position.

Left channel audio exits the switcher on pins 2 and 3 while right channel audio exits on pins 9 and 10. Both signals go now to IC1402 which limits the level of either audio source to prevent excessive volume above normal audio programming (SoundRite audio processing). The microprocessor turns IC1402 on or off by toggling the voltage at pin 11. If the voltage is lower than 1.5 volts, the circuit is off. If it is higher than 2.0 volts, the circuit is on.

The next leg of the journey routes both audio signals back into IC1400. Left audio enters at pin 5, and right audio enters at pin 38. The signals leave IC1400 at pins 5 and 6 and make their way to IC804 where they are amplified and sent to the speakers. Note that volume and mute are bus-controlled functions. Q800 mutes the speakers at turn on and turn off. I believe I have mentioned it before, and if I have, it bears repeating. When you encounter a "no audio" problem, check the voltage on pin 3 of IC1400. If it is low, immediately turn your attention to Q800. That little devil has fooled me twice in the last three months.

Some GX chassis are equipped with non-MTS stereo audio boards that, in the interest of space, I am not going to reproduce here. In those models baseband audio leaves ICX2200 and goes to IC1601, the audio decoder. After it is processed by IC1601, the audio goes to IC1602, a voltage-controlled amplifier that processes tuner audio and has provisions for switching external audio into the circuit. Audio leaves it and goes to IC803 that amplifies it and sends it on to the speakers. Transistors Q802 and Q803 are in the circuit to keep the speakers from popping when the TV is turned on or off.

Repair History

Before getting into a very brief repair history, let me acquaint you with two service bulletins. The first alerts us to the fact that a small quantity of Z-Line Z27X31D models were built using the Y-Line 9-1629-01 main module. It is almost identical to the 9-1811 main module used in the Z-Line products except for the type of sockets used for the PIP and audio/jack pack modules. When you service these products, be sure that the PIP and audio/jack pack match the sockets used on the main board. If you have to replace either module, you must match the module with the correct socket. The 9-1629-01 Y-Line main module uses the 9-1586 PIP module and the 9-1592 audio/jack pack while the 9-1811 Z-Line main module uses the 9-1814 PIP module and the 9-1733 audio/jack pack.

The second service bulletin refers to model Z27A11G and concerns loss of color on channel 4 when the TV is in broadcast mode and on channel 6 when it is in the cable mode. Locate surface-mount resistor R2241M in the collector circuit of Q2241 and check its value. If it is labeled "391," remove it and replace it with Zenith part number 63-11059-25A. You are changing the resistor from 390 ohms to 100 ohm, 5%, surface-mount resistor. It seems a certain number of 9-1791 main modules left the plant with the wrong value resistor installed.

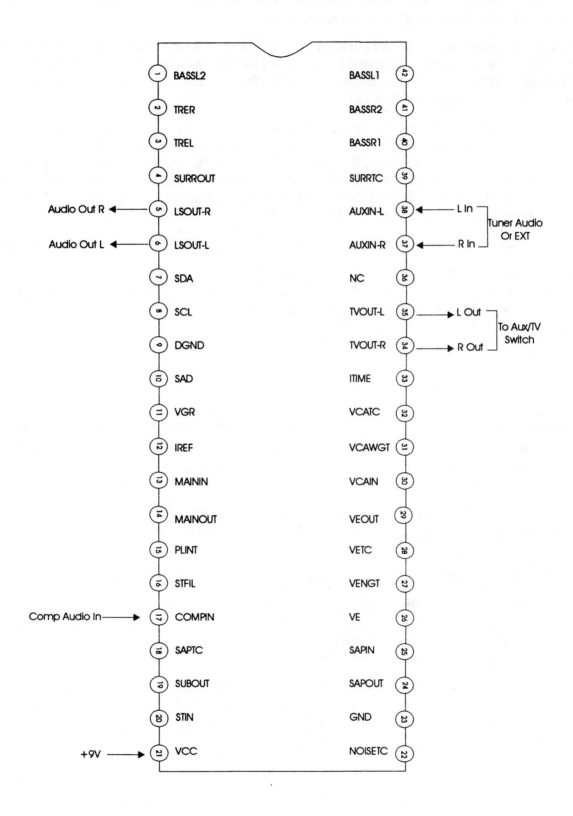

Figure 11-46 IC1400 MTS Stereo Decoder

Servicing Zenith Televisions

Just an aside comment. The "M" in the designation R2241M is Zenith's way of alerting you to fact that it is a surface-mount resistor. Now to the repair history.

I have carefully searched my memory and my database and concluded that I don't have much to offer you! I also checked Zenith's Z-Tips and found that it has quite a bit of information on the Z-Line. Let me once again recommend that you spend the few dollars it costs and add Z-Tips to your personal database. You won't be sorry when you do.

(1) Dead Set. When AC was applied, the power supply came up, but there was no +5 volts standby at IC3442. Do you recall that the +5 volts standby is derived from the +15-volt line and that the +15-volt line is fused by FX3402, a 3-amp, 250-volt pigtail (GJV) device? The fuse in this instance had opened. Since the biggest current drain on this line is the audio output IC, it is logical to suspect that it shorted. Replacing the audio output IC and FX3402 fixed the TV.

(2) Dead Set. There was no startup voltage at pin 2 of ICX3431 because R3404 had opened.

(3) Dead Set. You can take care of most dead set problems caused by a dead power supply by replacing ICX3431 (STR53041) and QX3401 (Zenith part number 121-1348. Be sure to check the diodes and resistors in the circuit and replace them as necessary. RX3407 (47 ohms at 1 watt) and RX3408 (22 ohms at ½ watt) are especially susceptible to damage. If the above parts check good, I suggest you check and replace, if necessary, CX3416 in the power supply.

(4) Dead Set. This TV had a properly operating power supply and system control, but it still wouldn't come on. I traced the problem to lack of horizontal drive and had to replace ICX2200 to fix it.

(5) Washed Out Picture. You might think the picture tube was defective. However, Q5104 (Remember this one?) on the CRT board had developed a collector-to-emitter leakage.

(6) Picture Too Wide (pincushion problems). I have known coil LX3201 to fail a couple of times. On one occasion the ferrite core loosened and fell to the bottom of the form. On the other occasion, the coil just burned out. Either fix the coil or replace it with part number 20-4530.

(7) Snowy Picture and Channels Won't Tune. Defective connections on the RF splitter or a defective splitter may cause the classic "snowy picture" syndrome.

(8) Shuts Off Soon After Turn-On. D6003 in the HEW circuit was breaking down under load and placing excessive voltage on pin 28 of IC6000. Monitor the voltage on pin 28 with a sample and hold meter to isolate the problem. I have also know CX3416 in the power supply to cause the same problem.

(9) No Audio. Q800 developed a collector-to-emitter leakage that resulted in a lowered voltage on the corresponding pin of audio output amplifier. The lowered voltage muted the output amplifier. I believe I used an ECG123AP to replace it.

CHAPTER 12
THE A-LINE CHASSIS

The A-Line series of televisions came out in 1998 for the model year 1999 and consists of the GA and GB chassis. Sounds simple, doesn't it? Before you reach that conclusion, let me tell you that each chassis has three versions! There are the GA-1, GA-2, and GA-3 as well as the GB-1, GB-2, and GB-3. Each version is slightly different, offering different features with the hardware and software to support those features.

I have struggled with the question of how to deal with these multiple chassis in a way that facilitates service and decided to treat all of the GA and GB variations under their respective headings and point out the differences as I go. If I don't do something like this, my presentation will become too long and too tedious to be helpful. I hope the slight variation in procedure won't be too confusing or taxing for you.

Available Literature

Let's start out by getting an overview of Zenith's literature for the A-Line. You notice that I have almost stopped referring to Sams PHOTOFACT for two reasons. First, Zenith's literature is ample for service needs, and second, a lot of these televisions are too new to have a PHOTOFACT yet. However, if and when I need a fresh perspective on a particular circuit and if it is available, I have no hesitation about pulling out a PHOTOFACT and taking a peek at how the technical folks at Sams have conceptualized the circuits.

If you work on commercial sets, you can obtain the necessary literature by calling Zenith and requesting their manual for the commercial A-Line products. I am currently looking at the A-Line commercial products addendum, a training manual that you can order by requesting part number 923-3354TRM.

The training manual for the GA chassis is number CM150TRG and may be ordered by requesting manual 923-3352TRM. The service manual, consisting of circuit descriptions, block diagrams, voltage and waveform charts, schematics, etc., belongs to the CM-150 series and is available by ordering manual number 923-3349. Zenith has even produced a videotape preview of the GA chassis (part number 892-75) that you might find interesting.

The GB chassis literature follows suit. The training manual belongs to the CM150TRG series (part number 923-3351TRG). The part number for the service manual is 923-3363, and the part number for the video training tape is 892-76.

I suggest that if you are not familiar with the way Zenith is now doing things you consider purchasing at least one of these videotapes. I have had the benefit of attending the training sessions and seen the chassis from that perspective. If I hadn't been able to attend the live training session, I would consider purchasing both tapes and watching them several times. There are benefits to listening to someone else explain the various features of a particular chassis.

GA Chassis

Let's begin our venture into the A-Line with the GA chassis. Before getting to the "how it works" section, I need to tell you about an interesting feature incorporated into the service manual. Pages 2-1 and 2-2 have a "model parts list." These pages list each model manufactured as a GA chassis, and under each model is a list of parts that you might have trouble locating if you don't know where to look. The list contains the part numbers for the module, speaker, cabinet, line cord, etc. Just make a note of it, and you won't have much trouble locating the number for a part that isn't listed with the resistors, capacitors, solid state devices, and so forth.

The GA chassis was developed for Zenith's "Sentry 2" series and is used in screen sizes 19" through 27" mono and stereo models. The chassis itself is built onto a single-sided printed circuit like previous chassis (figure 12-1). Three ICs handle signal, sync, and sweep processing: (1) ICX2200 processes the audio and video information and develops vertical and horizontal drive; (2) IC6000 is the system microcontroller, and IC6001 is the EAROM; (3) and IC2100/2101 develops vertical drive. There are, of course, other ICs on the chassis, but these three are the center of attention.

Each GA chassis category represents a different screen size and feature level. For example, the GA-1 category developed for 19" and 20" sets has these module variations (as of March 1998):

- 9-1869   mono without a jack pack
- 9-1870   non-MTS stereo without a jack pack
- 9-1871   MTS stereo with three jacks
- 9-1950   mono with a two-jack assembly

The GA-1 chassis uses the 221-1164 microcontroller and has a line-connected power supply.

The GA-2 chassis has these module variations:

- 9-1789   mono without a jack pack
- 9-1790   stereo with three jacks
- 9-1791   non-MTS stereo with three jacks
- 9-1831   stereo with three jacks

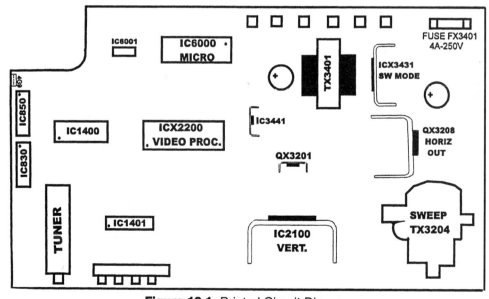

Figure 12-1 Printed Circuit Diagram

It uses the 221-1164 microcontroller and has a line-isolated, or switching, power supply.

The GA-3 chassis has these module variations:

9-1996	MTS stereo with five jacks
9-1997	MTS stereo with five jacks
9-1998	MTS stereo with five jacks

It uses the 221-1305 microcontroller and has a line-isolated power supply.

If you need a model-to-GA chassis list, take a look at table 12-1. It isn't exhaustive, but it should give you an idea about the number of models in the GA line.

Customer Menu

The customer menus are displayed over video and appear at the left of the screen. It can be set to scroll up vertically from the bottom of the screen or appear instantly. The method of display depends on the setting of item 26 in the service menu. The color of the menu depends on the level of the chassis, the level having been set at the factory during manufacturing. Zenith uses level 1, meaning the customer menu is blue on the GA-1 and GA-2 chassis and black on the GA-3. Private label sets are level 0 and have a dark red background.

A19A02D
A19A11D
A20A22D
LGA20A02GM
LGA20A04DM
LGA20A05DM
LGA20A11DM
LGA21A22DM
A25A02D
A25A11D
A25A12D
A25A74R
A25A76R
A27A11D
A27A12D
LGA26A02GM
LGA26A11DM
LGA26A12DM
LGA29A11GM
LGA29A12DM
A25A23W
A25A23W9
A27A23W
A27A23W9
A27A74R
A27A76R
LGA26A23WM
LGA29A23WM

Table 12-1 Model to GA Chassis List

The customer menu has a maximum of five "pages." Since features vary from set to set, not all of the items appear on all models. The "Setup Menu" includes auto program, add/del/surf, clock set, captions, caption/text, audio mode, and language. The "Special Features" menu has timer setup, channel labels, parental control, and auto demo in it. The "Audio Mode" permits the customer to control bass, treble, balance, audio mode, front surround, sound rite, and speakers. The "Video Menu" has contrast, brightness, color, tint, sharpness, color temperature, and picture preference in it. The "Source Menu" has the single listing of "main source" on it.

I will limit my comments to the relatively new features.

First, you see two headings for captions in the setup menu. "Captions" has three selections, on, off, and captions when muted. The first two are self-explanatory. The latter, however, is brand-new. If it is turned on, captions appear on the screen when the audio has been muted, letting the user read the captions and watch the picture while, for example, talking on the phone. The second heading, "caption/text," is the one we are accustomed to seeing.

Second, you have already seen "parental control" under special features. Even though I discussed the parental control feature in the last chapter, because it is a new feature, I will deal with it one more time. The feature has four subcategories: (1) "Block Channel/Block Video" permits the user to block one or more channels or block the video input. (2) "Set Hours" permits the user to set the hours the ports are be disabled. The setting is from one hour to 99 hours. (3) "Set Password" lets the user set a password that consists of four digits selected via the numerical keys on the remote control. (4) "Lock/On Off" turns the feature on or off.

NOTE. If the password is forgotten, the user must wait the prescribed time for the port or channel to unlock. Don't try to clear parental control by removing AC power to the TV because that merely prolongs the run-out time. However – and there may be a service call involved – entering and exiting the service menu will defeat parental control and cause the timer to reset to "0." Don't even consider telling the customer how to gain access to the service menu to turn the function off even if he (or she) is a dear friend. That kind of information simply must be kept private.

Third, two items in the video menu need to be mentioned. "Color Temp" gives the viewer a choice between "warm" and "cool." The latter adds a bit of blue to the picture. "Picture Pref" lets the customer choose between "preset" and "custom." The former sets brightness, contrast, color, etc. to the factory default settings while the latter permits the viewer to set those controls according to personal preference. Everything else in the customer menu is vintage Zenith and requires no additional comment.

Service Menu

I have discussed entry and exit of the service menu too many times to repeat the procedure. But there is one exception. The GA chassis comes equipped with a 10-key and a six-key keyboard. If you deal with the six-key keyboard, press and hold the menu button until the customer menu disappears and then without delay press at the same time volume up and channel down.

The service menu comes up just as it did, for example, in the Z-Line products. The number in the black bar at the top of the screen contains the part number for the software (figure 11-3a). Put a "121" in front of it, and you have the part number for the microprocessor, "121-1164." The numbers in the black bar at the bottom tell you when the set went through the factory and that it has been tested. The service menu comes up on item 03 (figure 11-3b) with one exception. If the set has a six-key keyboard, the service menu pops up on item 04 (Level).

Since the GA chassis makes use of two, different microcontrollers, the servicer must contend with two different service menus. Table 12-2 is the service menu for the GA-1 and GA-2 chassis (microprocessor 121-1164-01 or 121-1164-02), and table 12-3 is the menu for the GA-3 chassis (microprocessor 121-1305). The number of items in each chassis varies depending on the part number of the microcontroller and the feature level of the set. Since we have discussed many of these items in the past, I intend to limit my comments and depend on you to read the service menu discussions in previous chapters.

First, be sure to set item 04 (Level) correctly if you have to change modules or if the customer complains of loss of features. Incidentally, you can order a prewired jumper for connector 4G9 by requesting the "4G9 Level Set Jumper" from Zenith parts. Fabricate your own if you choose, but be careful to get the jumper connected to the correct pins.

Second, items 20 through 24 in table 12-2 are used to set the black and white tracking and correspond to items 22 through 26 in table 12-3.

Third, some of the items in both menus are not currently being used. For example, the information I have points out that item 28 in table 12-2 as well as items 34 through 37 and 38 through 41 in table 12-3 are not being used. Remember, Zenith designs a microprocessor that can be installed in more than one chassis rather than a specific one for each chassis. Their engineers are able to do this by controlling the features of the chassis via the service menu. Therefore, the unused items in the service menu have more or less been reserved for future use.

Fourth, make a note of item 25 (chassis type) in figure 12-2. It has one of five settings: 0 for mono, 1 for mono with auxiliary inputs, 2 for non-MTS stereo, 3 for non-MTS stereo with auxiliary inputs, 4 for MTS stereo, and 5 for MTS stereo with auxiliary inputs. If you don't set it correctly, the TV won't have the features it should have.

The GA-3 chassis has the more complicated service menu. If you need additional information about its features, please consult table 12-4.

1164-01 & 02 MICROCONTROLLER FACTORY RECOMMENDED SETTINGS

ITEM	RANGE	19/20"	25"	27"
Factory Mode 0 (Blue)				
00 F Mode	0-1	0	0	0
01 Pre Px	0-1	1	1	1
02 V Pos	0-24	10	7	7
03 H Pos	0-13	10	10	10
04 Level	0-2	1*	1*	1*
05 Band	0-7	0	0	0
06 AC On	01	0	0	0
Factory Mode 1 (Black)				
07 RF Bpf	0-1	1		1
08 3.58T	0-1	1	1	1
09 RF Brt	0-63	32	32	32
10 Aux Brt	0-63	32	32	32
11 V. Size	0-63	36	36	36
12 V. Phase	0-7	5	2	2
13 H. Phase	0-31	19	19	18
14 Aud Lvl	0-63	46	46	46
15 RF Agc	0-63	31	31	33
16 H Afc	0-1	1	1	1
17 WhComp	0-1	0	0	0
18 60 HzSw	0-3	2	2	2
19 PifVco	0-127	31	31	31
20 R Cut	0-254	0	10	5
21 G Cut	0-254	0	10	5
22 B Cut	0-254	0	10	5
23 G Gain	0-254	90	90	90
24 B Gain	0-254	90	90	90
25 C Type	0-5	2	2	2
26 Scroll	0-1	1	1	1
27 6 Keys	0-1	1	1	1
28 SpkrSw ***	0-1	0	0	0
29 5 Jack ***	0-1	0	0	0
30 St & Sap ***	0-1	0	0	0
28 SpkrSw ****	0-1	0	0	0
29 Surf ****	0-1	0	0	0
30 Vcurve ****	0-1	0	0	0

* 04 Level 0 and 1 are used for Zenith sets. Level 0 is Private Label.

** 05 Band. There are eight selections.

Items 28, 29, and 30 are not being used. Leave these set to zero which is off.

*** 1164-01 Microcontroller only

**** 1164-02 Microcontroller only

Table 12-2 Service Menu

Service Adjustments

New technology requires us to learn new service techniques. You can't perform adjustments "by the seat of your pants" anymore. If you think you can, you are setting yourself up for a magnificent fall! Since these new chassis are set up differently than Zenith's older ones, I am going to take the time to acquaint you with how Zenith recommends you do the set up. Let me emphasize the fact that if you want the television to be at its best, you will follow their suggestions.

Mechanical Adjustments

There are several "mechanical" adjustments on the GA chassis. R9738 adjusts the video gain for the 19/20" models. Set it for one volt peak-to-peak at the emitter of Q9706. R9745 sets the audio balance for the same chassis. Adjust it for equal outputs at the collectors of Q9701 and Q9703.

To set the focus, connect a signal generator to the RF port and inject a color bar pattern into the tuner. Adjust the focus for a good picture. Then call up the customer menu and continue to adjust the focus until there is no distortion in the letters of the menu.

Finally, set the G-2 voltage. If you haven't read the material in the last two chapters about setting the G-2 control, do so now. Make certain to set the G-2 control before you proceed with the gray scale or black and white tracking adjustment. This is a must!

RGB Cutoff Adjustment

If you are doing the GA-1 or GA-2, follow the procedure outlined in the last chapter. If you are doing the GA-3 adjustments, read on.

When you replace the picture tube or the main module, you must adjust the gray scale (or black and white tracking). Do you remember the discussion of the color temperature settings in the video menu that offers the customer a choice between "warm" and "cool?" You guessed it. The gray scale has to be set for each choice, or the TV won't display the proper picture for the selected color temperature.

Begin by selecting "cool" in the user's video menu, and set the color level to minimum (off) and tint to midrange. Second, access the service menu and set items 25 and 26 (table 12-3) to default settings for the screen size of the TV on which you are working. Third, fire up your signal generator set to inject the proper video pattern, and adjust items 22, 23, and 24 according to the procedure outlined in the previous chapter. Finally, return controls to normal and check your work. If the black level is too high, readjust item 9 (RF brightness) in the service menu.

Now let's set the gray scale for the "warm" selection in the video menu. Begin by selecting "warm" in the viewer's video menu. Then, access the service menu and set the following items to the specified value:

red cutoff 2 warm to the value of red cutoff cool 1 plus 3
green cutoff 2 warm to the value of green cutoff cool plus 3
blue cutoff 2 warm to the value of blue cutoff 1 cool plus 4
green gain 2 warm to the value of green gain 1 cool minus 25
blue gain 2 warm to the value of blue gain cool minus 48
Finally, exit the service menu and select "cool" in the viewer's video menu.

If the instructions don't make sense, take a look at table 12-5 because it gives you an idea about what to expect when you access the service menu to adjust the gray scale for the "warm" picture setting. When you understand that items 25 and 26 change according to the color temperature setting in the customer's video menu, then the instructions begin to make sense.

Servicing the GA Chassis

I am changing my format just a bit because I want to call your attention to the relatively new and brand-new features about these chassis that might cause you trouble if you aren't aware of them. I think the best way to do this lets you shake hands with the new features, stepping outside the normal way I have been presenting information.

1305 MICROCONTROLLER FACTORY RECOMMENDED SETTINGS

ITEM	RANGE	25"	27"
General Setting (Blue) Factory Mode 0			
00 F Mode	0-1	0	0
01 Pre Px	0-1	1	1
02 V Pos	0-24	14	17
03 H Pos	0-13	6	8
04 Level	0-2	0*	0*
05 Band	0-7	0**	0**
06 AC On	01	0	0
Technical Settings (Black) Factory Mode 1			
07 RF Bpf	0-1	1	1
08 3.58T	0-1	1	1
09 RF Brt	0-63	32	32
10 Aux Brt	0-63	32	32
11 MaxCon	0-63	63	63
12 V. Phase	0-254	86	53
13 H. Phase	0-254	110	104
14 V. Phase	0-7	3	1
15 H. Phase	0-31	15	20
16 Aud Lvl	0-63	46	46
17 RF Agc	0-63	31	31
18 H Afc	0-1	1	1
19 WhComp	0-1	0	0
20 60hz Sw	0-3	2	2
21 PifVco	0-127	32	32
Color Temp Cool Starting Values			
22 R Cut	0-254	5	5
23 G Cut	0-254	0	0
24 B Cut	0-254	10	10
25 G Gain	0-254	66	66
26 B Gain	0-254	14	14
27 Scroll	0-1	1	1
28 6 keys	0-1	1	1
29 A Att	0-1	9	9
30 A VCoc	0-15	31	31
31 A Fltr	0-63	31	31
32 Spctrl	0-63	31	31
33 W Band	0-63	31	31
34 PiP X1	0-63	7	7
35 PiP Y1	0-63	5	5
36 PiP X2	0-63	49	49
37 PiP Y2	0-63	32	32
38 PiP Ras	0-254	68	68
39 PiP Sw	0-15	8	8
40 PiP Lud	0-7	2	2
41 PiP TOF	0-63	3	3
42 C. In OSDC	0-1	0	0
43 OSD FR	0-1	0	0

Table 12-3 Service Menu

1305 MICROCONTROLLER

07 Rf Bpf: (RF bandpass) Range is 0 - 1.

08 3.58T: 3.58 Mhz trap. Range is 0 - 1.

09 Rf Brt: (RF Brightness) Set the adjustment range of customer control for brightness in RF mode. Range is 0 - 63.

10 Ax Brt: (Aux Brightness) Sets adjustment range of customer control for brightness in the AUX mode. Range is 0 - 63.

11 MaxCon: (Max Contrast) Set adjustment range of customer control for contrast. Range is 0 - 63.

12 V. Size: (Vertical Size) Adjusts size of picture vertically. Range is 0 - 254.

13 H. Size: (Horizontal Size) Adjusts size of picture horizontally. Range is 0 - 254.

14 V. Phase: (Vertical Phase) Shifts picture vertically. Range is 0 - 7.

15 H. Phase: (Horizontal Phase) shits picture vertically. Range is 0 - 7.

16 AudLvl: (Audio Level) Set gain for Composite Audio from Video processor. Range is 0 - 63.

17 RF Agc: Range is 0 - 63.

18 H Afc: There are two setting 0 and 1. Setting is usually 1.

19 WhComp: (White Compression) There are two settings 0 and 1. Setting is 0.

20 60hzSw: (60 Hertz compression) There are two settings 0 and 1. Setting is 0.

21 PifVco: (PIF Voltage Controlled Oscillator) Range is 0 - 127.

Items # 22 through #26 are for B&W tracking

22 R Cut: Range is 0- 254.

23 G Cut: Range is 0 - 254.

24 B Cut: Range is 0 - 254.

25 G Gain: (Green Gain) Range is 0 - 254.

26 B Gain: (Blue Gain) Range is 0 - 254.

27 Scroll: Selects the method the User Menus that will appear on the screen. Scroll Off is 1, 1 is On.

28 6 Keys: Set to 1 for the 6 button keyboard.

29 A Att: (Audio Attenuator) Range is 0 - 15.

30 A VCO: (Audio Voltage Controlled Oscillator) Range is 0 - 63.

31 A Fltr: (Audio Filter) Range is 0 - 63.

32 Spctrl: High Frequency separation. Range is 0 - 63.

33 W Band: (Wide Band Low Frequency Separation) Range is 0 - 63.

34 PiP Z1: Adjusts horizontal position of the insert picture on left side.

35 PiP Y1: Adjusts vertical position of the insert picture on left side.

36 PiP X2: Adjusts horizontal position of the insert on right side.

37 PiP Y2: Adjust vertical position of the insert picture on right side.

38 PiP Ras: (Picture in Picture Raster) Range is 0 - 255.

39 PiP Sw: (Pip Switch Delay) Used to center PIP Boarder and PIP picture in the horizontal direction. Range is 0 - 15.

40 PiPLud: (PIP Luminance Delay) Used to match Luma and Chroma of inset picture. Range is 0 - 7.

41 PIP Tof: (PIP Tint level register) Range is 0 - 63.

42 C. In OSDC: On screen dispalay internal oscillator. Range is 0 - 1.

43 OSD FR: On screen display frame. Range is 0 - 1.

Table 12-4 GA-3 Service Menu

First, if the set works fine except that certain features are missing from the customer's menu, check the setting for "level" in the service menu to make sure it corresponds to the level called for on the sticker located on the back of the set. A glitch in the AC power or a spike caused by lightning or a power surge might do no damage except to cause a change or two in the settings of the service menu.

Second, if the set shuts down after turnon and cannot be turned back on unless it is disconnected and reconnected to AC (resetting the microprocessor) and if the power supply voltages are up to par, investigate the XRP and vertical circuits. The literature says the x-ray protection circuit turns the set off just 1.6 seconds after it starts while the vertical protection software turns it off 3 seconds after it starts (booklet 923- 3352TRM, page 16). There isn't much difference between 1.6 and 3 seconds, but

The A-Line Chassis

	GA3	
Tube Size	25"	27"
Customer Contrast	0	0
RF Bright/Aux Bright	40	40
Customer Brightness	32	32
Color Temp	Cool	Cool
Red Cut 1	15	15
Green Cut 1	8	5
Blue Cut 1	12	12
Green Gain 1	95	100
Blue Gain 1	70	80
Color Temp	Warm	Warm
Red Cut 2	15	16
Green Cut 2	10	8
Blue Cut 2	17	16
Green Gain 2	70	75
Blue Gain 2	25	32

Table 12-5 Accessing Service Menu to Adjust Gray Scale

if you pay attention you should be able to tell the difference. Moreover, the vertical circuit is more apt to cause trouble than the XRP circuit. Well, I ought to say that it does on the sets on which I have worked.

The vertical deflection circuit must work and the vertical pulses must get to the microprocessor. If the vertical circuit fails or the vertical pulses don't get to the microcontroller because of other component failures, the microcontroller implements its CRT protection software and turns the set off in three seconds. I know I have made this point lots of times, but I repeat it because it is a new feature and will fool you simply because you aren't thinking about it when shutdown occurs. And, I know from firsthand experience!

The X-ray protection circuit is the second circuit you should check. If the +130-volt source is too high or the retrace pulse is too narrow, the TV will shut down. If the former is too high, you should troubleshoot the power supply. If the latter is the fault, then you should suspect the retrace capacitor in the collector circuit of the horizontal output transistor. But then you probably already know that, don't you?

Third, the GA-1 and GA-2 chassis shut down if you turn them on while the video output module is removed from the picture tube base. Because R3212 is used in the heater and the shutdown circuit, removing the video output module from the CRT permits the voltage at pin 29 of ICX220 to go high and activate the shutdown circuit. Should you need to troubleshoot the set with this module unplugged, add a load of 9 ohms between pins 9 and 10 of the module. The service manual instructs you to use a 10-watt, 5% tolerance, resistor for the 9-ohm load.

Fourth, don't get excited when you encounter a GA chassis that has a black raster (no video, no snow, just black) and no audio but displays a perfect customer menu. Check the clock and data pulses before you do anything else. If they are not at a 5-volt peak-to-peak level, then something connected to the bus line is pulling them low. Experience leads me to suspect the tuner first. The best way to proceed is simply to unsolder it. With the tuner out of the circuit, chances are you will get a snowy raster and a rush of white noise out of the speakers. It's the same problem we servicers have been seeing since the Y-Line came on the market and is caused by the tuner controller IC failing and pulling the clock and data lines low.

I just fixed a GA-1 chassis that added a bit of a twist to this problem. The set came on with perfect audio and displayed a perfect menu but had a black raster. Since the set had audio, I began to probe around ICX2200 and found the luminance signal everywhere I looked except at the output. I changed the chip, thinking the problem had to be internal to it. But the new chip didn't fix the problem. I scratched my bald head and wondered what I had missed.

Remembering a thing or two about previous repairs, I checked the clock and data pulses at the tuner and found one at 2.65 volts peak-to-peak and the other at the correct value. I removed the tuner and soldered a new one (175-2720) into the circuit then sent the set home. As I said, a new twist on an old problem.

Power Supplies

The GA chassis makes use of two, different power supplies. The GA-1 uses a line-connected power supply (figure 12-2) while the GA-2 and GA-3 (figure 12-4) use a line-isolated switching supply. However, both require that you use an isolation transformer when you service them. We have seen both power supplies in other chassis and neither really require extended comment except to give you points to check the various voltages each power supply produces.

Let's begin with figure 12-2. The power supply produces three standby voltages: +13, +12, and +5. The +13 volts is developed from DX3405 (half-wave rectification) and CX3411. The +12 volts is derived from the +13 volt supply, and the +5 volts is derived from the +12 volt supply via ICX3402.

The main power supply comes on-line when an on command from the microprocessor turns on QX3402 causing the relay (KX3401) to energize and supply AC to the bridge rectifiers. The bridge produces about +150 volts that is supplied to the linear regulator to develop +130 volts for the horizontal deflection circuit. The +130 volts is also used to develop +9 volts for the horizontal drive section of ICX2200 and +33 volts for the tuner. The rest, as they say, is "old hat." If you need to know more about it, reread the discussion about the line-connected power supply in the last chapter.

You may check the +13 volts at CX3411, the +12 volts at pin 1 of ICX3402, and the +5 volts at pin 3 of ICX3402. Look for the switched voltages at these points: +150 volts at CX3406 or the input to the linear regulator and +130 volts at FX3402. Check the sweep-derived voltages at these test points:

+5 volts at the cathode of ZD3206
+9 volts at pin 3 of IC3201
+12 volts at + side of C3222
+25 volts at the cathode of D3202
+180 volts at the + side of C3207

The GA-2 and GA-3 chassis use the same power supply we have been seeing from page one, though with certain variations. For example, the standby voltages marked "standby" are switched voltages when the TV comes equipped with "Energy Star," a topic for later discussion. The power supply develops four standby voltages: +124 volts for horizontal deflection; +18 volts for the audio output circuit; +15 volts for audio and low-level video processing, and +5 volts for the microcontroller and the IR receiver via RX3416 and ZD3401. The +18-volt and +15-volt lines are fuse protected. The +124-volt line isn't fuse protected because an overcurrent protection circuit monitors it. The rest of the power supply, like the on/off scenario, follows Zenith's usual way of doing things.

Check the standby voltages by looking for

+5 volts at the cathode of ZD3401
+15 volts at the emitter of Q3403
+123 volts at the + side of CX3420
+150 volts at RX3404

Check the switched voltages by checking for

+ 5 volts at the cathode of ZD3402
+9 volts at the pin 3 of IC3431
+15 volts at the collectors of Q3404 and Q3403

When it comes on-line, horizontal sweep develops three voltages besides the EHT, focus, and G2. Check for the +25 volts at RX3242, the +35 volts at the positive terminal of CX3272, and the +215

The A-Line Chassis

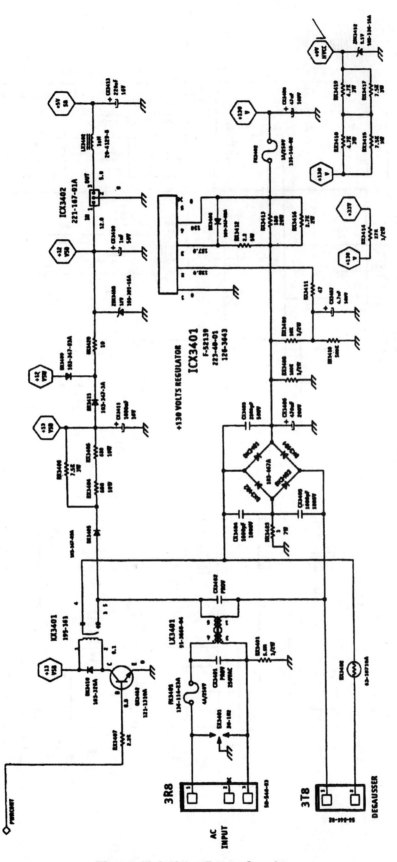

Figure 12-2 GA-1 Power Supply

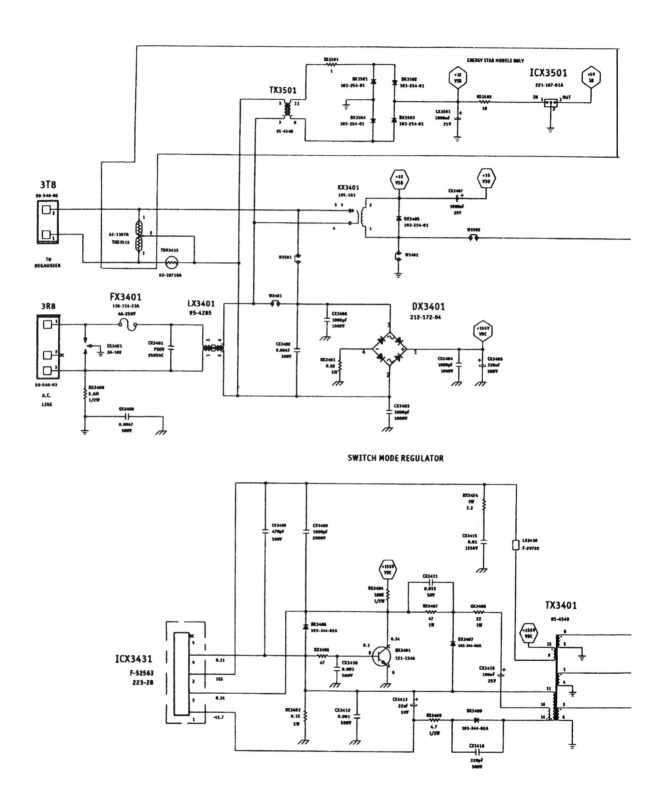

Figure 12-4 GA-2 Power Supply

The A-Line Chassis

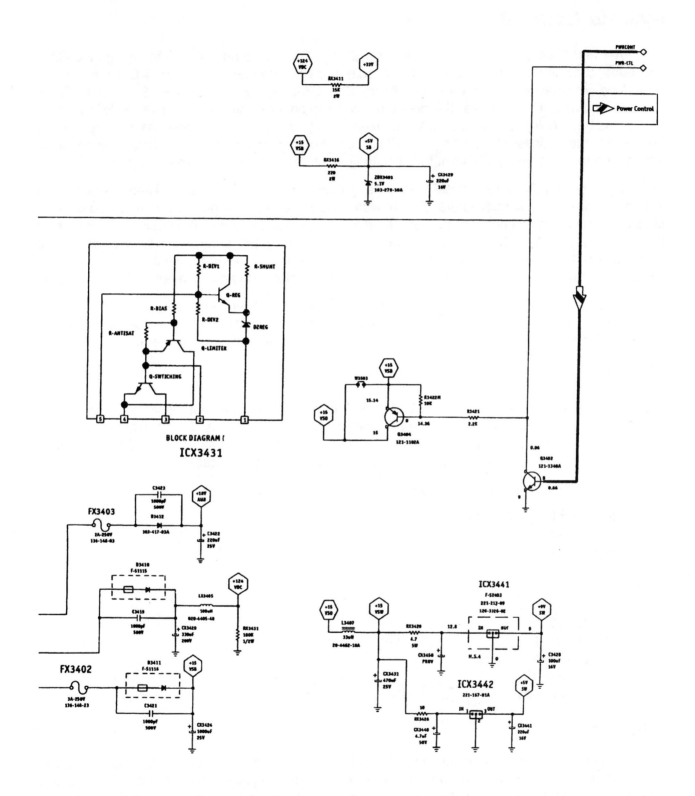

Figure 12-4 GA-2 Power Supply (Continued)

volts at the junction of CX3296 and RX3277. You may check for filament voltage at pins 1 and 2 of connector 2F5 using either an AC RMS meter or your scope.

System Control

The GA-1 and GA-2 use one microprocessor while the GA-3 uses another. Even though the GA-1 and GA-2 chassis use the same one, the circuits are not identical. Therefore, I feel the need to reproduce schematics for all three (GA-1, figure 12-5; GA-2, figure 12-6; and GA-3, figure 12-7). If you take a moment to look over the diagrams, you should see that the configuration follows what Zenith has been doing for the last few years. Therefore, I am going to skip the usual description and get down to the business of providing you with troubleshooting information. If you find the schematics a little hard to follow, you might want to check the pinout descriptions in figures 12-8 and 12-9.

Begin by checking the "must haves": +5 volts, ground, reset, and oscillator. Then check for bus activity. Unlike other manufacturers, Zenith microcontrollers use two buses, meaning you need to check two clock and two data lines for proper activity. I cannot overstress the importance of making those checks, especially for problems you might not think are system control related. Remember the discussion of the defective tuner that pulled the clock line low? The tuner caused the problem by swamping one of the bus signals. Other components connected to the bus can (and will) do the same.

Also, remember that the microprocessor must have horizontal and vertical pulses to operate properly. It uses the horizontal pulse to time its activity and the vertical pulse to position the OSD in the picture. The set will shut down about three seconds after turnon if the vertical pulse is missing.

Tuner

The GA chassis use two tuners, the 175-2720 and the 175-2721. They are electrically identical but mechanically dissimilar. 175-2720 is used in the line-connected chassis (GA-1) and must be used with an antenna isolation block. 175-2721 is used in line-isolated chassis (GA-2 and GA-3) and does not require the isolation block. It also has the RF terminal affixed to its case.

The pin functions of the tuner are as follows:

pin 1, AGC input from ICX2200
pin 2, not connected
pin 3, not connected
pin 4, serial clock
pin 5, serial data
pin 6, +9 volts switched
pin 7, +5 volts switched
pin 8, not connected
pin 9, +33 volts tuning voltage
pin 10, not connected
pin 11, IF output to SAW filter and IXC2200

Keyboard

The GA chassis supports the same kind of A/D keyboard Zenith has been using for several years (figure 12-10). As we have seen in past discussions, such a keyboard produces a distinct voltage level for each key depression. In the GA chassis, these voltage levels are input to pins 7 and 8 of the microcontroller.

Deflection Circuits

Because there are differences among the GA chassis, I will give you schematics for each of the three (GA-1, figure 12-11; GA-2, figure 12-13; and GA-3, figure 12-14). However, because most of the circuits are now familiar to you, I am going to limit my comments. Most of the illustrations come from the training manual for the GA chassis as opposed to the service manual simply because they are easier to reproduce.

Vertical Deflection

Vertical drive comes from pin 22 of ICX2200 and goes to IC2101 of the GA-1 chassis and to IC2100 of the GA-2 chassis. Replace either with a 221-903 (Zenith) or an LA7833. Use the appropriate item in the service menu if you need to make an adjustment to vertical size.

The GA-3 circuit is totally different. It uses the 221-992-01 vertical deflection IC that Zenith has been putting in some of its newer chassis. The signal path from ICX2200 to the vertical deflection chip is even different. Vertical drive comes from the collector of Q6002 instead of pin 22 of ICX2200, and the drive that feeds Q6002 – which is configured as an inverter – originates at pin 24 of ICXC2200 instead of pin 22!

The GA-3 uses a circuit consisting of C2107, DX2102, DX2103, R2105M, R2151M, R2152M, and R2153M to prevent failure of vertical scan from damaging the CRT. It works by tapping a pulsing waveform off pin 11 of the vertical deflection IC, rectifying it, and applying the resulting DC voltage to a voltage divider network. The output of the voltage divider network connects to pin 13 of IC6000 where the voltage is monitored by the CRT protection software. During normal operation, the voltage falls between approximately 3.6 and 5.1 volts. If it falls outside the specified range (like to 0 volts), then the microcontroller shuts the TV down about three seconds after turnon.

A shutdown will occur when vertical deflection fails or when a component like CX2109 in the CRT protection circuit fails. The question naturally crops up, "How can I troubleshoot a TV for a vertical deflection problem if it won't stay on?" Well, you can apply about +5 volts to pin 13 of the microprocessor and force the set to stay on. If you do, turn the G2 voltage down to prevent phosphor burns on the face of the picture tube. Get the +5 volts by using an external power supply or use a small ohm resistor to connect pin 13 to one of the TV's existing +5-volt supplies. Either scheme works well, but the latter might be easier because it eliminates a set of test leads connected to the chassis.

Horizontal Deflection

The horizontal deflection circuits are vintage Zenith and just like those we have seen in previous chapters. Since I have examined the schematics and read circuit descriptions looking for something "out of the ordinary" and found nothing, I'm going to pass over them without further comment. However, the shutdown circuit does need to be mentioned. It monitors the horizontal sweep pulse at pins 7 and 9 of the IFT (TX3204) in the GA-1 chassis and at pins 5 and 6 of the GA-2 and GA-3 chassis. The circuit rectifies the monitored pulse and applies it to a 9.1 zener diode (DX3003 or ZD3004, depending on the chassis). When the high voltage reaches its maximum permitted value, the zener conducts and pushes the voltage at pin 29 of ICX2200 beyond its threshold of 3.5 volts, causing the television to shut down. The microprocessor will keep the TV turned off until it has been reset by disconnecting and reconnecting AC.

Servicing Zenith Televisions

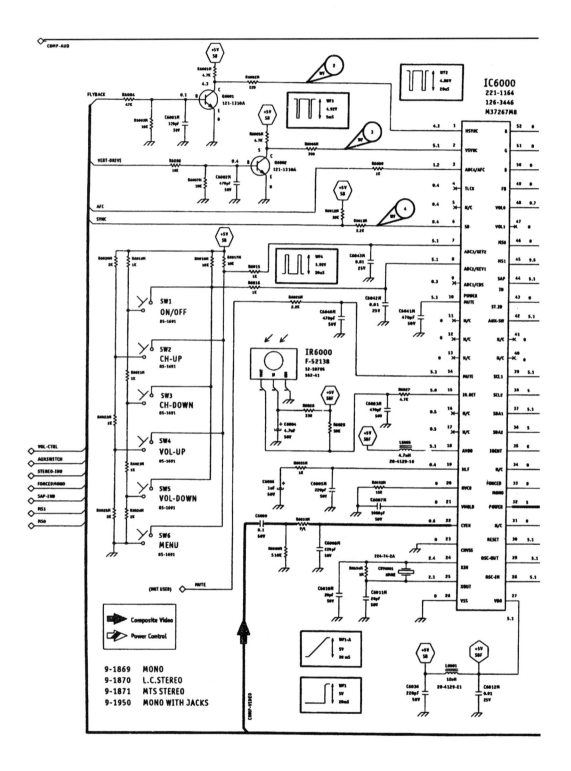

Figure 12-5 GA-1 Microcontroller

The A-Line Chassis

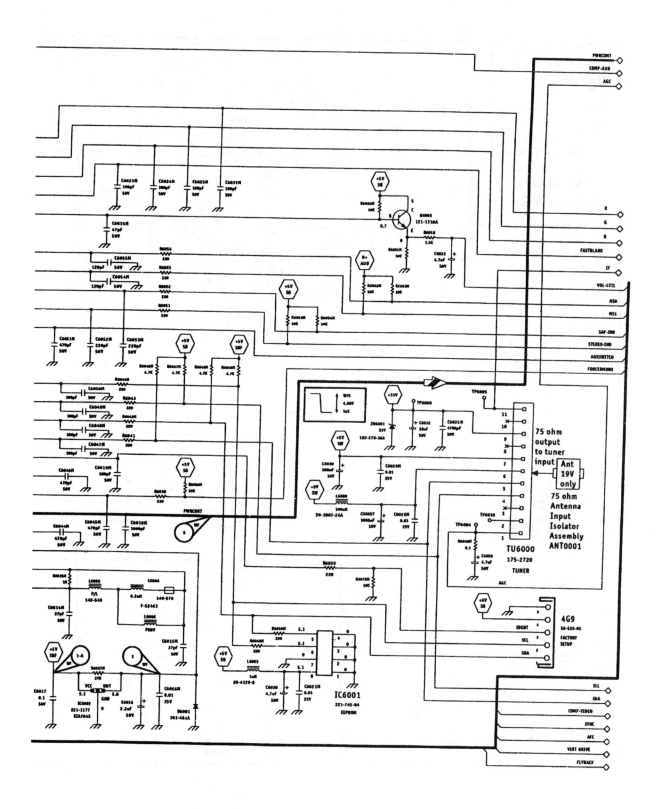

Figure 12-5 GA-1 Microcontroller (Continued)

341

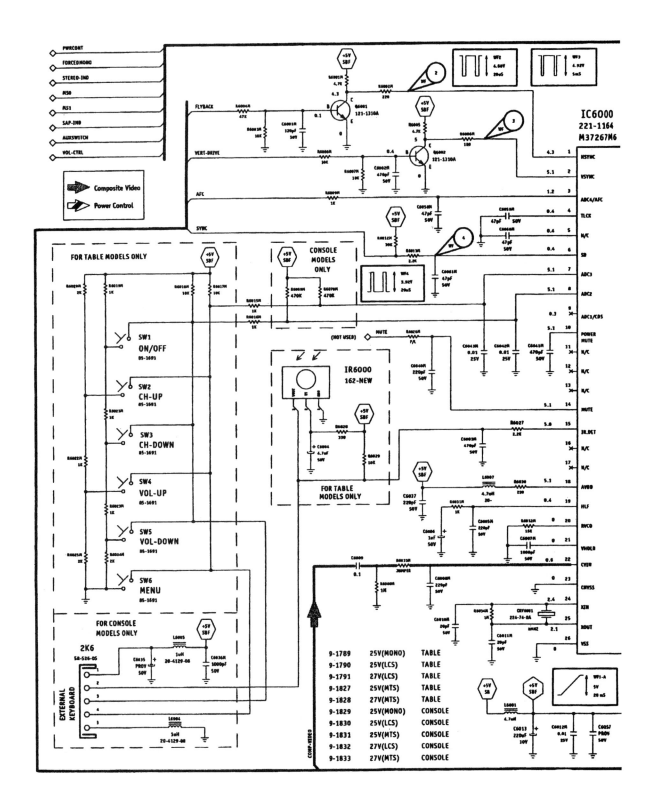

Figure 12-6 GA-2 Microcontroller

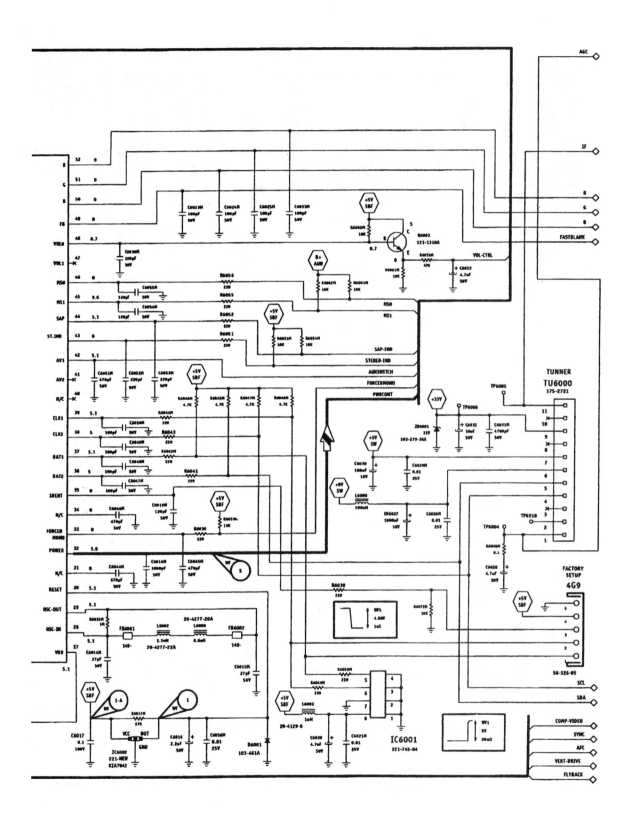

Figure 12-6 GA-2 Microcontroller (Continued)

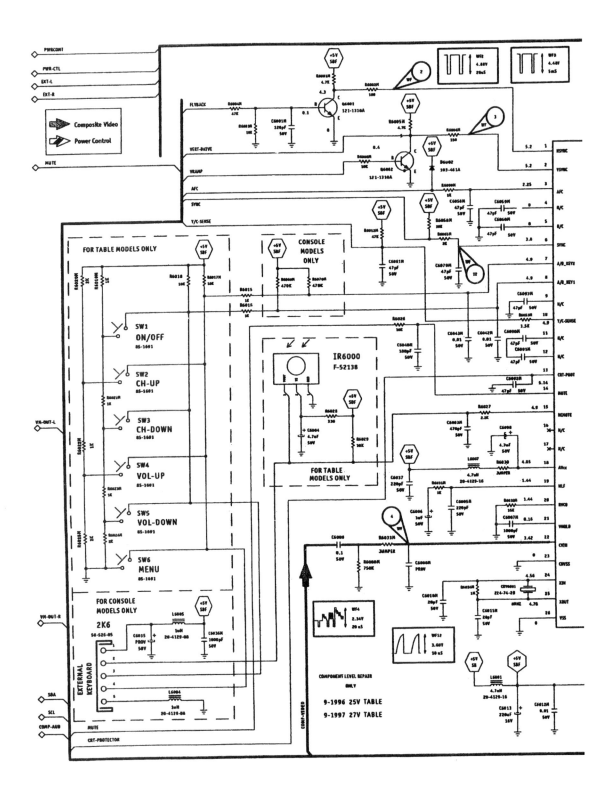

Figure 12-7 GA-3 Microcontroller

The A-Line Chassis

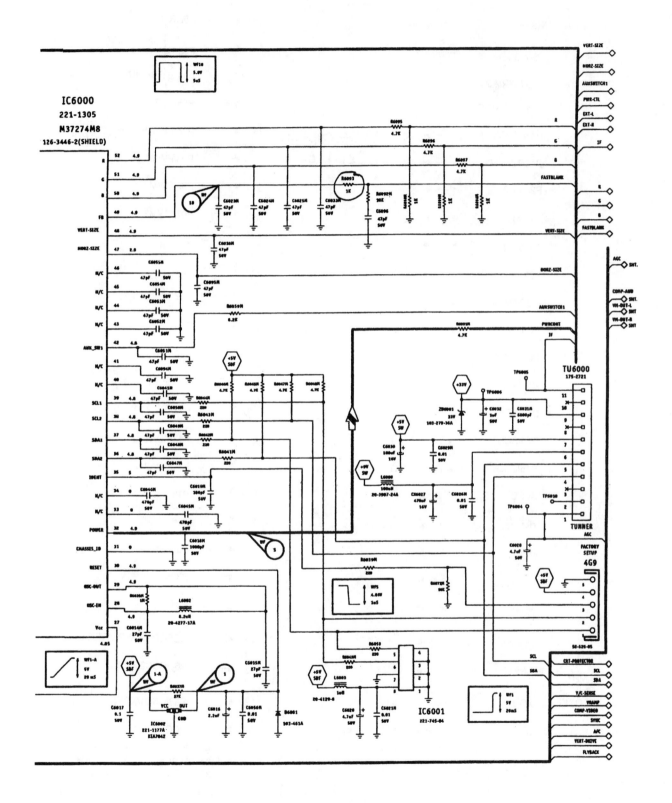

Figure 12-7 GA-3 Microcontroller (Continued)

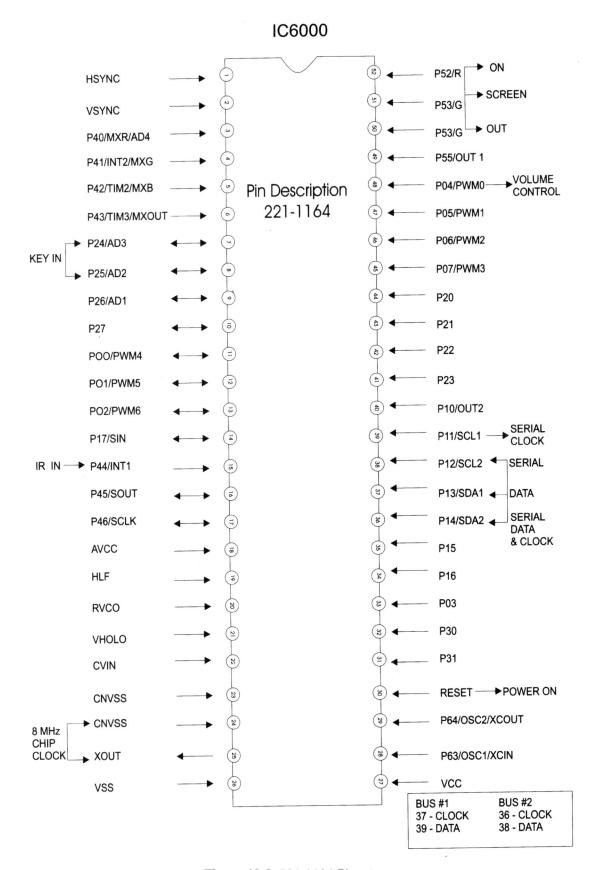

Figure 12-8 221-1164 Pinouts

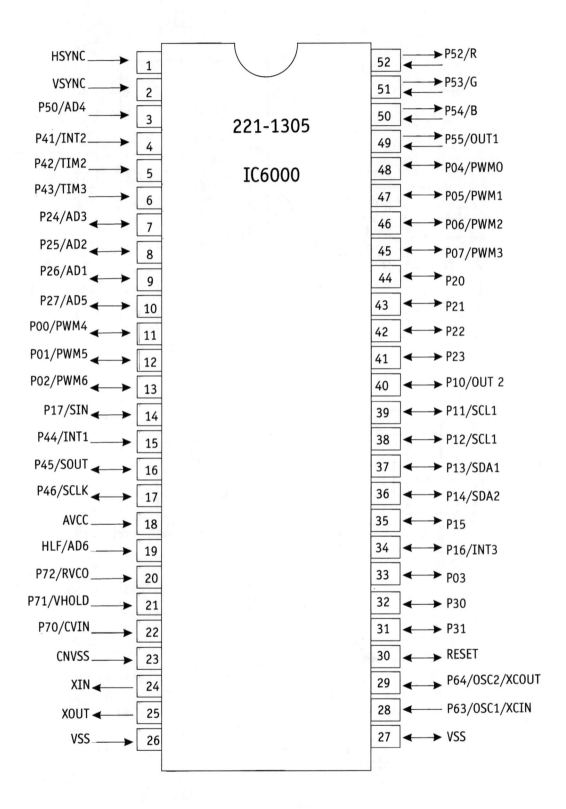

Figure 12-9 221-1305 Pinouts

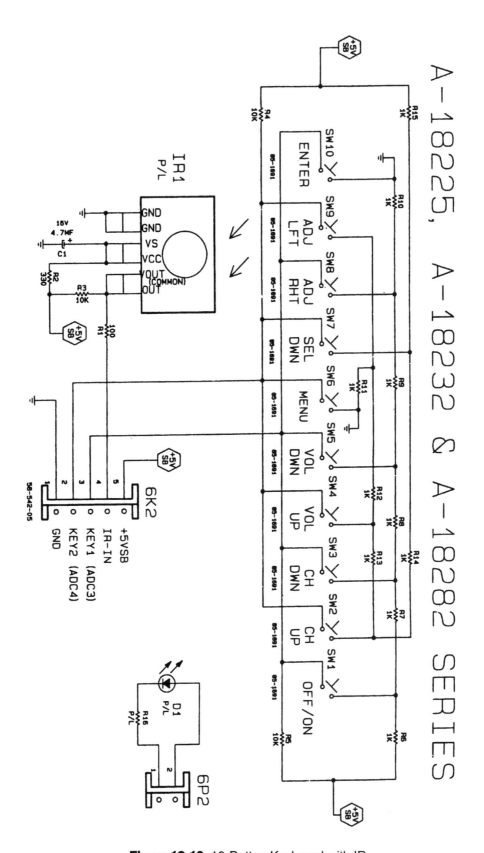

Figure 12-10 10-Button Keyboard with IR

The A-Line Chassis

Incidentally, the GA-1 should generate about 27kv maximum while the GA-2 and GA-3 should generate 29kv maximum. These measurements should be taken with brightness and contrast controls set for minimum screen illumination, resulting in zero beam current.

Video Circuits

Each of the GA chassis use the 221-1165 video processing IC that Zenith has been using for a while and continues to use. The parts list describes it as an "integrated circuit I2C bus control NTSC" component. A very complicated product, it is the heart of the video processing circuit (figure 12-15). Since there are minor differences among the chassis, I will give you the diagrams for each (GA-1, figure 12-16; GA-2, figure 12-17; GA-3, figure 12-18). Because the horizontal, vertical, and video muting circuits work exactly like those sections in chassis we have previously discussed, I see no need to deal with them here.

Video IF enters ICX2200 on pins 7 and 8, is processed, exits at pin 47, and reenters at pin 37. Auxiliary video from the jack pack appears at pin 39. (The chip switches between the two internally.) The selected video exits at pin 41 and is processed by transistor(s) Q2201 (GA-1) or Q2201 and Q2202 (GA-2) or Q2201 and DL2200 (GA-3) and their associated components to separate luma from the chroma. Luminance reenters at pin 43 and chroma at pin 45. OSD information from the microcontroller makes its appearance at pins 15, 16, and 17. RGB signals leave at pins 19, 20, and 21 on their way to the CRT module.

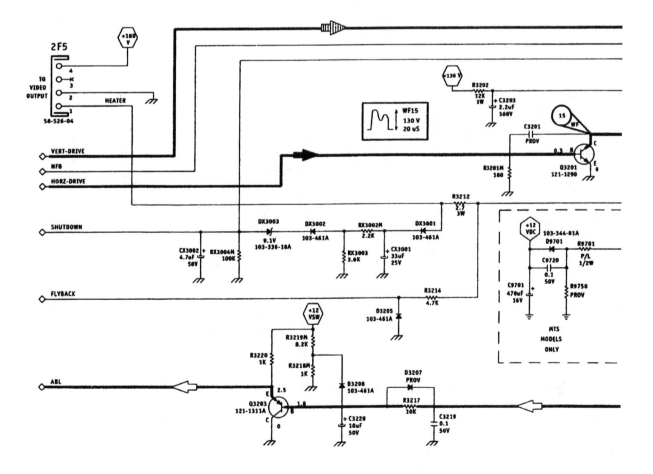

Figure 12-11 GA-1 Deflection Circuits

Servicing Zenith Televisions

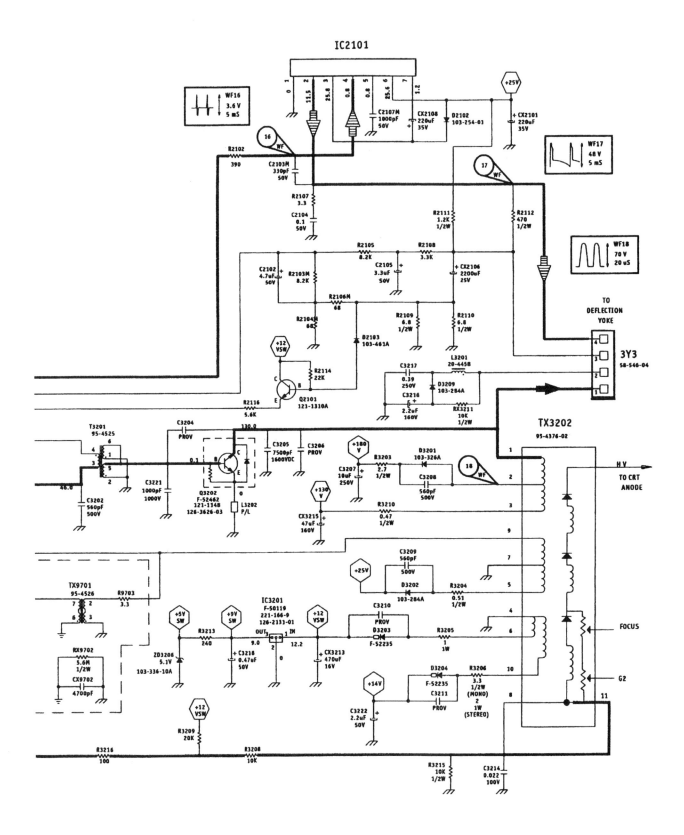

Figure 12-11 GA-1 Deflection Circuits (Continued)

The A-Line Chassis

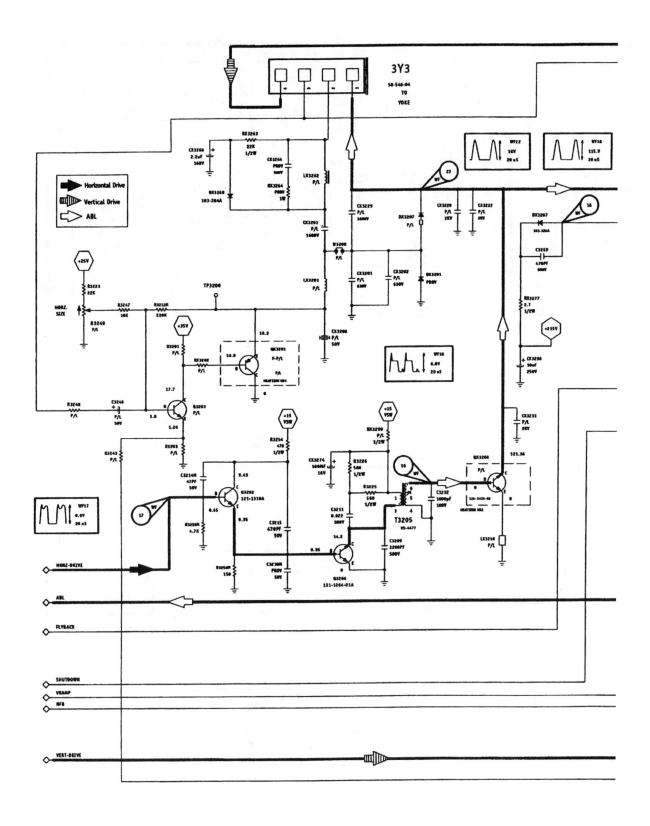

Figure 12-13 GA-2 Deflection Circuits

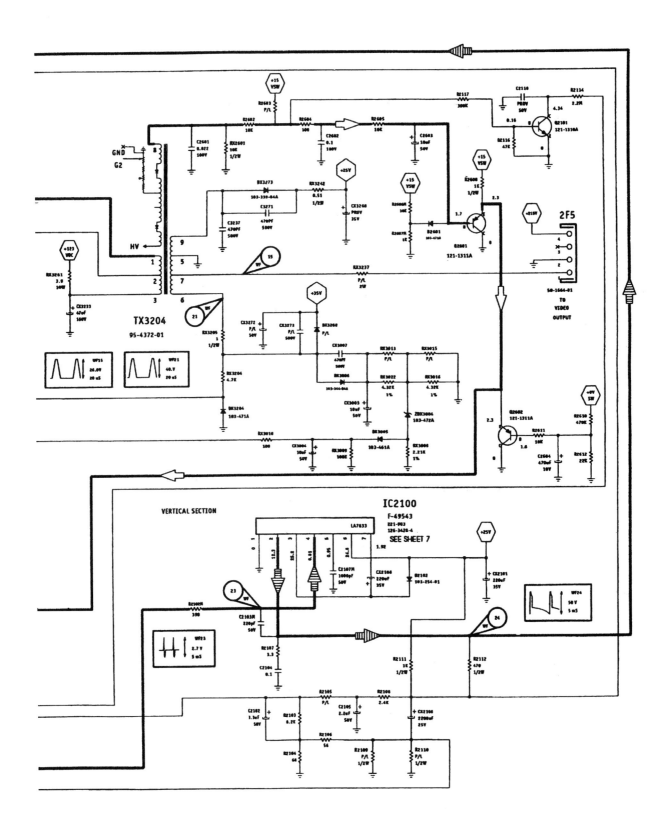

Figure 12-13 GA-2 Deflection Circuits (Continued)

The A-Line Chassis

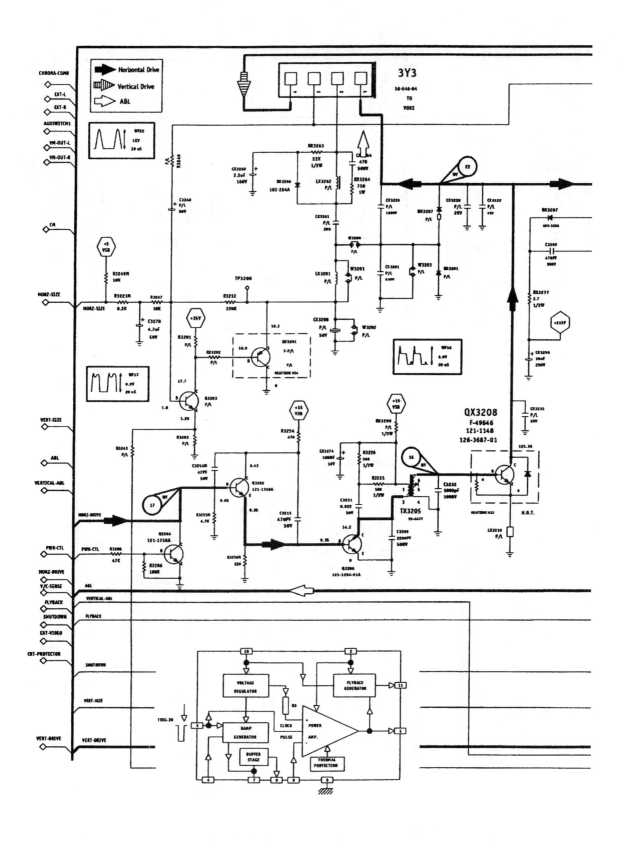

Figure 12-14 GA-3 Deflection Circuits

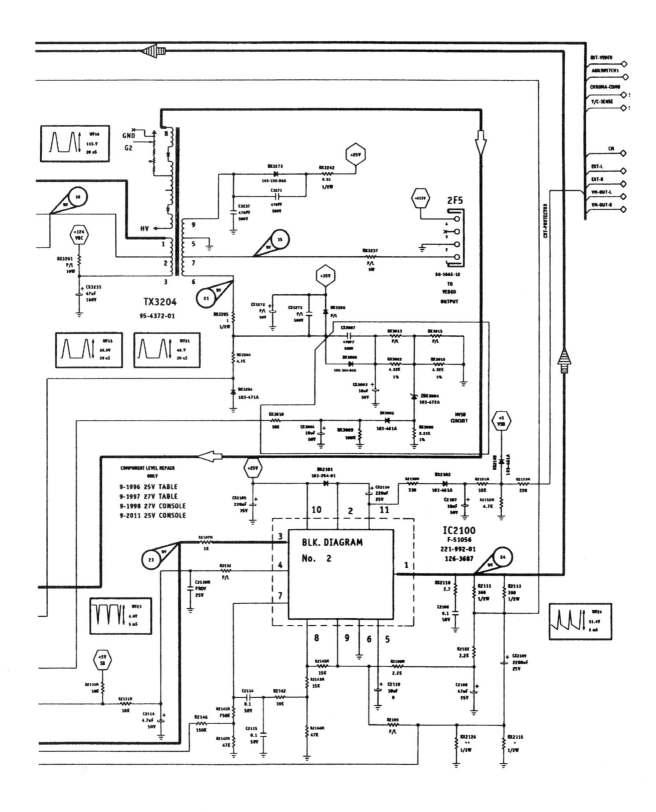

Figure 12-14 GA-3 Deflection Circuits (Continued)

The 4.5 MHz audio IF signal is extracted from the video signal by filter U1200 and coupled to pin 52 of ICX2200 where it is processed. The detector coil, L1205, is connected to pin 4. Composite audio exits at pin 2 on its way to the audio circuitry.

Troubleshoot ICX200 by looking for these key signals:

switched video out to the microcontroller at pin 47
IF from the tuner at pins 7 and 8
tuner composite video at pin 37
luminance in at pin 43
chroma in at pin 45
composite audio out at pin 2
horizontal drive at pin 32
vertical drive at pin 24
blue video out at pin 3 of 2C5
green video out at pin 2 of 2C5
red video out at pin 1 of 2C5
B+ of 9 volts at pin 9, 23, 46, and 48
clock and serial data at pins 27 and 28

If you have a problem with the video circuit, don't automatically assume that ICX2200 causes the problem. I told you about a tuner that pulled the I2C bus low and led me to believe the luminance section inside ICX2200 had failed. This chip is a "slave" of the microcontroller. If it doesn't receive instructions (or the proper instructions!), the IC won't work. Here is an area we techs must grasp if we are going to service the new electronic products on the market. Things are not the way they used to be, and we have to keep up with the times or go out of business. The microcontroller simply has to communicate with the other chips on the chassis, or the TV just won't work. Using an oscilloscope is about the only way to find such a problem. So, if you don't have one, you ought to make arrangements to purchase one now. If you have one that you almost never use, dust it off because you're going to need it.

Video Output Modules

Check figures 12-19, 12-20, and 12-21 for the video output module schematics. Remember that the GA-1 and GA-2 shut down when you attempt to operate them with the module off the CRT. If you have forgotten or just started reading at this point, refer to the section titled "Servicing The GA Chassis" for the instructions about how to keep the set on with the module off the CRT. As far as I know, the warning about not firing up the larger TVs while the video output module is off the CRT does not apply to the GA chassis.

Audio Circuit

Following recent trends, Zenith offers three audio systems options for its A-Line televisions: monophonic, non-MTS stereo (right and left speakers), and MTS-stereo. The last option is available only in the GA-3 chassis.

Monophonic audio

The mono audio option is available in the GA-1 and GA-2 chassis (figures 12-22 and 12-23.). On the models equipped with the mono option, audio from pin 2 of ICX2200 enters pin 4 of IC801, is amplified, and coupled to the speaker at connector 9M4. Pins 5, 6, and 7 are tied to ground providing a bit of heat sinking for the IC. The microprocessor controls volume and mute functions inside ICX2200 via the I2C bus. The part number for audio amplifier (IC801) is 221-981. I am not aware of a generic substitute as of this writing.

Non-MTS Stereo

This option is available for the GA-1 and GA-2 chassis (figures 12-22 and 12-23) only. I assume you know that it provides for left and right audio out, including audio from two speakers, but isn't in any sense true stereo. The electronics are located on the stereo decoder module and make use of IC830 and IC850 as audio amplifiers for the left and right channels. These chips also have built-in "auxiliary/ TV" switches that permit the viewer to select between tuner audio and auxiliary audio. They also have voltage-controlled volume controls.

Let's assume the viewer has the TV in the TV mode (audio and video from the tuner). Composite audio exits pin 2 of ICX2200 and goes to pin 12 of the MTS module. Left and right audio exit at pins 3 and 4 after processing and are routed to IC830 and IC850 where they are amplified and sent on their way to the speakers via connector 9S4. Now let's assume the viewer wants to use audio from the jack pack. Auxiliary audio enters on pin 3 of IC830 and IC850 where it is amplified and output on pin 9 of each chip. It goes from there to connector 9S4 and on to the speakers.

The GA-1 chassis needs extra circuitry to isolate the auxiliary inputs from the chassis because the chassis is line-connected (a "hot" chassis). The isolation circuitry consists of Q9702, Q9703, and optoisolator IC9701 for the left channel and Q9700, Q9701, and optoisolator IC9700 for the right channel.

MTS Stereo

The MTS stereo option is available on the GA-3 chassis only and consists of two ICs, five transistors, and their associated components (figure 12-24). IC1400 (figure 12-23a) does the brunt of the work, serving as the MTS stereo decoder and audio processor. It also controls tone, volume, and balance, switches between tuner and auxiliary inputs, and has built-in audio enhancement and SoundRite volume limiter circuits. It is in a "slave" configuration with respect to the microprocessor because each of these audio features is controlled over the I2C bus. IC804 (figure 12-24a) and five transistors complete the MTS stereo circuit. IC804 is the two-channel audio amplifier. It boosts the audio signal and outputs it to connector 9S4 that sends it on its way to the speakers. Transistors Q1401 through Q1404 are variable audio monitor outputs. These transistors take left and right audio out signals from IC1400, amplify them, and make them available as left and right audio out signals at the jack pack. The last component on the list is Q800 whose job is to turn the speakers on or off via the menu and to keep them from making a popping sound when the TV is turned on or off. If Q800 develops, let us say, a collector-to-emitter short, it effectively inhibits audio. Be sure to check it when the customer complains of a no-audio condition.

The MTS stereo-equipped chassis has six audio adjustments, all of which are located in the service menu. Five directly affect IC1400. The remaining adjustment sets the composite audio output level of ICX2200.

Auxiliary Video Switch

You never really know what you're going to encounter when you put one of these new televisions on your bench. In the event that you need it, I'm including a schematic of the auxiliary video switching arrangement (figure 12-25). IC2001 is the workhorse of the circuit and is the logical place to begin troubleshooting when you suspect a problem in the video circuit. You may be familiar with it by its other name, LA7222.

The A-Line Chassis

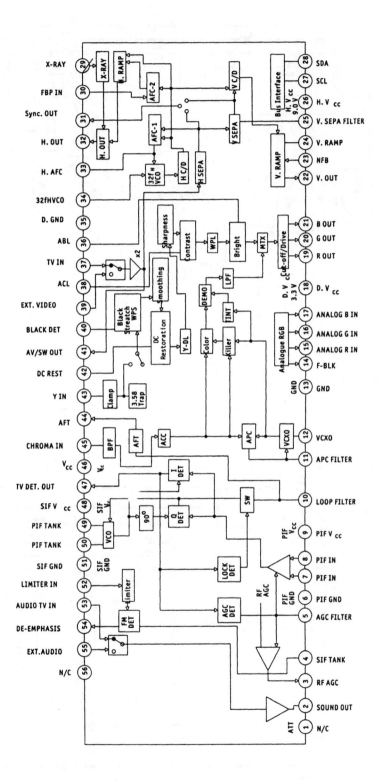

Figure 12-15 221-1165 Block Diagram

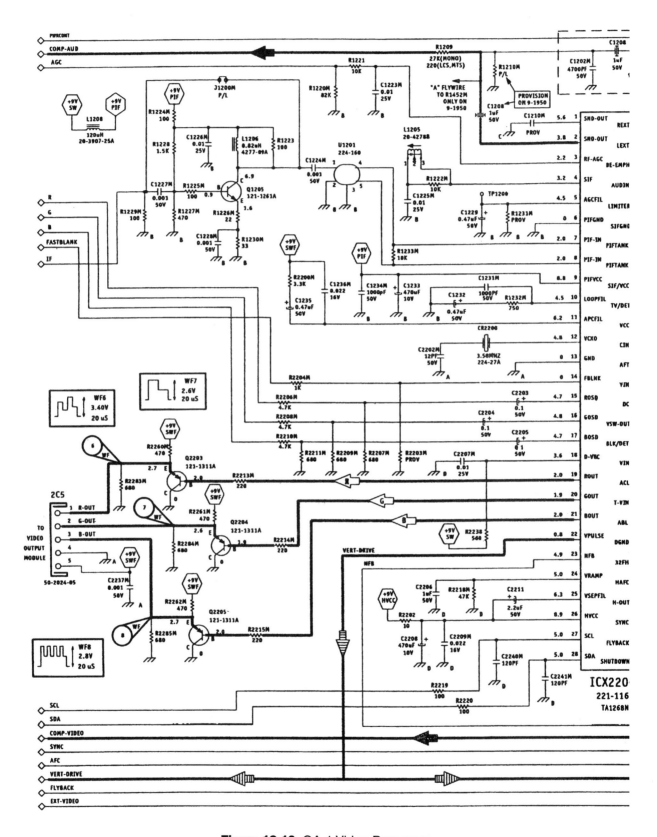

Figure 12-16 GA-1 Video Processor

The A-Line Chassis

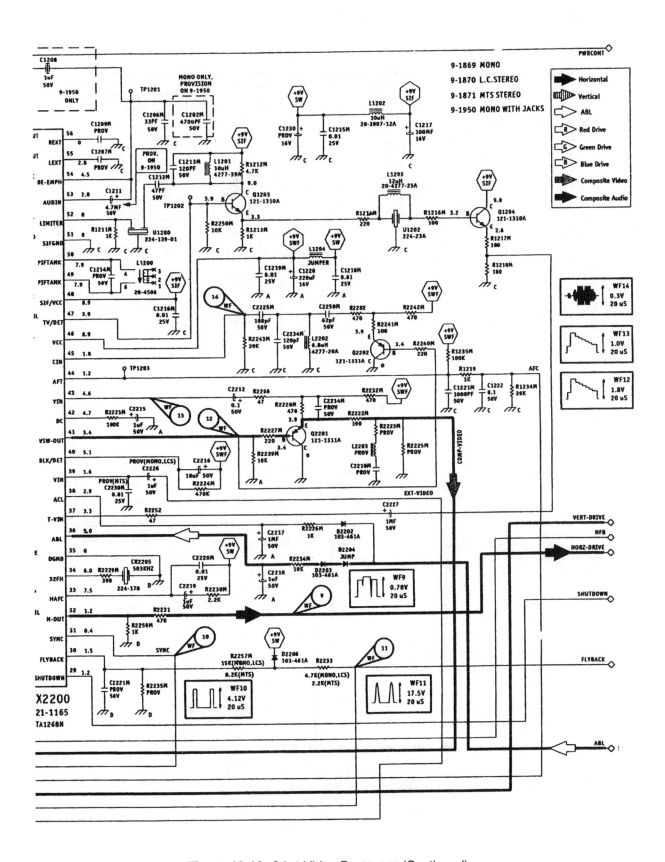

Figure 12-16 GA-1 Video Processor (Continued)

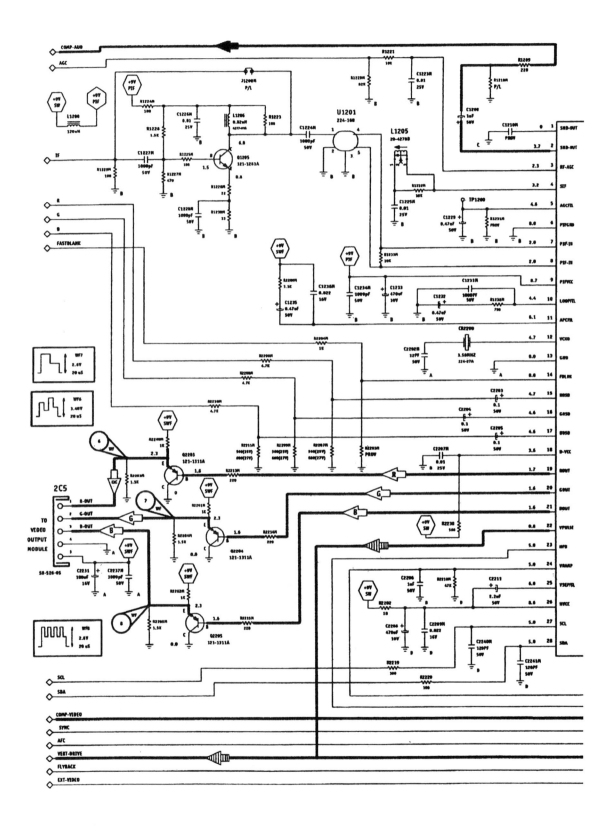

Figure 12-17 GA-2 Video Processor

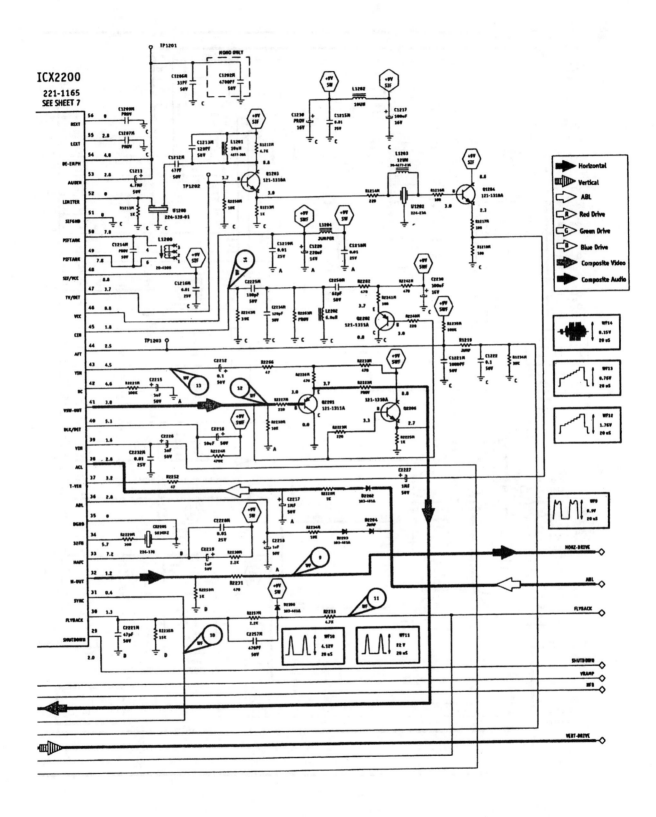

Figure 12-17 GA-2 Video Processor (Continued)

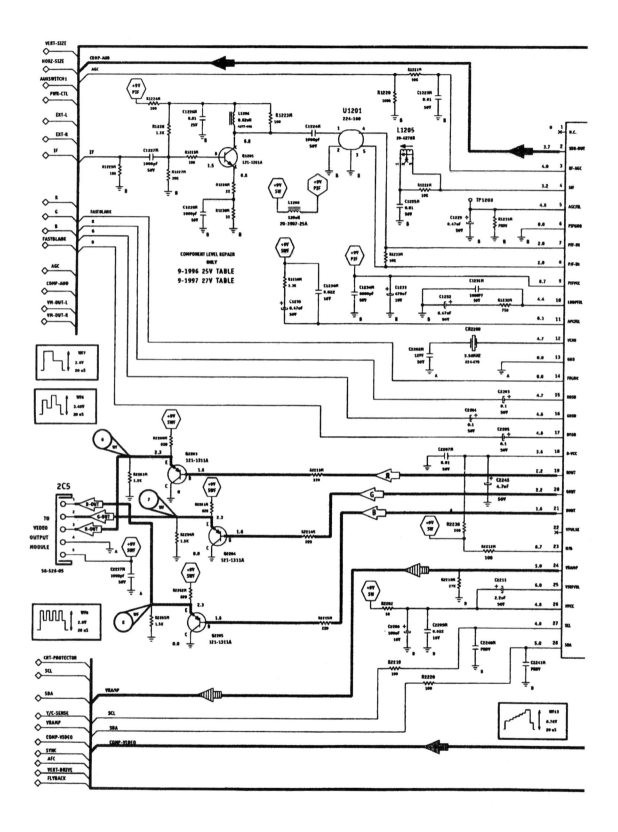

Figure 12-18 GA-3 Video Processor

The A-Line Chassis

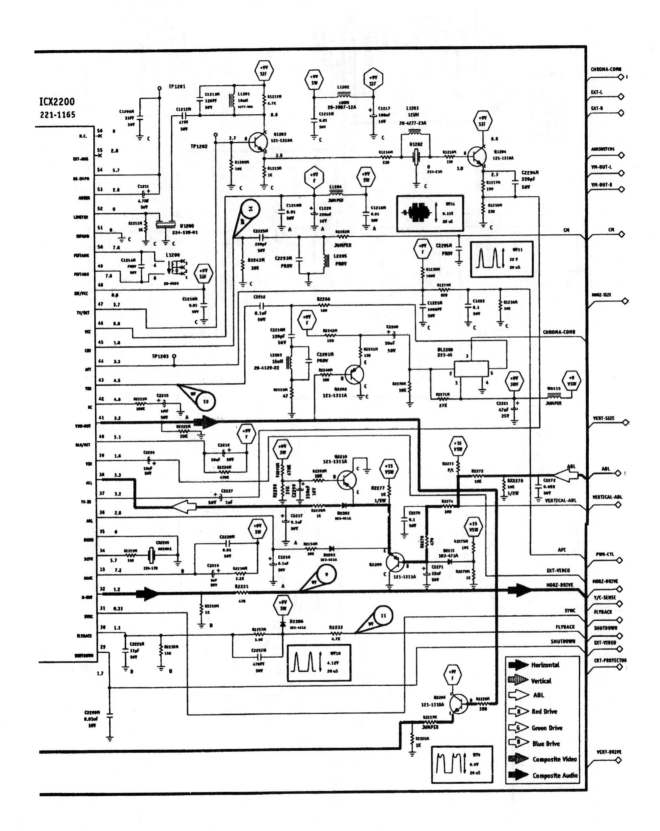

Figure 12-18 GA-3 Video Processor (Continued)

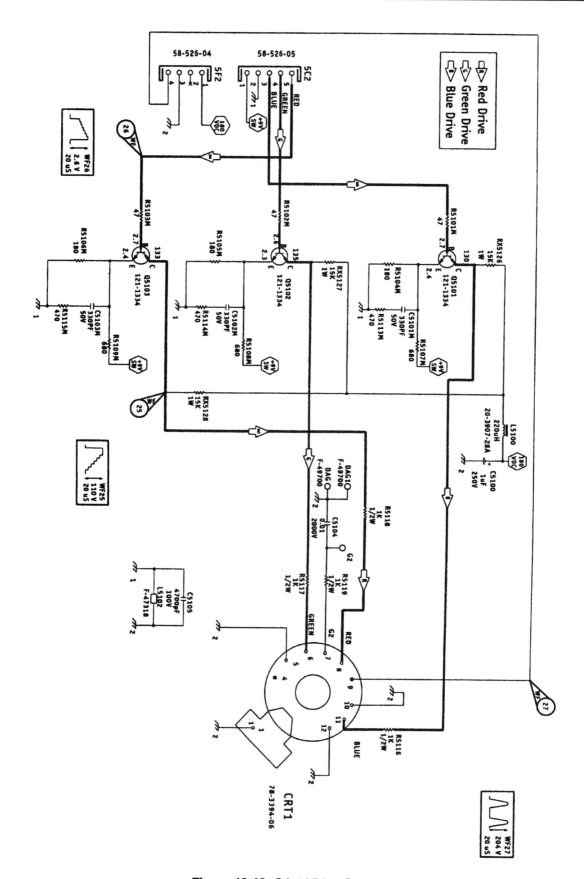

Figure 12-19 GA-1 Video Output

The A-Line Chassis

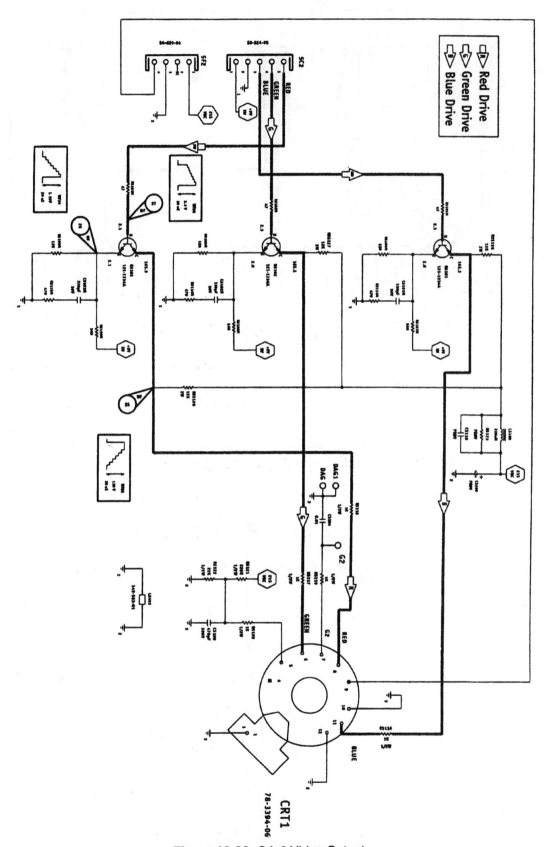

Figure 12-20 GA-2 Video Output

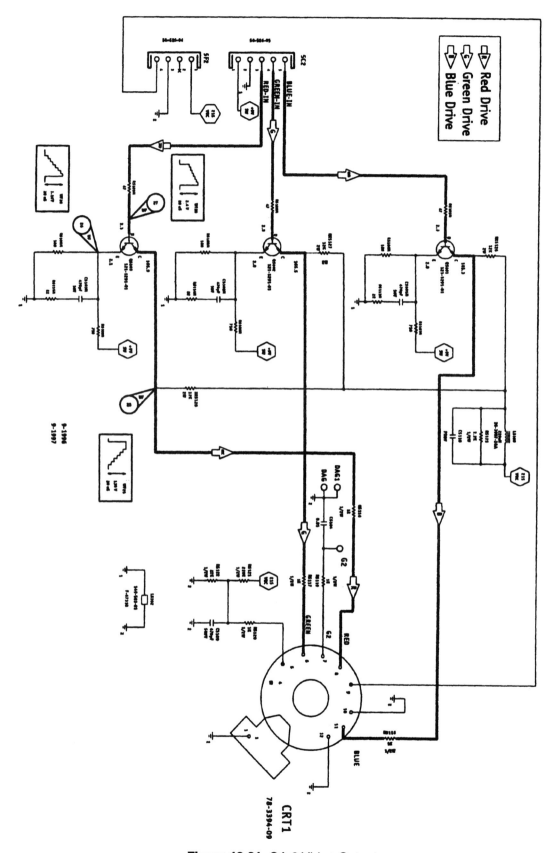

Figure 12-21 GA-3 Video Output

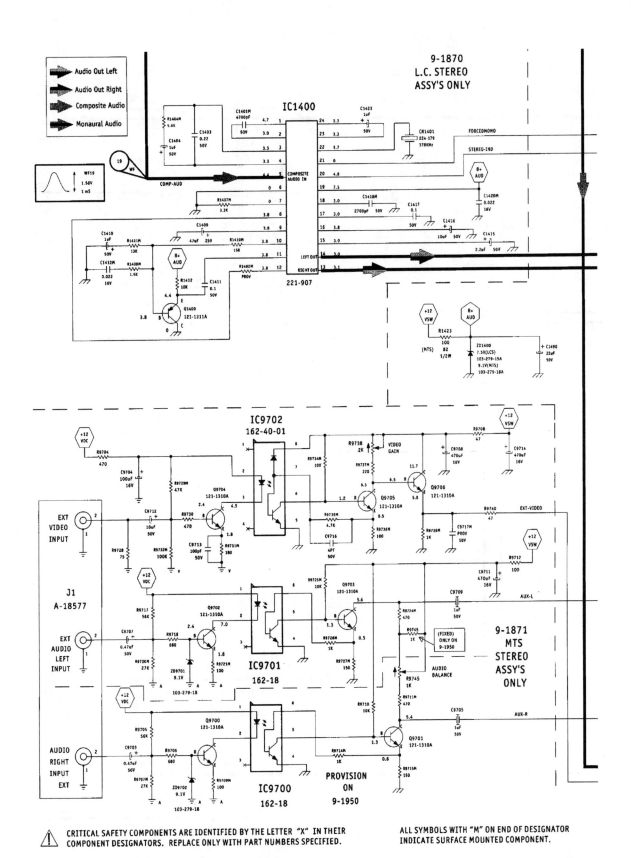

Figure 12-22 GA-1 Audio Circuit

Servicing Zenith Televisions

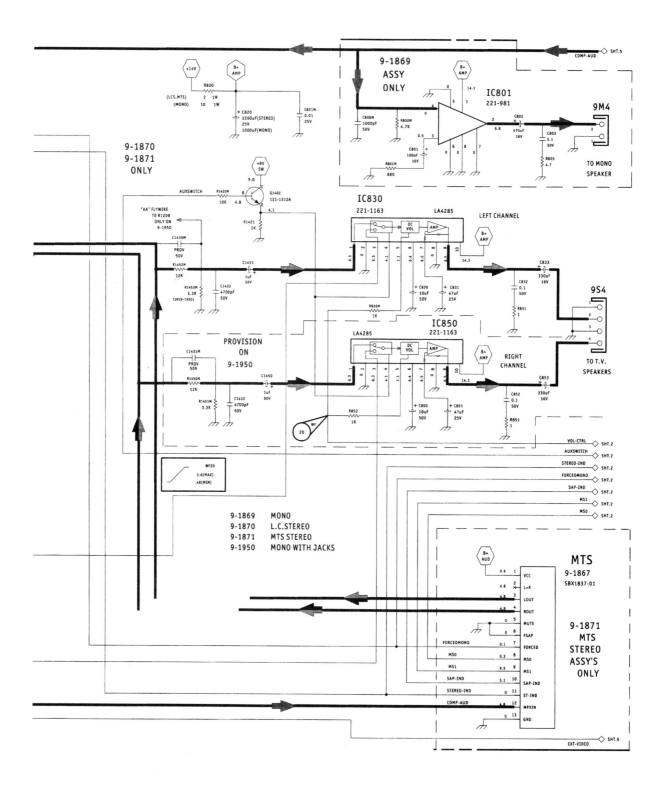

Figure 12-22 GA-1 Audio Circuit (Continued)

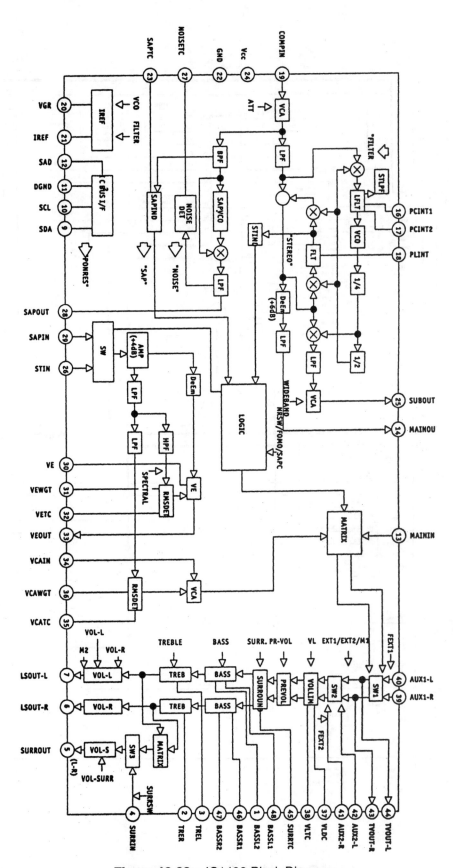

Figure 12-23a IC1400 Block Diagram

Servicing Zenith Televisions

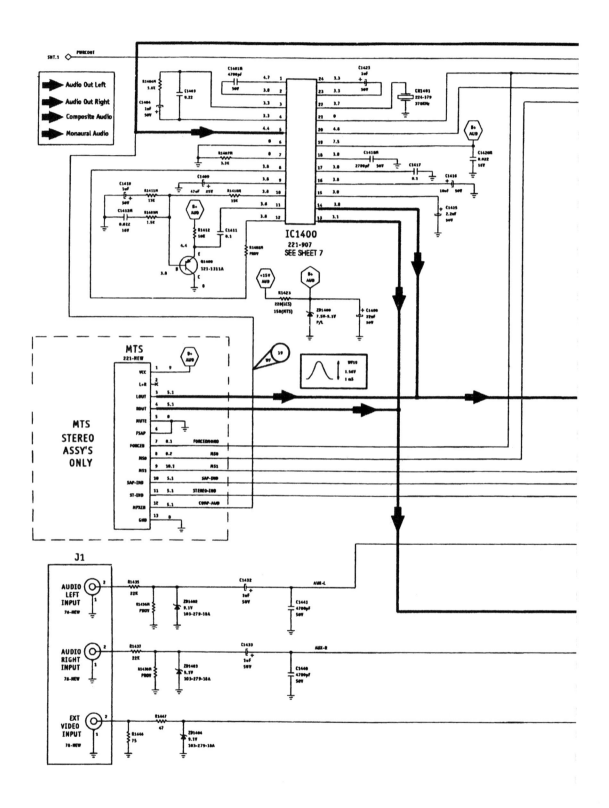

Figure 12-23 GA-2 Audio Circuit

The A-Line Chassis

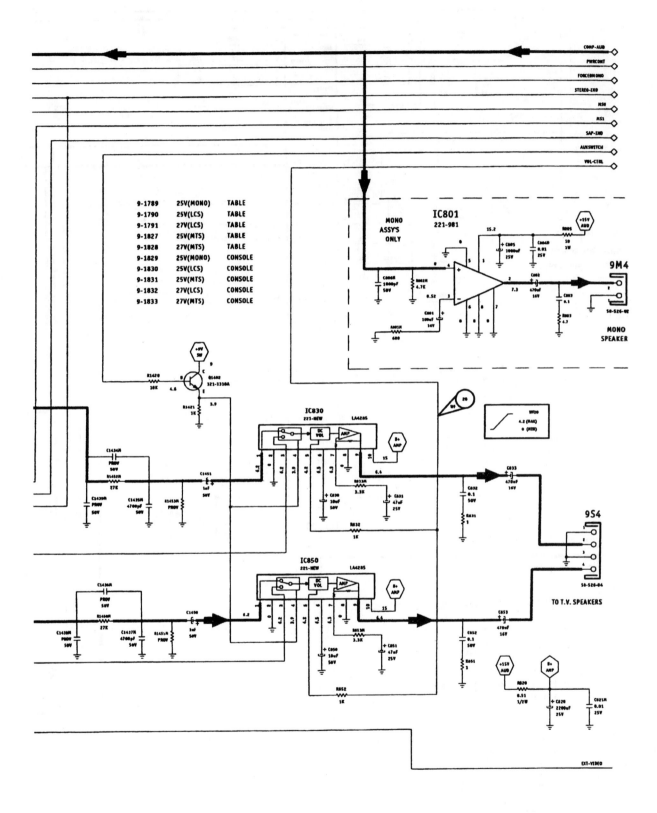

Figure 12-23 GA-2 Audio Circuit (Continued)

Servicing Zenith Televisions

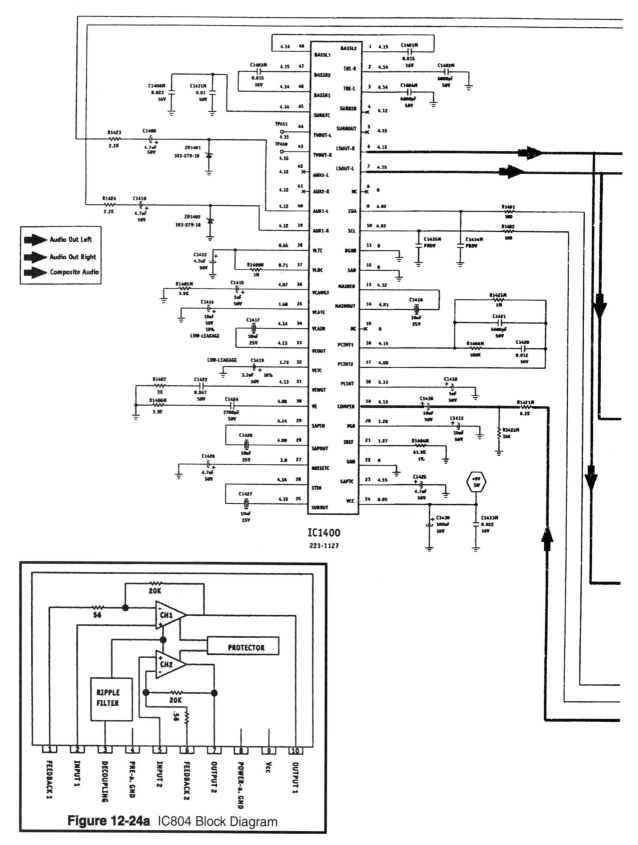

Figure 12-24a IC804 Block Diagram

Figure 12-24 GA-3 Audio Circuit

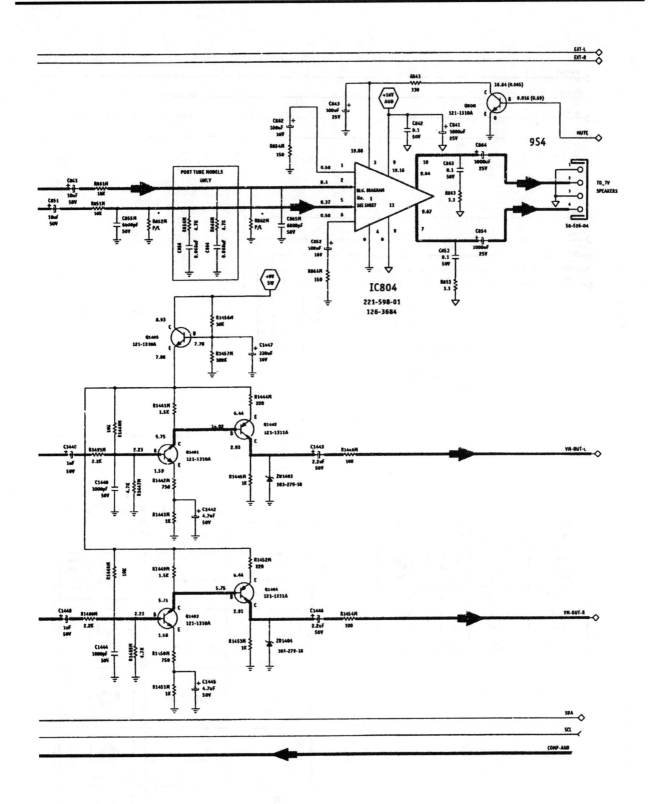

Figure 12-24 GA-3 Audio Circuit (Continued)

Servicing Zenith Televisions

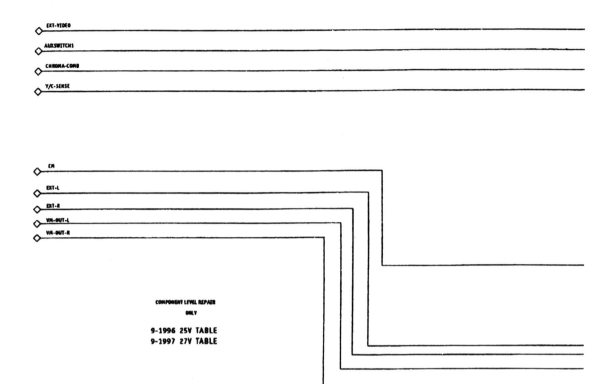

Figure 12-25 Auxiliary Video Switching Arrangement

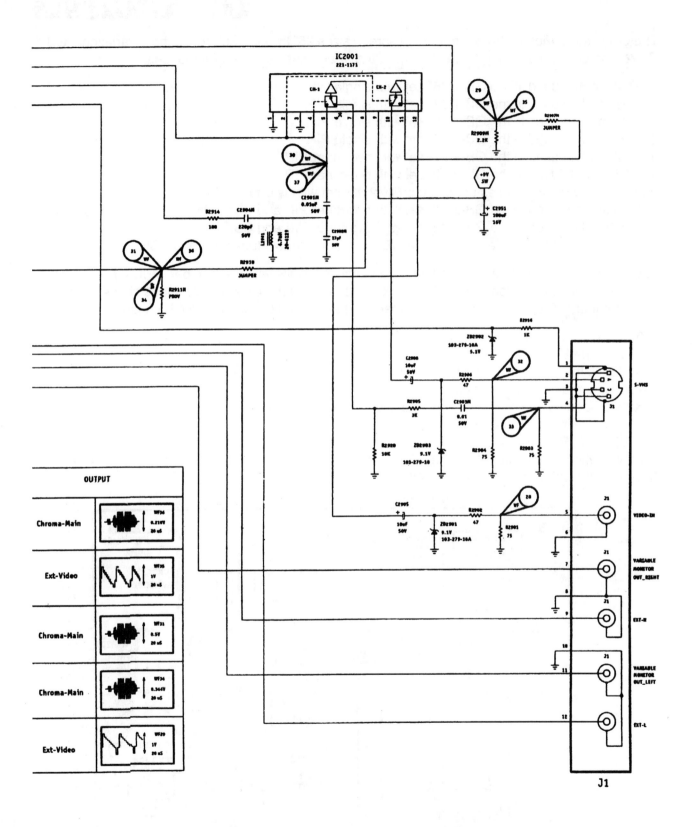

Figure 12-25 Auxiliary Video Switching Arrangement (Continued)

Servicing Zenith Televisions

GB Chassis

The GB chassis (figure 12-26) appears in screen sizes 25" through 36" and in three variations, GB-1, GB-2, and GB-3. According to the literature, the three chassis have 11 variations:

9-2022 module used in the GB-1 25" set with single PIP
9-1999 module used in the GB-1 27" set with single PIP
9-2000 module used in the GB-1 32" set with single PIP
9-2001 module used in the GB-1 36" set with single PIP
9-2002 module used in the GB-2 27" set with two-tuner PIP
9-2003 module used in the GB-2 32" set with two-tuner PIP
9-2005 and 9-2023 modules used in the GB-2 36" set with two-tuner PIP
9-2006 module used in the GB-3 27" set with two-tuner PIP and with BBE
9-2007 module used in the GB-3 32" set with two-tuner PIP and BBE
9-2008 and 9-2024 modules used in the GB-3 36" set with two-tuner PIP and BBE

These modules make use of nine integrated circuits. However, the number of ICs in a particular chassis depends on its feature level. For example, IC2150 appears only on those chassis that have the two-tuner PIP option. Here are the other integrated circuits used in the GB chassis.

IC6000 is the microcontroller
IC6001 is the EAROM
ICX3431 is the controller IC for the power supply
IC2900 is the audio/video switch
ICX2200 handles audio, video, sync, and sweep drive processing
IC2100 develops vertical sweep
IC2000 is the PIP processor
IC2150 is the second tuner PIP video IF processor
IC1401 is the BBE sound enhancement audio circuit

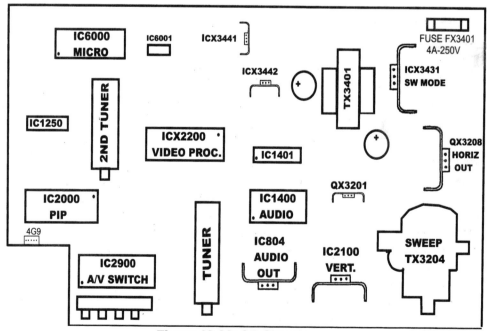

Figure 12-26 GB Chassis

The A-Line Chassis

I know of 15 models (as of July 1998) that belong to the A-Line GB series of televisions. There are surely others by now, but these are the ones with which I am familiar:

A25B33W	A27B33W	A32B41W	A36B41W	LGA29B33W
A25B33W8	A27B41W	A32B43W	A36B41W8	
	A27B43W	A32B84R	A36B43W	
		A32B86R	A36B43W8	
			A36B86R	

As far as I can tell, the part number of the microprocessor is the best way to distinguish among the GB chassis variations. The GB-1 uses the 221-1305-01, the GB-2 uses the 221-1307, and the GB-3 uses 221-1308. Each variation has it own features that have been programmed into the EAROM at the factory and supported by additional "daughter" boards like the one-tuner PIP, the two-tuner PIP, etc.

Customer Menu

Like the previous chassis, the GB customer menu appears over video and at the left of the screen and consists of six pages: Setup, Special Features, Audio, Video, PIP, and Source. Most of the menu features are "old hat," but there are a few new additions. For example, there is an entry on the Special Features page labeled "Data Source." According to my notes, it is not being currently used but has been set aside to accommodate future features like connecting the TV to a computer. The Audio Menu page has a feature called "BBE." If it is turned on, the BBE feature provides for enhanced audio. The Audio Menu also has a feature that allows the customer to turn the internal speakers on or off. If the speakers have been turned off, the volume bar at the bottom of the screen reads "EXT Volume" instead of showing the level to which the audio has been set.

The Source Menu has two options. The "main source" option permits the viewer to select from antenna cable, video 1, video 2, or F-video (front video). The "PIP source" option offers the customer a choice of antenna cable, video 1, video 2, or F-video as the source for the PIP.

Make a note that when "main source" is in the video mode, the setup menu shows six items: clock set, timer setup, parental control, captions/text, language, and auto demo.

Service Menu

Each GB variation has its own service menu. These menus have a lot in common but enough differences that I felt it necessary to provide all three (GB-1, table 12-6; GB-2, table 12-7; GB-3, table 12-8). You should know by now how to access the service menu, what you see when it comes up, and how to gain access to all of the items rather than the first few. If you don't, read (or reread) the appropriate sections in this book.

GB-1 Service Menu

Position 04 (level) for the GB-1 chassis has settings of 1 and 2. Unlike previous designs, the factory sets the level, and it cannot be changed in the field. That makes one less item to be concerned about.

- Items 22 through 26 permit the servicer to set the black and white tracking.
- Be aware that items 25 and 26 change depending on the color temperature setting in the video menu.
- Items 34 through 41 for the PIP only.
- Items 42 and 43 let you adjust the position of the on-screen display and don't affect anything else on the screen.

GB-2 Service Menu

Note that there are two audio level adjustments (items 16 and 17). The first is for the main tuner audio, and the second is for the second tuner (or PIP) audio. Items 23 through 27 are for black and white tracking and, like the GB-1, items 26 and 27 change when the color temperature setting in the video menu changes. Items 35-42 deal just with the PIP option.

GB-3 Service Menu

Most of the nomenclature for the items remains the same as they were in preceding chassis, but the numbers that designate the items change slightly. Item 45 (front jack option) is a brand new addition and has two settings. A setting of 1 enables the front audio/video Y/C inputs while a setting of 0 turns them off.

Servicing the GB Chassis

If you replace the main module – and you probably will at some point if you do Zenith warranty work – remember that an out-of-the-box module might not be set correctly for the model on which you are working. When you get the new module installed and fired up, go into the service menu and check the setting for each of the items. You don't have to set the level (item 4) because it has been preset at the factory and cannot be altered.

You must reset the black and white tracking when you replace the picture tube or the video processor. Follow the procedure I outlined for the A-Line GA chassis when the video menu has been set for "cool." Don't forget to set the affected items to their default settings for the screen size on which you are working before you begin the adjustments. After you have completed that part of the gray scale adjustment, select "warm" in the video menu. Then return to the service menu, and make the following adjustments to red, blue, and green cutoffs and green and blue gain 2:

red cutoff 2 warm equals red cutoff 1 cool plus 25
green cutoff 2 warm equals green cutoff cool plus 2
green cutoff warm equals blue cutoff 1 cool minus 3
green gain 2 warm equals green gain 1 cool minus 25
blue gain 2 warm equals blue gain cool minus 45

Finally, select "cool" in the user's video menu and adjust red, green, and blue cutoffs and green and blue gain 1 according to the same formula.

If you service a GB chassis that has no viewable picture, begin troubleshooting by accessing the service menu and checking the default settings of items 05, 18, and 22 (band, RF AGC, and PIF VCO) before you do anything else. Then tune in a good off-the-air signal, place a high impedance meter at pin 44 of ICX2200 or R1219 and check for a reading of 2.5 volts. If the voltage is high or low, adjust item 22 until you get a reading of 2.5 volts DC on the meter. This is the AFC crossover point.

There may come a time when you need to adjust the AGC delay. In that event, input a strong, noise-free signal into the RF port and reduce the setting of item 18 until the signal gets noisier. Then increase the setting until the noise disappears. A setting much higher than 40 might permit the tuner to overload and cause "beats" to appear in the picture.

The time may come when you have to adjust the audio detector, especially after you have replaced ICX2200. Input a signal from your generator into the tuner, making sure the generator is also inputting an audio signal. Connect a meter to pin 54 of ICX2200 and adjust coil L1205 for a reading of 4.0 volts

GB-1 221-1305-01 MICRO
GENERAL SETTINGS

ITEM	RANGE	25"	27"	KEY ABBREVIATIONS
00 F Mode	0 - 1	0	0	Factory Mode (refer to page 2-1)
01 Pre Px	0 - 1	1	1	Used to store customer menu adjustments
02 V Pos	0 - 24	16	17	Moves on screen displays vertically
03 H Pos	0 - 75	6	23	Moves on screen displays horizontally
04 Level	1 - 2	1	1	Level is fixed at the factory
05 Band	0 - 7	1	1	Broadcast band adjust
06 AC On	0 - 1	0	0	Will turn on and off set when AC power is applied and removed
TECHNICAL SETTINGS				
07 RFBpf	0 - 1	1	1	RF bandpass filter
08 3.58T	0 - 1	1	1	3.58 trap
09 RF Brt	0 - 63	21	29	RF Brightness
10 Ax Brt	0 - 63	21	29	Auxiliary Brightness
11 Max Con	0 - 63	63	63	Max Contrast
12 V. Size	0 - 254	56	120	Vertical Size, adjusts size of picture
13 H. Size	0 - 254	107	80	Horizontal size, adjusts size of picture
14 V Phase	0 - 7	2	0	Vertical Phase, shifts picture vertically
15 H Phase	0 - 31	19	18	Horizontal Phase, shifts picture horizontally
16 AudLv1	0 - 63	46	46	Audio level
17 RF Agc	0 - 63	31	31	Automatic gain control
18 H Afc	0 - 1	1	1	Automatic frequency control
19 WhComp	0 - 1	0	0	White compression
20 60 hzSw	0 - 1	2	2	60 Hertz switched
21 PifVco	0 - 127	31	31	PIF Voltage controlled Oscillator
COLOR TEMP COOL STARTING VALUES				
22 R Cut	0 - 254	0	0	B&W tracking
23 G Cut	0 - 254	0	0	B&W tracking
24 B Cut	0 - 254	0	0	B&W tracking
25 G Gain	0 - 254	95	100	Green Gain Color Temp
26 B Gain	0 - 254	70	80	Blue Gain Color Temp
COLOR TEMP WARM STARTING VALUES				
22 R Cut	0 - 254	0	0	B&W tracking
23 G Cut	0 - 254	0	0	B&W tracking
24 B Cut	0 - 254	0	0	B&W tracking
25 G Gain	0 - 254	70	70	Green Gain Color Temp
26 B Gain	0 - 254	40	40	Blue Gain Color Temp
27 Scroll	0 - 1	1	1	Selects method for User Menu
28 6 Keys	0 - 1	1	1	Keyboard, set to 1 for 6 keyboard, set to 0 for 10 key
AUDIO SETTINGS				
29 A Att	0 - 15	9	9	Audio input attenuator
30 A Vco	0 - 63	31	31	Audio Voltage Controlled Oscillator
31 A Fltr	0 - 63	3	31	Audio Filter
32 Spctrl	0 - 63	31	31	High frequency separation
33 W Band	0 - 63	31	31	Wide band low frequency separation
PIP SETTINGS				
34 PIPX1	0 - 63	9	8	Sets horizontal position on left side
35 PIPY1	0 - 63	5	5	Sets vertical position on left side
36 PIP X2	0 - 63	49	51	Sets horizontal position on right side
37 PIP Y2	0 - 63	32	33	Sets vertical position on right side
38 PIP Ras	0 - 254	69	69	Picture in Picture raster
39 PIP Sw	0 - 15	9	9	Picture in Picture switch delay
40 PIP Lud	0 - 7	1	1	Picture in Picture luminance delay
41 PIP TOF	0 - 63	18	18	Picture in Picture level register
42 InOSDC	0 - 1	1	1	OSD internal oscillator
43 OSD FR	0 - 1	0	0	OSD Frame

Table 12-6 GB-1 Service Menu General Settings

GB-2 221-1307 MICRO
GENERAL SETTINGS

ITEM	RANGE	27"	32"	36"	36"	KEY ABBREVIATIONS
00 F Mode	0 - 1	0	0	0	0	Factory Mode (refer to page 2-1)
01 Pre Px	0 - 1	1	1	1	1	Used to store customer menu adjustments
02 V Pos	0 - 31	17	18	18	18	Moves on screen displays vertically
03 H Pos	0 - 75	23	20	23	23	Moves on screen displays horizontally
04 Level	1 - 2	1	1	1	1	Level is fixed at factory
05 Band	0 - 7	1	1	1	1	Broadcast band adjust
06 AC On	0 - 1	0	0	0	0	Will turn on and off set when AC power is applied and removed

TECHNICAL SETTINGS

ITEM	RANGE	27"	32"	36"	36"	KEY ABBREVIATIONS
07 RFBpf	0 - 1	1	1	1	1	RF bandpass filter
08 3.58T	0 - 1	1	1	1	1	3.58 trap
09 RF Brt	0 - 63	29	29	29	45	RF Brightness
10 Ax Brt	0 - 63	29	29	29	45	Auxiliary Brightness
11 Max Con	0 - 63	63	63	63	63	Max Contrast
12 V. Size	0 - 254	110	132	140	188	Vertical Size, adjusts size of picture
13 H. Size	0 - 254	20	90	140	80	Vertical Size, adjusts size of picture
14 V Phase	0 - 7	2	0	2	0	Vertical Phase, shifts picture vertically
15 H Phase	0 - 31	17	17	18	18	Horizontal Phase, shifts picture horizontally
16 AudLv1	0 - 63	46	46	46	46	Audio level
17 AudLv2	0 - 63	46	46	46	46	Audio level
18 RF Agc	0 - 63	31	31	31	31	Automatic frequency control
19 H Afc	0 - 1	1	1	1	1	Horizontal AFC
20 WhComp	0 - 1	0	0	0	0	White compression
21 60 hzSw	0 - 1	2	2	2	2	60 Hertz switched
22 PifVco	0 -127	31	31	31	31	PIF voltage controlled oscillator

COLOR TEMP COOL STARTING VALUE

ITEM	RANGE	27"	32"	36"	36"	KEY ABBREVIATIONS
23 R Cut	0 -254	0	0	0	0	B&W tracking
24 G Cut	0 -254	0	0	0	0	B&W tracking
25 B Cut	0 -254	0	0	0	0	B&W tracking
26 G Gain	0 -254	100	100	100	100	Green Gain Color Temp
27 B Gain	0 - 254	80	80	80	80	Blue Gain Color Temp

COLOR TEMP WARM STARTING VALUES

ITEM	RANGE	27"	32"	36"	36"	KEY ABBREVIATIONS
23 R Cut	0 - 254	0	0	0	0	B&W tracking
24 G Cut	0 - 254	0	0	0	0	B&W tracking
25 B Cut	0 - 254	0	0	0	0	B&W tracking
26 G Gain	0 - 254	70	70	70	40	Green Gain Color Temp
27 B Gain	0 - 254	70	70	70	40	Blue Gain Color Temp
28 Scroll	0 - 1	1	1	1	1	Selects method for User Menu
29 6 Keys	0 - 1	1	1	1	1	Keyboard, set to 1 for 6 keyboard, set to 0 for 10 key

AUDIO SETTINGS

ITEM	RANGE	27"	32"	36"	36"	KEY ABBREVIATIONS
30 A Att	0 - 15	9	9	9	9	Audio input attenuator
31 A Vco	0 - 63	31	31	31	31	Audio voltage controlled oscillator
32 A Fltr	0 - 63	31	31	31	31	Audio filter
33 Spctrl	0 - 63	31	31	31	31	High frequency separation
34 W Band	0 - 63	31	31	31	31	Wide band low frequency separation

PIP SETTINGS

ITEM	RANGE	27"	32"	36"	36"	KEY ABBREVIATIONS
35 PIPX1	0 - 63	8	8	8	8	Set horizontal position on left side
36 PIPY1	0 - 63	5	5	5	5	Sets vertical position on left side
37 PIP X2	0 - 63	52	50	51	51	Sets horizontal position on right side
38 PIP Y2	0 - 63	33	33	33	33	Sets vertical position on right side
39 PIP Ras	0 - 254	69	69	69	69	Picture in Picture raster
40 PIP Sw	0 - 15	9	9	9	9	Picture in Picture delay
41 PIP Lud	0 - 7	1	1	1	1	Picture in Picture luminance delay
42 PIP TOF	0 - 63	33	31	31	31	Pip in Picture tine level register
43 InOSDC	0 - 1	1	1	1	1	OSD internal oscillator
44 OSD FR	0 - 1	0	0	0	0	OSD frame

Table 12-7 GB-2 Service Menu General Settings

GB-3 221-1308 MICRO
GENERAL SETTINGS

ITEM	RANGE	27"	32"	36"	36"	KEY ABBREVIATIONS
00 F Mode	0 - 1	0	0	0	0	Factory Mode (refer to page 2-1)
01 Pre Px	0 - 1	1	1	1	1	Used to store customer menu adjustments
02 V Pos	0 - 24	17	18	14	18	Moves on screen displays vertically
03 H Pos	0 - 13	23	20	23	23	Moves on screen displays horizontally
04 Level	1 - 2	1	1	1	1	Level is fixed at factory
05 Band	0 - 7	1	1	1	1	Broadcast band adjust
06 AC On	0 - 1	0	0	0	0	Will turn on and off set when AC power is applied and removed

TECHNICAL SETTINGS

ITEM	RANGE	27"	32"	36"	36"	KEY ABBREVIATIONS
07 RFBpf	0 - 1	1	1	1	1	RF bandpass filter
08 3.58T	0 - 1	1	1	1	1	3.58 trap
09 RF Brt	0 - 63	29	29	29	45	RF Brightness
10 Ax Brt	0 - 63	29	29	29	45	Auxiliary Brightness
11 Max Con	0- 63	63	63	63	63	Max Contrast
12 V. Size	0 - 254	104	132	130	188	Vertical size
13 H. Size	0 - 254	145	90	140	80	Horizontal size
14 V Phase	0 - 7	0	0	2	0	Vertical phase
15 H. Phase	0-31	18	18	18	18	Horizontal phase
16 AudLv1	0 - 63	19	17	18	18	Audio level
17 AudLv2	0 - 63	46	46	46	46	Audio level 2
18 RF Agc	0 -63	31	31	31	31	Automatic Gain control
19 H Afc	0 - 1	1	1	1	1	Horizontal AFC
20 WhComp	0 - 1	0	0	0	0	White compression
21 60 hzSw	0 - 1	2	2	2	2	60 Hertz switched
22 PifVco	0 - 127	31	31	31	31	PIF voltage controlled oscillator

COLOR TEMP COOL STARTING VALUES

ITEM	RANGE	27"	32"	36"	36"	KEY ABBREVIATIONS
23 R Cut	0 - 254	0	0	0	0	B&W tracking
24 G Cut	0 - 254	0	0	0	0	B&W tracking
25 B Cut	0 - 254	0	0	0	0	B&W tracking
26 G Gain	0 - 254	100	100	100	100	Green Gain Color Temp
27 B Gain	0 - 254	80	80	80	80	Blue Gain Color Temp

COLOR TEMP WARM STARTING VALUES

ITEM	RANGE	27"	32"	36"	36"	KEY ABBREVIATIONS
23 R Cut	0 - 254	0	0	0	0	B&W tracking
24 G Cut	0 - 254	0	0	0	0	B&W tracking
25 B Cut	0 - 254	0	0	0	0	B&W tracking
26 G Gain	0 - 254	70	70	70	70	Green Gain Color Temp
27 B Gain	0 - 254	40	40	40	40	Blue Gain Color Temp
28 Scroll	0 - 1	1	1	1	1	Selects method for User Men
29 6 Keys	0 - 1	1	1	1	1	Keyboard, set to 1 for 6 keyboard, set to 0 for 10 key

AUDIO SETTINGS

ITEM	RANGE	27"	32"	36"	36"	KEY ABBREVIATIONS
30 A Att	0 - 15	9	9	9	9	Audio input attenuator
31 A Vco	0 - 63	31	31	31	31	Audio voltage controlled oscillator
32 A Fltr	0 - 63	31	31	31	31	Audio filter
33 Spctrl	0 - 63	31	31	31	31	High frequency separation
34 W Band	0 - 63	31	31	31	31	Wide band low frequency separation

PIP SETTINGS

ITEM	RANGE	27"	32"	36"	36"	KEY ABBREVIATIONS
35 PIPX1	0 - 63	8	8	8	8	Sets horizontal position on left side
36 PIPY1	0 - 63	5	5	5	5	Sets vertical position on left side
37 PIP X2	0 - 63	51	50	51	51	Sets horizontal position on right side
38 PIP Y2	0 - 63	33	33	33	33	Sets vertical position on right side
39 PIP Ras	0 - 254	69	69	69	69	Picture in Picture raster
40 PIP Sw	0 - 15	9	9	9	9	Picture in Picture switch delay
41 PIP Lud	0 - 7	1	1	1	1	Picture in Picture luminance delay
42 PIP TOF	0 - 63	31	31	31	31	Picture in Picture tint level register
43 InOSDC	0- 1	1	1	1	1	OSD internal oscillator
44 NC Ready	0- 1	0	0	0	0	OSD frame
45 F-Jack	0- 1	0	0	0	0	Front Jack option

Table 12-8 GB-3 Service Menu General Settings

DC on the meter. When you achieve this reading, you will have adjusted the circuit for maximum audio output from ICX2200.

The stereo level adjustment (item 16 in the service menu) should also be checked when ICX2200 has been replaced. Connect a high impedance AC meter between jumpers W1611 (+ lead) and W1603 (- lead). Place a 4700pf capacitor from the + side of C1211 to the ground lead jumper W49 on the main board. Turn the AC off and back on to reset the microprocessor if you haven't already done so. Apply an RF signal with a good video and an audio signal of 400 hertz at 100% modulation to the tuner, and adjust item 16 for a reading between 490 and 500 millivolts on the meter.

If you have replaced the audio IC or the main module, access the service menu and confirm that the five stereo adjustments correspond to the data on the bar code label. I know I keep on repeating myself, but it's important folks!

When you face a video problem, don't forget that all video travels through the PIP IC and video/audio switch. If either fails, the video will suffer. checking the PIP and the video switch is as simple as using a scope to confirm that "what goes in" is also "what comes out."

WARNING:
Never disconnect the video output module from the CRT when you are troubleshooting the GB chassis. If you remove it when the set is on or turn the set on when the module has been disconnected, the CRT will arc and probably be damaged. Moreover, you will expose yourself to dangerous voltages. If you recall our discussion of the Z-Line products, you remember there is a termination circuit (R5110 in this instance) on the video output module for a voltage divider circuit inside the CRT. The warning applies only to the 32" and 36" sets.

Power Supply

Figure 12-27 is the power supply schematic for the GB chassis. It works exactly like the power supplies we have already discussed. If you suspect it is causing the problem you are servicing, make these checks when the set is in the standby mode. I do not list these voltages in any particular order. Let your common sense determine which you check first:

+150 volts DC at CX3404
+130 volts DC at CX3420 or the collector of the horizontal output transistor
+18 volts DC at C3422
+15 volts DC at the emitter of Q3404
+5 volts DC at pin 3 of IC3442

After you have confirmed the presence of the standby voltages, check the power on sequence by looking for keyboard input at pins 19 and 20 of IC6000 or an IR input at pin 2. If the readings are correct, check to see if pin 52 responds to the power on request by going high. Then check to see if the power-on signal translates to about 0.7 volts on the base of Q3402 and that it turns on. Finally, confirm the presence of the switched voltages by confirming that +9 volts appears at pin 3 of IC3441 and +15 volts at the emitter of Q3404.

All standby voltages marked as "standby" will be switched voltages when the Energy Star power supply becomes a part of the TV's circuitry. I don't have any information right now about when Energy Star will be integrated into the circuitry, but you should begin to expect it at any time.

System Control

Figure 12-28 depicts the system control circuit. Troubleshoot it using the guidelines we have been over several times. Note that the vertical input to IC6000 is clearly labeled "CRT protect." If the voltage is missing, the microcontroller responds by turning the TV off about three seconds after it turns on.

Deflection Circuits

The deflection circuits correspond to those we have looked at several times and do not require extended comment.

Vertical Deflection

The GB chassis uses the 221-992-01 vertical output IC that the GA-3 chassis uses and is configured the same way. For example, vertical drive comes from the collector of Q6002 instead of pin 22 of ICX2200. The CRT protection circuit consists of C2107, DX2102, DX2103, R2105M, R2151M, and R2153M. It works by tapping a pulsing voltage off pin 11 of IC2100, rectifying it, and applying the resulting DC voltage to a voltage divider network, the output of which goes to the CRT protection input of IC6000.

Horizontal Deflection

The basic horizontal deflection circuit is vintage Zenith stuff. Horizontal drive exits pin 32 of ICX2200, goes to the base of Q3202 that is configured as an emitter follower to give the signal current gain, and then goes to the base of the horizontal driver, etc.

Horizontal deflection produces +23 volts for the vertical circuit, +215 volts for the video output transistors, drive for the filaments of the CRT, and the usual EHT, G2, and focus voltages.

The larger CRTs require pincushion correction circuitry. Therefore, the GB chassis make use of Q3201, Q3203, and associated components to form a diode modulator circuit that corrects for horizontal pincushion.

The XRP circuit consists of DX3005, ZDX3004, CX3004 and a resistor network to monitor high-voltage production. When the resulting voltage reaches 3.5 volts DC or higher at pin 29 of ICX2200, horizontal drive stops, and the set shuts down just about 1.6 seconds after turnon. In correctly working televisions, expect to find a maximum of 29 kV on the 27" sets, 31 kV on the 32" sets, and 33 kV on the 36" sets except for the GB-2 and GB-3 36" sets, which usually have a maximum of 31.5 kV.

Video Circuits

The video processor is the now familiar 221-1165 IC (figure 12-30). The difference between its configuration in the GB chassis and previous chassis lies in the signal path, not in the way the circuit works. For example, composite video leaves ICX2200 at pin 47, is processed by Q1203 and Q1204, and coupled to pin 50 of IC2900, the audio/video switch. The signal leaves IC2900 at pin 36 and goes to the comb filter where the composite video is separated into its luminance and chrominance parts. The separated signals reenter IC2900 at pins 30 and 38 and exit at pins 34 and 32.

The signals then go into pins 29 (luma) and 32 (chroma) of IC2000, the PIP IC. Luma exits at pin 49 and passes through Q2009 on its way back into pin 43 of ICX2200. Chroma exits at pin 51 and reenters ICX2200 at pin 43.

Servicing Zenith Televisions

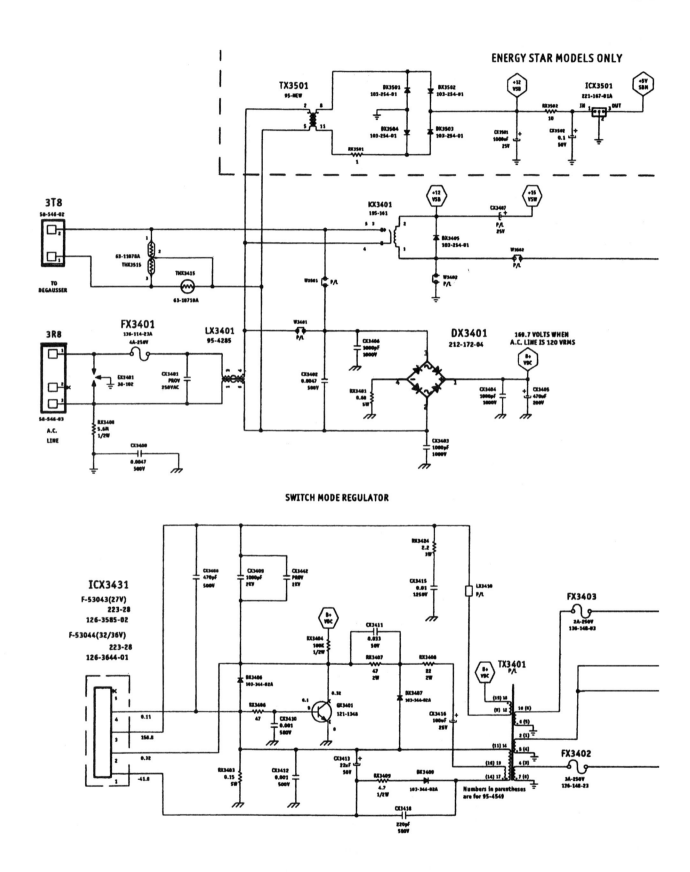

Figure 12-27 Switch-Mode Power Supply

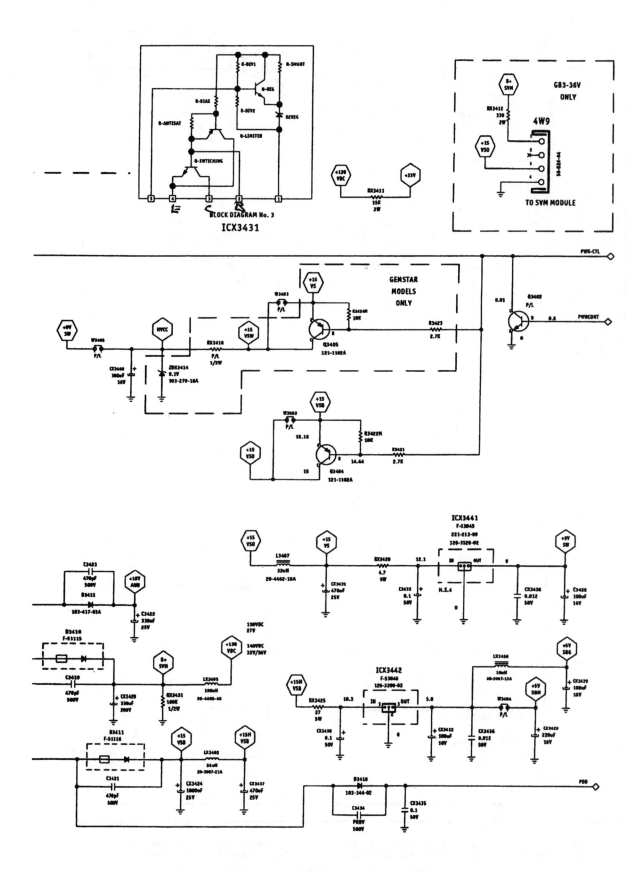

Figure 12-27 Switch-Mode Power Supply (Continued)

Servicing Zenith Televisions

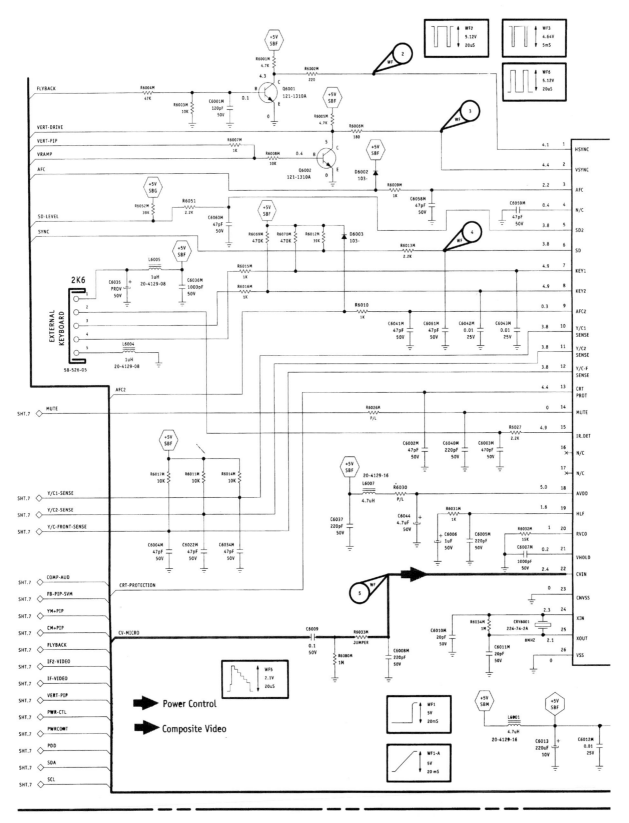

Figure 12-28 System Control Circuit

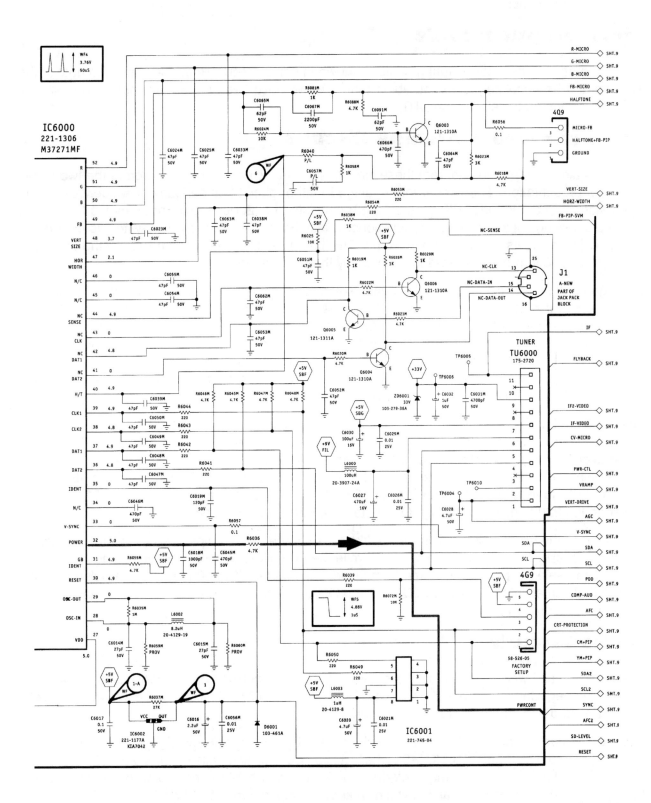

Figure 12-28 System Control Circuit (Continued)

Troubleshoot ICX2200 by using the information given when I discussed how to troubleshoot the video circuit in the GA chassis.

Video Output Module

The schematic for the video output module is shown in figure 12-31. Read the important note in the diagram and be absolutely sure you abide by it. As we have seen in past examples, the value of R5110 depends on the size of the CRT.

PIP and Video Switching Circuits

Since it is almost impossible to treat these circuits separately, I am going to integrate the two in this very brief presentation. Remember that a failure in either circuit has serious consequences for the video signal. The one-tuner PIP (figure 12-33) consists of IC2000, a chip that handles the analog signal processing, logic, and memory functions. It also uses IC2900, the video switch, to control the routing of signals to its circuitry. The two-tuner PIP (figure 12-34) uses the same ICs as the one-tuner PIP plus IC1250 to process the video supplied by the second tuner.

The video switch (figure 12-32) lets the viewer select among the various inputs for his or her viewing pleasure. For instance, main tuner video from ICX2200 goes to pin 50 of IC2900. If the set has a second tuner, the video from it goes to pin 53 of IC2900. External video from source one comes into pin 7 while left and right audio appear at pins 8 and 10. External video from the second source comes into pin 13 while its left and right audio come into pins 14 and 16. S-VHS luminance from the front jack appears at pin 21 and chroma at pin 23. External video from source three (the front jacks) comes into pin 19 while left and right audio come into pins 20 and 22.

The video switch has two outputs. PIP luma exits on pin 44, passes through Q2001, and enters pin 36 of IC2000. Chroma exits at pin 42, passes through Q2002, and enters pin 34 of IC2000. Main luma out exits on pin 34, passes through Q2901, and enters IC2000 at pin 29. Main chroma out exits at 32, passes through Q2906, and enters pin 32 of IC2000 PIP output from IC2000 is at pins 49 and 51. Main and PIP luminance signals are on pin 49 while main and PIP chroma are on pin 51. Both are routed to ICX2200, the luma through Q2009 to pin 43 and chroma to pin 45.

In addition to the usual circuit requirements, like a B+ source, these chips need a few other signals to function properly. For example, the microcontroller sends its instructions over the I2C bus to IC2900, causing it to select a particular source from its several inputs. IC2000 must have a pulse from the flyback that comes into pin 1 through Q2003 and Q2004 a vertical pulse that comes into the IC at pin 2 for it to work properly.

Audio Circuits

The audio circuit I want to talk about develops MTS stereo (figure 12-35). It consists of IC1400, the stereo decoder, IC804, the SoundRite volume limiter, dual power amplifiers, and may include IC1401 the BBE sound enhancement audio processor.

The BBE feature is the new kid on the block. When it turns on, left and right channel audio from IC1400 goes to pins 3 and 17 of IC1401 where it is processed and exits on pins 9 and 11 on its way back into IC1400. I wish I had more information to give you about the new feature especially about how it enhances the audio signal, but you now know as much about it as I do.

The A-Line Chassis

You probably see similarities between this circuit and the one used in the GA chassis. For example, left channel audio out goes to Q1401 and Q1402 to provide variable left channel audio out. Right channel audio goes to Q1403 and Q1406 to provide variable right channel audio out.

Repair History

These chassis are so new that I haven't accumulated much of a repair history for them. But such as I have, I offer. By the way, I see no need to reproduce the information about the power supply that I have given you in several places in this book. The same applies for other circuits. If you don't find the information you are looking for, you might thumb through a few of the previous discussions beginning with the Y-Line.

(1) "Aux" appears on the screen and won't go off. Every feature works fine except that "aux" won't disappear from the screen. Remember the discussion about R9738, the adjustment for the video gain? Adjust it slightly until "aux" disappears.

(2) Dead set. Follow previous tips about how to repair a nonfunctioning power supply, especially the discussion of the power supply in the chapter on the Y-Line. If all components check okay, become suspicious of CX3416.

(3) No raster, no picture, no sound, but menu works. Immediately suspect the tuner. Make a quick check by unsoldering clock and data lines to see if raster appears and white noise is heard in the speakers. If it does, replace the tuner.

(4) Dead set but power supply is okay. Check to see if horizontal drive is available out of ICX2200. If it isn't, replace the chip. Don't forget to check CX3416 in the power supply if the voltage at the collector of the horizontal output transistor falls substantially when you try to turn the set on.

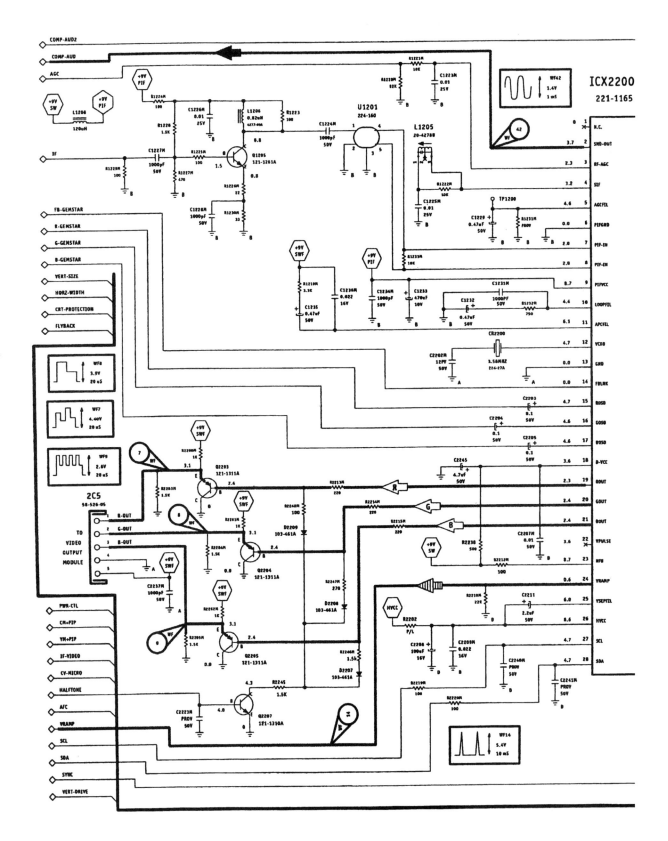

Figure 12-30 Video Processor

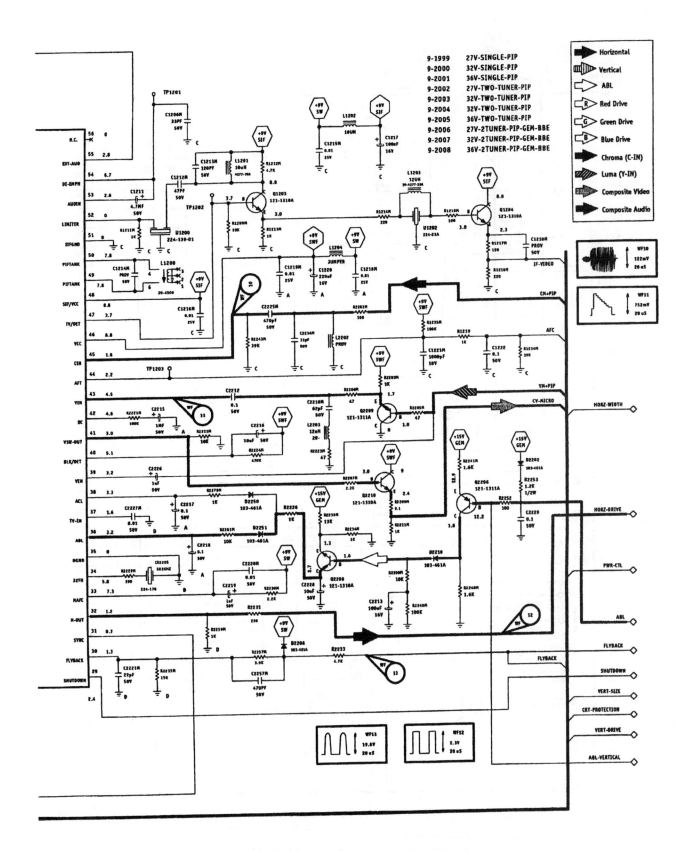

Figure 12-30 Video Processor (Continued)

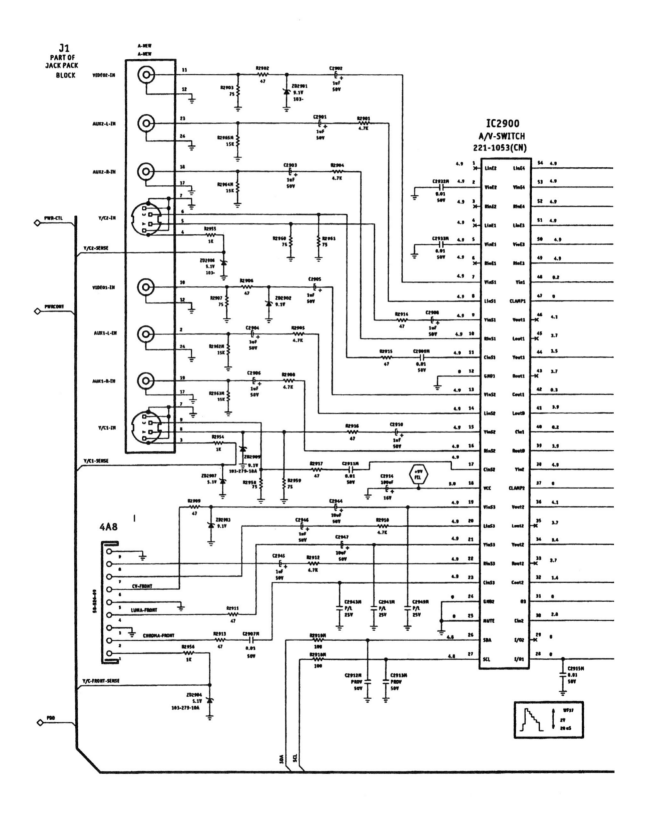

Figure 12-32 Audio/Video Switch

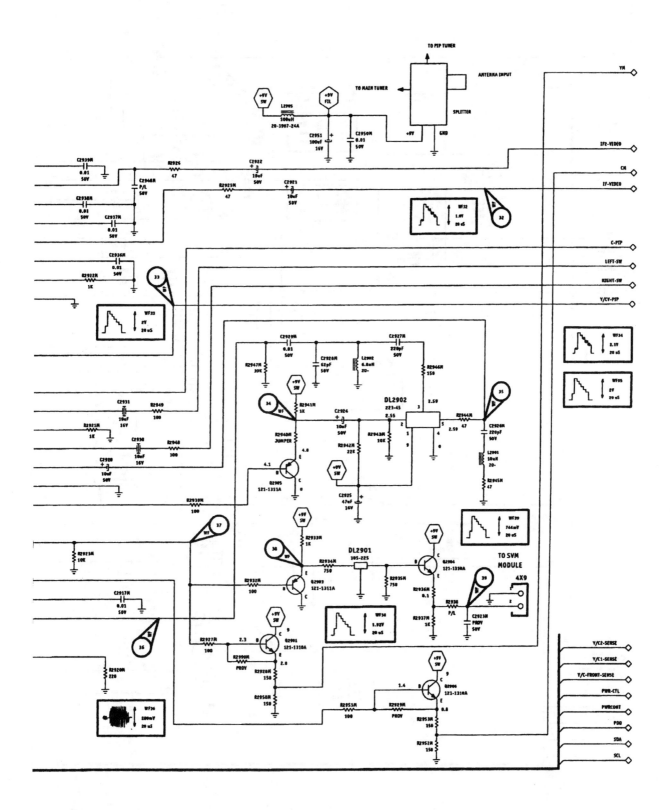

Figure 12-32 Audio/Video Switch (Continued)

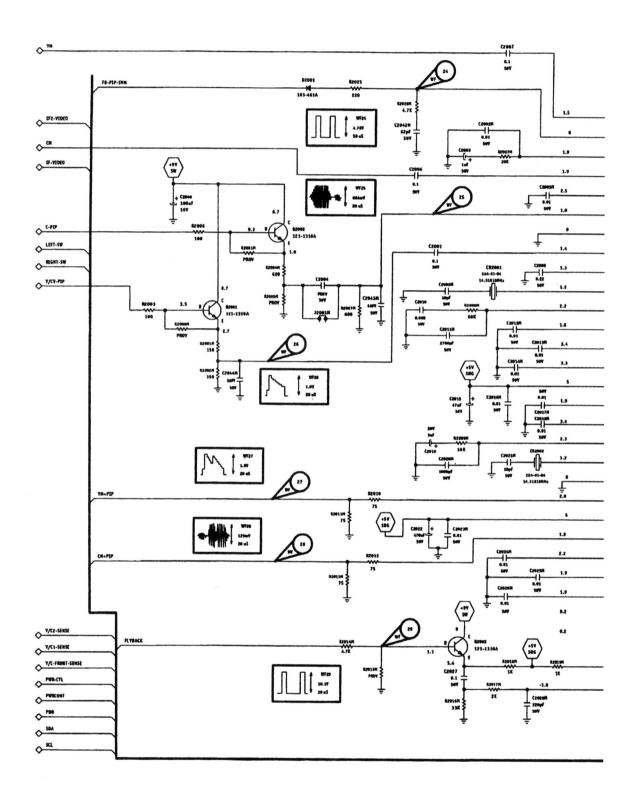

Figure 12-33 PIP

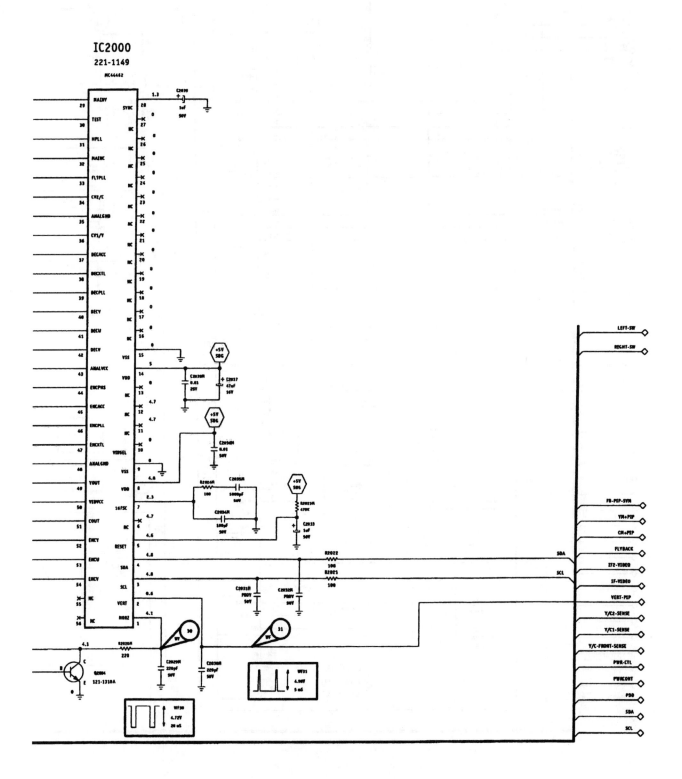

Figure 12-33 PIP (Continued)

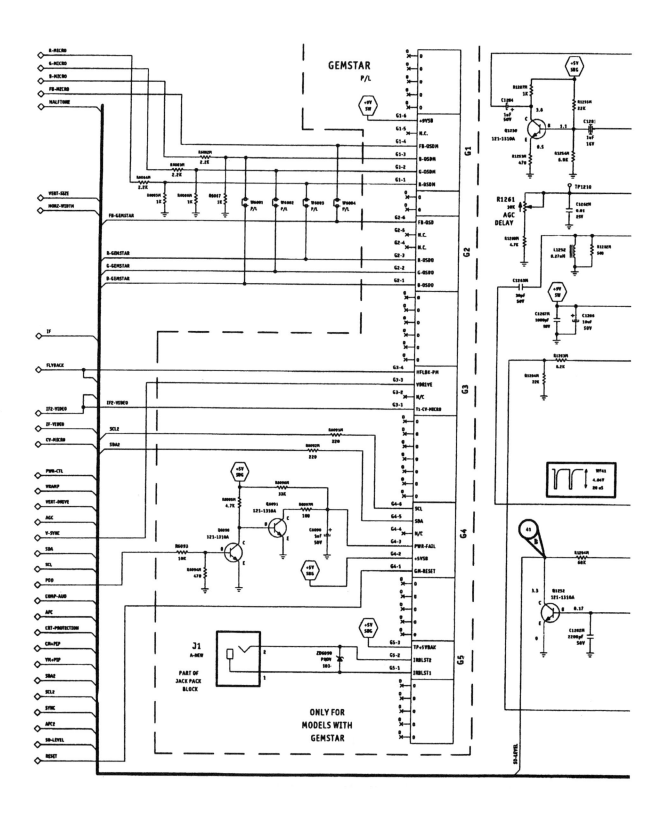

Figure 12-34 Two-Tuner Video IF

The A-Line Chassis

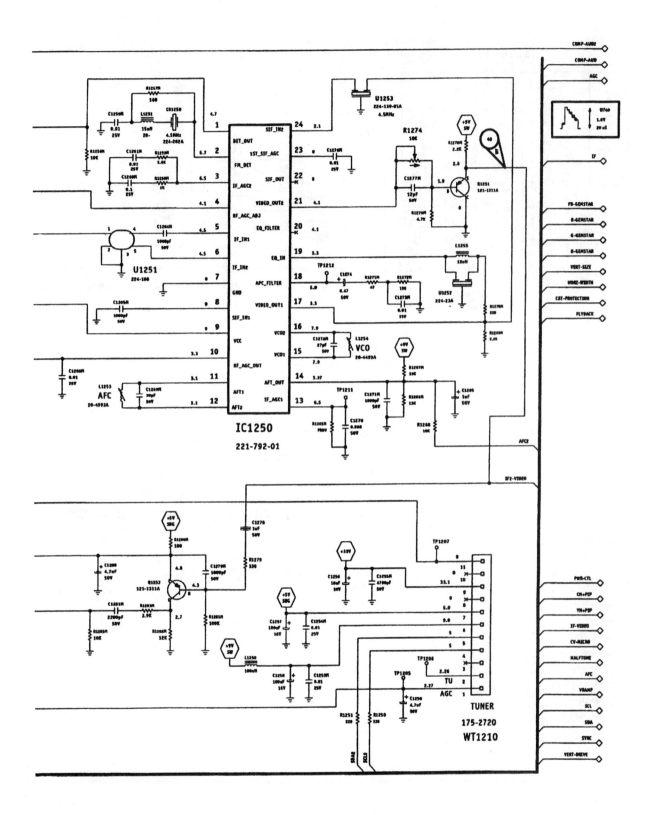

Figure 12-34 Two-Tuner Video IF (Continued)

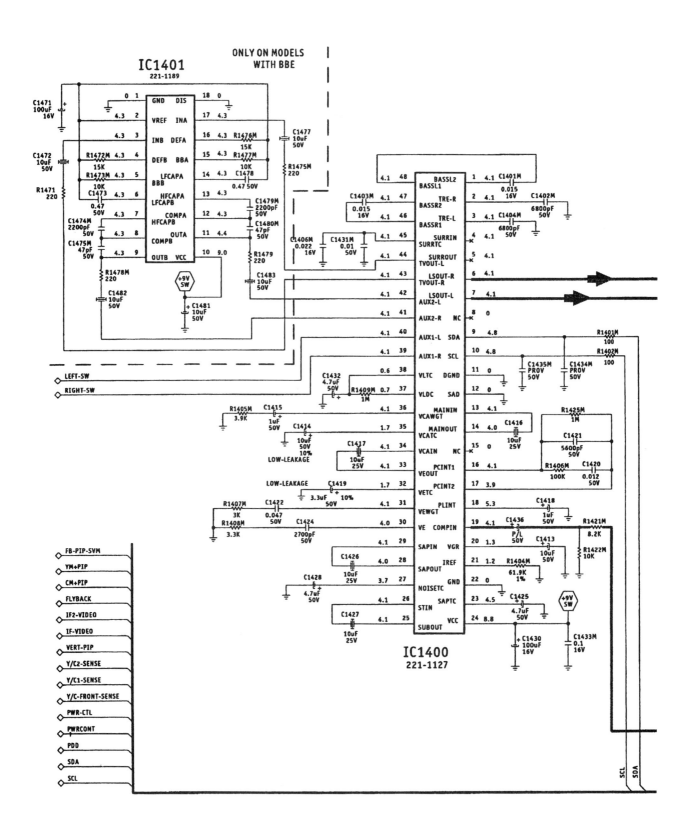

Figure 12-35 MTS Stereo Audio

The A-Line Chassis

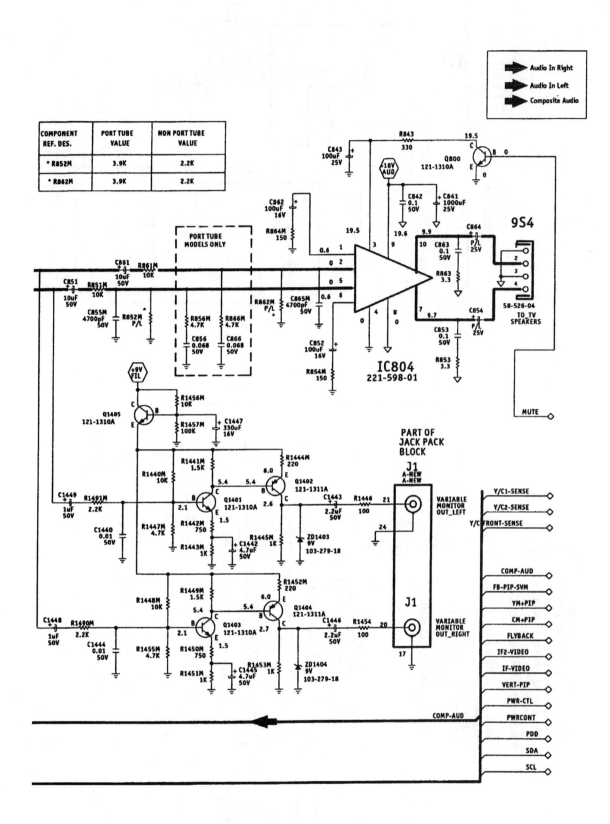

Figure 12-35 MTS Stereo Audio (Continued)

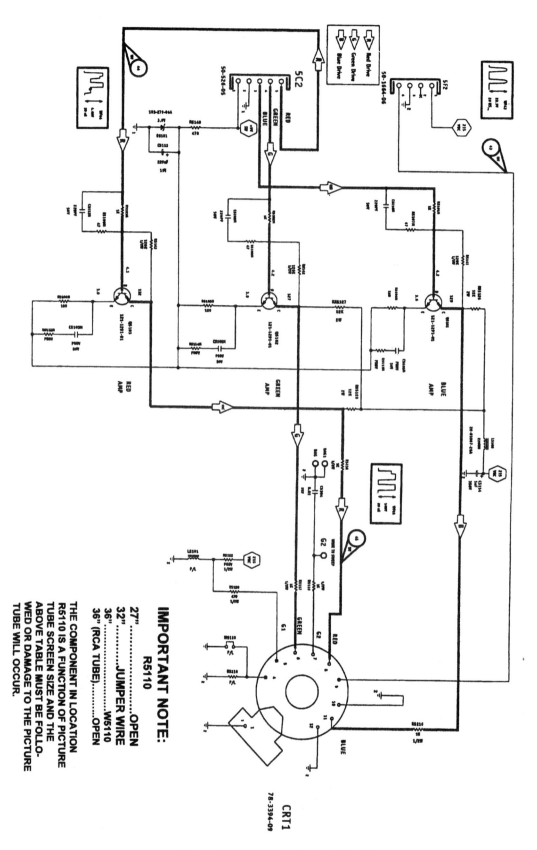

Figure 12-31 Video Output Module

CHAPTER 13

THE B-LINE CHASSIS

Zenith's B-Line, which became available in 1999, is like the previous line in that it is made up of two chassis, the CA and CB. Table 13-1 contains a list of the models that make up both chassis as of this writing. The list provides cabinet style, gives you an idea about the feature level, which microcontroller the chassis uses, and whether Zenith requires module level or component level repair. Following a trail blazed by the Y-Line, Zenith expects component level repair for the 27-inch and smaller sets and module level repair (module replacement) for the 32-inch and larger sets, including the projection televisions. Of course, these requirements pertain only to warranty work.

The CA variation (figure 13-1) appears in 25-inch and 36-inch screen sizes under the Sentry 2 rubric. The microprocessor, like the modules in previous lines, is programmed for added features during manufacturing, and plug-in daughter boards support those features. Even a cursory look leaves you thinking that the CA chassis uses quite a few ICs. However, the same faithful few do the bulk of the work. ICX2200 processes the audio, video, sync, and sweep drive. IC6000 is the main microprocessor, and IC6001 is the EAROM. IC2100 or IC2101 develops vertical drive.

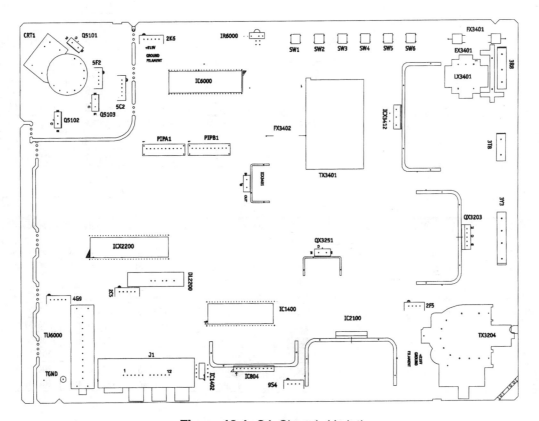

Figure 13-1 CA-Chassis Variation

CA Chassis Model Information

MODEL	SCR	CABINET	AUDIO	PIP	JACKS	COMB	MICRO	OP GUIDE
COMPONENT LEVEL REPAIR									
B25A10Z	25	TableTop	Stereo	None	0	Yes	221-01384	206-03476
B25A10ZC	25	TableTop	Stereo	None	0	No	221-01384	206-03476
B25A11Z	25	TableTop	Stereo	None	3	No	221-01384	206-03477
B25A24Z	25	TableTop	MTS/SAP	None	6	Yes	221-01385	206-03479
B25A30ZC	25	TableTop	Stereo	1 Tuner	6	No	221-01385	206-03480
B25A74R	25	Console	MTS/SAP	None	6	Yes	221-01385	206-03450
B25A76R	25	Console	MTS/SAP	None	6	Yes	221-01385	206-03450
B27A11Z	27	TableTop	Stereo	None	3	No	221-01384	206-03477
B27A11ZC	27	TableTop	Stereo	None	3	No	221-01384	206-03477
B27A13Z	27	TableTop	Stereo	None	0	No	221-01384	206-03478
B27A24Z	27	TableTop	MTS/SAP	None	0	Yes	221-01385	206-03479
B27A30ZC	27	TableTop	Stereo	1 Tuner	6	No	221-01386	206-03480
B27A34Z	27	TableTop	MTS/SAP	1 Tuner	6	Yes	221-01387	206-03481
B27A74R	27	Console	MTS/SAP	None	6	Yes	221-01385	206-03372
B27A76R	27	Console	MTS/SAP	None	6	Yes	221-01385	206-03372
CB27A10Z	27	TableTop	Stereo	None	0	No	221-01384	206-03476
CB27A10ZC	27	TableTop	Stereo	None	0	No	221-01384	206-03476
LGB26A11ZM	25	TableTop	Stereo	None	3	No	221-01384	206-03494
LGB29A10ZM	27	TableTop	Stereo	None	0	No	221-01384	206-03495
LGB29A11ZM	27	TableTop	Stereo	None	3	No	221-01384	206-03498
LGB29A13ZM	27	TableTop	Stereo	None	0	No	221-01384	206-03498
LGB29A24ZM	27	TableTop	MTS/SAP	None	6	Yes	221-01385	206-03497
LGB29A30ZM	27	TableTop	Stereo	1 Tuner	3	No	221-01386	206-03496
MODULE LEVEL REPAIR ONLY									
B32A24Z	32	TableTop	MTS/SAP	None	6	Yes	221-01385	206-03479
B32A30Z	32	TableTop	Stereo	1 Tuner	3	No	221-01386	206-03480
B32A34Z	32	TableTop	MTS/SAP	1 Tuner	3	Yes	221-01387	206-03481
B36A24Z	36	TableTop	MTS/SAP	1 Tuner	6	No	221-01385	206-03479
B36A30Z	36	TableTop	Stereo	1 Tuner	3	No	221-01386	206-03480
B36A34Z	36	TableTop	MTS/SAP	1 Tuner	6	Yes	221-01387	206-03481

CB Chassis Model Information

MODEL	SCR	CABINET	JACKS	FILTER	RESOLUTION	EXTRAS	REMOTE	MICRO	OP GUIDE
COMPONENT LEVEL REPAIR									
IQB27B42W	27	TableTop	15	Comb	650	V-Chip+	MBR3464Z	221-01388	206-03482
IQB27B44W	27	TableTop	15	Digital Comb	800	Guide Plus (GemStar)	MBR3466Z	221-01389	206-03483
MODULE LEVEL REPAIR ONLY									
IQB32B42W	32	TableTop	15	Comb	650	V-Chip+	MBR3464Z	221-01388	206-03482
IQB32B44W	32	TableTop	15	Digital Comb	800	Guide Plus (GemStar)	MBR3466Z	221-01389	206-03483
IQB32B84R	32	Console	15	Comb	650	V-Chip+	MBR3464Z	221-01388	206-03482
IQB32B86R	32	Console	15	Comb	650	V-Chip+	MBR3464Z	221-01388	206-03482
IQB36B42W	36	TableTop	15	Comb	650	V-Chip+	MBR3464Z	221-01388	206-03482
IQB36B44W	36	TableTop	15	Digital Comb	800	Guide Plus (GemStar)	MBR3466Z	221-01389	206-03483
IQB36B86R	36	Console	15	Comb	650	V-Chip+	MBR3464Z	221-01388	206-03482

Table 13-1 B-Line Models

The CB variation (figure 13-2) has also been developed for use in 25-inch through 36-inch televisions for the Sentry 2 line, making it very similar to its CA sibling. The CA and CB chassis are not the only televisions in the B-Line. I am aware of the CN-001C and CN220A chassis that are found in certain 13-inch and 19-inch sets. I am also aware of the CTV-6091 that Zenith simply calls a "small-screen color TV." If my information is correct, at least some of these chassis are "purchased" ones, that is, manufactured by someone other than Zenith to be sold under the Zenith name. I consider them beyond the scope of this book and won't spend time dealing with them. However, Zenith's technical information about these TVs is remarkably clear and a more-than-adequate servicing guide.

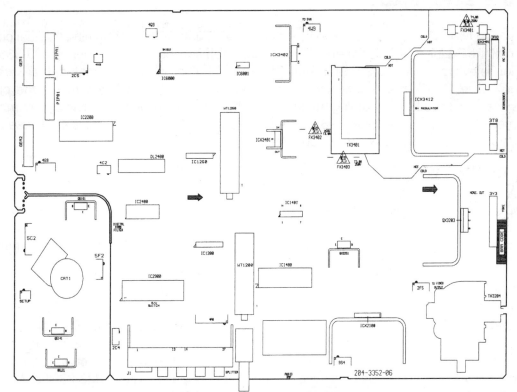

Figure 13-2 CB-Chassis Variation

Moreover, the CA and CB chassis have a number of variations. If you look at the service literature, you will find discussions for four members of the CA family, each based on the microprocessor used in the particular module. Those microprocessors are 221-1384, 221-1385, 221-1386, and 221-1387. The CB chassis makes use of two, 221-1388 and 221-1389. As you might expect, the customer menus, service menus, and features also vary depending on the microprocessor.

Since we are looking at a minimum of six microprocessors, I am going to take a cue from the B-Line training material and use a generic-style approach to the discussion of the B-Line products. That is, some of the circuits and features I talk about won't apply to every model in the B-Line but will apply to most.

Now let me conclude this introduction on a somewhat personal note. The B-Line products have not been out long. As a matter of fact, Zenith held its annual training meeting for my area in Memphis in the fall of 1999 at which point we were given the training material. The service manuals began to arrive at about the same time. This means I have very limited experience with B-Line products and must write on the basis of what I have learned "from the book" and my discussions with other techs as opposed to largely hands-on experience. I hope you keep this in mind as we proceed.

Since the CA chassis contains circuits that we have seen and discussed many times, I won't give you its schematics, saving the available space for the schematics of the CB chassis because Zenith has done some new things with it. As you read through the material for the CB chassis, you might have questions I haven't answered. I can't answer some of them even for myself because I don't have as much information on the new circuits as I would like. Certain voltages can't be provided because they are unavailable in the service literature, which I hope is an oversight Zenith corrects before the B-Line CB chassis becomes history.

Available Literature

I can point you to two training manuals and the usual service data. The first training manual contains a discussion of the B-Line's features, menus, and circuits. You may obtain it by ordering 923-3351TRM (B-Line Color Television: CA and CB Chassis). The second piece of training material, *B-Line Color Television: CA and CB Chassis Schematics* (part number 923-3401-TRM), contains the clearest schematics I have ever seen in a training manual, and I have seen lots of them. The part numbers for the factory service manuals are 923-3396 (CA chassis) and 923-3397 (CB chassis). These are the basic ones, but there are several smaller supplements.

Customer Menu

Please keep in mind the procedure I outlined in the opening paragraphs of this chapter and remember that menu features vary depending on the part number of the microcontroller. I am going to pass over the menus that are similar to those we have looked at and focus on the features that have either a new look or are brand-new.

Setup Menu

The Setup Menu page lists these options: EZ Program, Add/Del/Surf, Clock Set, Captions, Captions/Text, and Language. We have encountered each feature in previous sets except EZ Program. It has been described as a feature that "... automatically searches for all available channels and marks them as 'Added' so that they may be accessed via the channel Up/Down key" (923-3351TRM, page 2-5). EZ Program, it seems, is another name for "auto program," but there is more.

If you select EZ Program, a message at the top of the screen highlights the request while a message at the bottom informs you that need to use the Up/Down and Left/Right Arrow keys for auto program selections. You are also told whether the current RF input is off-the-air antenna or cable TV. Use the Up/Down keys to change the RF input if it doesn't suit you. When you are ready to proceed with auto program, press the Left/Right key, and auto program commences.

There are a few points about the EZ Program feature of which you should be aware: (1) It clears all surfing channels when you run it; (2) it disables all keys except the power key; (3) it clears the factory mode if it has been turned on; and (4) if it finds no channels, it gives you the message, "Make sure that the cable/ant. is connected and try again". If the EZ Program line in the setup menu happens to be red, the module has not passed one of factory tests and might have to be replaced. Just make a note of this oddball fact in the event you need it.

Add/Del/Surf has also been around for a while, but you might need your memory refreshed. After you run EZ Program (auto program), the available channels appear marked by the word "added." However, some of the channels might be scrambled or undesirable for other reasons. The customer may then choose to use Add/Del/Surf to delete the scrambled and-or undesirable ones. Of course, the surf feature permits the user to set up a list of favorite channels for quick scan.

Clock Set is certainly not a new feature, but it has a new twist because it contains a clock set and a date set option. Let's look first at the clock set feature. When you enter it, you must choose between "manual" and "auto." The manual option lets you set the time yourself while the auto option lets the TV set the time using information taken from the XDS packet that rides on line 21 of field 2 of the video information. If you choose the auto option, you also have the choice of setting the time zone and selecting Daylight Saving Time (turning it on or off).

The Captions/Text option has been expanded to include captions 1 through 4 and text 1 through 4. There are three things about this feature of which you should be aware: (1) If you select the text option and a text box appears, pressing the CC button on the remote control causes the caption box to appear on the screen; (2) if you press one of the volume control keys while you are using the captions, you won't get the familiar volume bar; and (3) if you aren't using the captions feature and press the CC button, the captions box appears offering you the choice of turning on or off "Captions When Muted". The last option displays the caption text on the screen when the TV has been muted permitting the user to read the captions and watching the picture while talking on the phone.

Special Features Menu

Special Features contains EZ Timer, Channel Labels, XDS Display, Parental Control, EZ Help, and EZ Demo. Some are new; some are not. For instance, EZ Timer is the access point for the sleep timer and on/off timer. Channel Label has also been around for a while, but Zenith has expanded the list of available labels and added a feature that permits the TV itself to add the labels from the broadcast signal. Selecting this option activates the auto channel option that means the program automatically takes the channel labels from the XDS (Extended Data Service) feature. If you don't want the channel labels added automatically, select the option labeled "None." The XDS choice has two settings, on and off.

Parental Control

The much-touted V Chip technology makes parental control possible. I assume you know that the microprocessor incorporates those features that let it perform the tasks assigned to the V Chip, meaning there is no chip on the chassis that is labeled V Chip. You should also know setting up the feature is rather involved. Figure 13-3 gives you an idea about how Zenith has configured the technology for use in its products.

V Chip technology takes advantage of information transmitted via the XDS (Extended Data Service) choice. Part of the XDS broadcast consists of ratings for each program being broadcast. If you expect to use V Chip technology, you must become familiar with the rating systems on which the technology depends. The system uses two ratings, the MPAA and the TV Parental Rating System.

The rating system of the MPAA (Motion Picture Association of America) looks like this:

G	General Audience	Content not offensive to most viewers
PGP	Parental Guidance	Content is such that parents may not want their children to view it
PG-13	Parental Guidance Suggested	Program is inappropriate for preteens, with a greater degree of offensive material than PG-rated programs
R	Restricted Viewing	Not for children under age 17. Strong elements of sex and/or violence

NC-17	Restricted Viewing	Not for children under age 17 under any circumstances. Strong sexual content
X	Hard-Core Films	Same as NC-17 rating

The TV Parental Guideline Rating System looks like this:

G	General Audience	Content not offensive to most viewers
TV-G	General Audience	Suitable for all audiences; children may watch unattended
TV-PG	Parental Guidance Suggested	Unsuitable for younger children because it may contain suggestive dialogue, bad language, sex, and violence scenes
TV-14	Parents Strongly Cautioned	Unsuitable for children under 14 because it may contain strong dialogue, bad language, sex, and violence scenes
TV-MA	Mature Audience Only	Adults only. May contain strong dialogue, bad language, sex, and violence scenes
TV-Y	Children	Considered suitable for all children under seven years old
TV-Y7	Children seven and up	Considered suitable for children over seven and may contain fantasy violence scenes

Aux Block, MPAA, Age Block, Content Blk, Set Hours, Set Password, and Lock On/Off are found in the Parental Control submenu. Since these categories are brand-new, let me give you a brief explanation for each one.

Auxiliary Block lets the user block use of the auxiliary TV inputs.

MPAA permits the customer to select having the channel unblocked G and above, PG and above, R and above, or X.

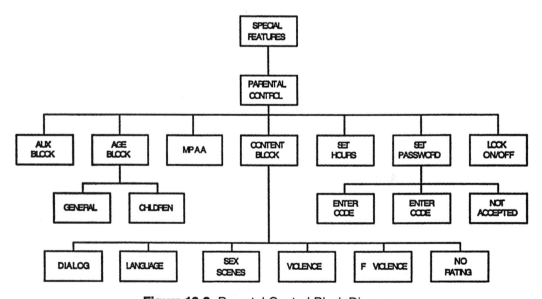

Figure 13-3 Parental Control Block Diagram

Age Block has two options, General and Children. Using the General option, the parent may select Unblocked, TV-G and Above, TV-PG and Above, TV-14 and Above, or TV-MA. Or the parent may enter the "Children" option and select among Unblocked, TV-Y and Above or TV-Y7.

Content Block has these categories under it: Dialogue Block, Language Block, Sex Scenes Block, Violence Block, F Violence Block, and No Content Block. When the parent enters this submenu, he or she must choose " block" or "unblock" for each of the categories. If "block" is selected, the parent must then use the TV Parental Guideline Rating System to decide what the children may see and hear.

Set Hours determines how many hours the Parental Control feature will be active. The range is 1 hour to 99 hours. The timer must time itself out once it has been set and activated unless it has been turned off using another feature in the menu.

Set Password permits the user to select a four-digit code that he or she must use to activate or deactivate Parental Control.

Lock On/Off is the final submenu category. It turns the Advanced TV feature on or off. It can be turned on only if the hours are greater than zero and the password has been set. If you try to turn it on without one or both of these features having been activated, you will see the error message, "Must Set Hours" or "Must Set Password."

If someone tries to access a blocked channel, the message, "Parental Lockout Is Active," flashes on the screen and is followed by the amount of time in hours the lockout remains effective. Once the parental control feature has been activated, the software won't permit immediate entry into the Advanced TV Control Menu. If you try it, you will be asked for the password. When you enter the correct password, "Lock On/Off" appears on the screen. If you don't enter the correct password, the error message, "Not Accepted!" flashes on the screen and stays there for five seconds before it disappears. During those five seconds, you will not be permitted to reenter a password.

Let me repeat myself for emphasis. There are just two ways to defeat parental control when the hours have been set and the feature turned on: (1) Use the password to turn the feature off, or (2) enter and exit the service menu. Otherwise, the feature has to time itself off. Unplugging the TV won't work! You may be tempted to tell your customer how to enter and exit the service menu, but I strongly urge against it. That kind of information MUST be kept out of consumer hands for their benefit. It's up to you to make a service call at which point you must decide whether to charge for your services or, in the case of a good customer, write it off as a pro bono exercise.

EZ Help gives the TV user a bit of assistance by giving brief explanations for each of the customer menu functions and a suggestion or two about how to solve certain common problems. For example, the "EZ Program" feature gives a brief explanation for the menu options. Its main screen, however, has a feature called "EZ Help" that offers solutions to the most common problems TV users have, like how to make the correct antenna connection to the tuner.

EZ Demo is a sales feature designed to show off the icon menus and other icon displays while the set is on the dealer's showroom floor. This feature has about 23 items in its display.

All in all, Zenith has designed and produced a smart television that offers a host of really neat features and should be fun both to play around with and to use. Based on my experience of the past three years, I think Zenith has also manufactured a reliable product that gives the buyer a good product in exchange for money spent. It is up to servicers like you and me to become as familiar with the product as we can, which is why you are reading a book that I have written.

Audio Menu

Most of the options in the audio menu are self-evident, like bass, treble, balance, audio mode, and speakers. Others probably need a clarifying statement. EZ SoundRite is another name for SoundRite. But, the feature itself remains unchanged. EZ Sound has six selections from which the customer may choose: custom, normal, stadium, news, music, and theater. Each selection makes changes in the audio to suit a particular audience and environment. For example, the stadium setting makes the audio sound as if it originates in a large, open-aired arena with tiered seats. It is suitable for watching sports and rock concerts. The theater setting mimics a movie theater. If you do audio work I am sure you are familiar with each of these.

Video Menu

The video menu has the usual listings plus one called EZ Picture. It allows the customer to maintain separate settings for contrast, brightness, color, tint, and sharpness. "Preset," the first choice, lets the customer use those setting that have been established at the factory while "custom" permits the user to establish his or her preferences.

PIP Menu

The PIP Menu contains adjustments for contrast, tint, and size.

Source Menu

The Source Menu has options for antenna, cable, and video (jack pack input) under the "main source" feature. The PIP Source lets the user select the PIP source for the main picture.

Service Menu

Beginning with the B-Line, Zenith changes terminology and calls the service menu the "Installer's Menu." It is now accessed only by the remote control (Menu-9-8-7-6-Enter). The "installer's menu" comes up the usual way displaying the software part number at the top, H Pos (item 3) in the center, and the date of manufacture and test status at the bottom. Once you get into the menu, turn the factory mode on to gain access to all of the items and proceed the way you usually do. Since they are important, I give you the complete service menus in tables 13-2a through 13-2f at the end of the chapter. Pay close attention to the note at the bottom of some of the pages telling you that when you replace an EAROM replace the microprocessor with an updated version. The level was set at the factory for some of the modules and cannot be changed but, you have to set the level on others.

Servicing Notes

Let's begin with a look at the CA chassis.

If you do warranty service, you are expected to repair to the component level the 19-inch through 27-inch chassis. However, you may get permission to replace the module if you encounter an extraordinarily difficult problem. You are expected to replace the module for the 32-inch and 36-inch sets, but you may repair it to the component level if that means giving quick service and satisfying the customer. I do suggest that you get permission before you do the unexpected, like repair a module that should be replaced, especially if you expect to get paid for the job.

When you replace the picture tube or the video processor IC, you need at least to check the black and white tracking and reset it if necessary. These adjustments are made through the service menu.

The B-Line Chassis

If you replace the MTS stereo module, be sure to unplug the TV and plug it back in to reset the microcontroller. The software has an auto detect feature that runs at reset. If it doesn't reset, the microprocessor won't recognize the new module, meaning the TV acts as if the new module wasn't present. After you have reset the microprocessor, enter the service menu and make sure the stereo adjustments are set properly by comparing them with the bar code on the module itself.

When you troubleshoot a video problem, remember that all video travels through the PIP (if the set has a PIP) and the video switching IC. A defect in either circuit has a profound effect on the video.

Finally, NEVER disconnect the video output module from the CRT when you are working on the 32-inch and 36-inch receivers. I know I discussed this point in the past, but it is so important that you need to be reminded of it again. Nobody wants to damage one of those large tubes and have to pay for a replacement! Nor do you want to create a situation where you could receive a dangerous electric shock.

That said, let's now shift our attention to the IF and audio alignment procedures.

Following recent procedures, if you encounter a set with no viewable picture, access the service menu and check the default settings for RFAGC and PIFVCO.

Remember, the item number varies according to the microprocessor installed in the module on which you are working. The preset value for the former is 42 while the preset value for the latter is 52. When you adjust the AGC delay, apply a strong signal to the tuner and adjust the RFAGC delay downward until the signal gets noisy. Then, increase the setting until the noise disappears. If the setting goes much above 50, the tuner might overload and cause beats to appear in the picture. When you adjust the PIFVCO, input a good off-the-air signal into the tuner, place a high impedance meter at pin 44 of ICX22200, and adjust the register for reading of 2.5 volts on the meter. You see, there is nothing complicated or time consuming about these adjustments. Once you have done one or two, they become "old hat."

If the set exhibits a no-audio symptom, access the service menu and check the default setting for Audio Level (AUD LEV). Tune in a good off-the-air signal, place a meter at pin 54 of ICX2200, and adjust LX202 for a reading of 4 volts on the meter. Remember that its preset level is 46.

The MTS stereo decoder alignment is a bit more involved, and the instructions Zenith gives are not entirely clear. Moreover, you need a very good signal generator to do the alignments correctly. If you need to make these adjustments, read – very carefully – the instructions on pages 3-6 and 3-7 of manual 923-3396. However, you probably won't ever have to do one.

The G2, focus, and black and white tracking adjustments are performed the same way they are done for the Z-Line and A-Line products. The differences are the register numbers of the items you are expected to adjust in the service menu.

Now let's take a look at the CB chassis.

The various IF alignments pertain to the main IF (IC1200) and the PIP IF (IC1250). Unlike previous chassis where you make alignments via the service menu, you make mechanical adjustments when you align these items in the CB chassis!

If there is no picture on the screen, check first the adjustments for the IF circuits. Let's begin by aligning the VCO. Select the "off-air" input in the Setup Menu. Next, place your high impedance meter at pin 18 of IC1200. Then without any signal present, short pin 13 of IC1200 to ground and make a note of the reading on your meter. Finally, remove the short, tune in a good signal, and adjust coil

L1202 until you get the same voltage reading at pin 18 as you did when you shorted pin 13 to ground. Now repeat the same procedure for the PIP IF, remembering that you are now working with pins 13 and 18 of IC1250 and adjusting coil L1254. It's a good idea to make the adjustments and repeat them at least once to confirm that they are correct. Now move from the IFVCO to the AFC. Adjusting it means that you are once again adjusting a coil and not a service menu item. Begin by monitoring the voltage at R1206 (not at pin 14 of IC1200) and adjust coil L1207 for a reading of 3.0 volts plus or minus 0.1volt. Repeat the procedure for the PIP IF using R1268 (not pin 14 of IC 1250) to monitor the voltage while you adjust coil L1253 for the reading of 3.0 volts plus or minus 0.1 volt. It is important that you get the adjustment as close to the stated reading as you can. I suggest that you put as little pressure on the core of the coil as possible because excessive pressure increases the chances of misalignment.

After you have completed the adjustments to the VCO and AFC circuits, move your meter to pin 1 of the tuner (WT1200/WT1210) and prepare to adjust the AGC delay. Begin by adjusting the AGC pot until you get the maximum possible voltage reading on the meter. Then, reverse the pot until the maximum voltage begins to decrease. Continue to adjust it until you get 0.5 volts less than the maximum reading.

The last alignment concerns the MTS decoder. Before you start it, let the TV warm up for at least 10 minutes. Begin by inputting an IF signal of 30 dbmy modulated at 100% with a 100 hertz tone at 25 kHz deviation. Complete the procedure by adjusting the A ATT register for a reading of 490 millivolts at pin 44 of IC1400.

The G2 adjustment procedure is unlike anything we have encountered so far. It requires special handling because the "auto tracking system" (AKB) won't work unless the G2 control has been properly adjusted. Make the adjustments in this order:

(1) Create a short circuit at the SETUP connector on the video output module by connecting a jumper between pins 1 and 2. If I read the schematic correctly, you are placing the emitters of the video output transistors at ground potential.

(2) Adjust the G2 control until retrace lines appear on the screen. Then decrease the G2 setting until those lines disappear or are just barely visible.

(3) Remove the jumper you installed at the SETUP connector.

(4) Set the color level at a minimum and the brightness and contrast to their midrange settings.

(5) Inject a color bar pattern into the tuner. The gray tones should stand out, the last white bar should not be saturated, and the first bar should be completely black. Remember, you have turned the color off via the customer menu.

(6) Enter the service mode with factory mode set to 01 and press the Swap key on the remote control to open a status window at the bottom of the screen. Check the second digit on the second data row for a reading of "1." The "1" indicates that the AKB (auto track system) is working.

(7) Press Swap key again to close the status window. Set Factory Mode to 00, and exit the service menu by pressing Enter on the remote control.

If you change the CRT or the chassis, you need at least to check the RGB cutoffs and adjust them if they aren't set correctly. The procedure is like those we have looked at in previous chassis, but there are important differences. I'll try to simplify the steps involved by putting them into a 1-2-3 order:

(1) Connect a signal generator to the antenna port and select a flat white signal.

(2) Access the customer menu and select the "Medium" color temperature. Set tint and brightness to midrange (about 31), and turn the color level to minimum.

(3) Select contrast and turn it down. Then increase the contrast level until the image on the screen has a grayish tone.

(4) Observe the color that predominates (red, green, or blue) and keep its setting in the cutoff register at 0. Keep the gain registers at the values specified in the service menu tables for the chassis on which you are working.

(5) Select the medium color temperature if you haven't already done so and enter the service menu. Set the cutoff and gain registers according to the settings given in the service menu. Check the picture to see if the gray color has a kind of smog appearance. If it doesn't, adjust the cutoff registers without moving the predominant color register to obtain the smoky gray color.

(6) Input a color bar pattern from your generator. The color should still be off making the color bars from the generator appear as bars of various shades of gray and black. If the brightness is too high, readjust the RF brightness register. According to the service menu (page 3-3), Aux Brightness is the same as RF Brightness register. Then readjust YUV brightness till it reads seven points (-7) less than the RF brightness register. If it reads less than 0, set the register to 0. By the way, "Y" is luminance; 'U" is"B-Y," and "V" is "R-Y" – a sample of the new terminology we must also learn.

(7) Go to EZ Picture and select "Normal Picture." Then select "Cool" from the Video menu and adjust the cutoff and gain registers by adding and offsetting values according to this table:

Register	27v	32v	36v
Red Cutoff 2	-1	-2	-1
Green Cutoff 2	-1	0	-1
Blue Cutoff 2	2	4	4
Red Gain 2	-10	-6	-6
Green Gain 2	-5	-5	-5
Blue Gain 2	0	0	0

(8) Select "Warm" as the color temperature from the Video menu and adjust the registers according to these values:

Register	27v	32v	36v
Red Cutoff 3	5	4	2
Green Cutoff 3	1	-1	-1
Blue Cutoff 3	0	-2	-2
Red Gain 3	1	10	9
Green Gain 3	6	2	1
Blue Gain 3	-4	-4	-4

(9) Finally (!), select "Medium" as the color temperature and adjust it according to this table:

Register	27v	32v	36v
Red Cutoff 1	9	11	12
Green Cutoff 1	2	2	3
Blue Cutoff 1	0	2	2
Red Gain 1	48	40	41
Green Gain 1	37	35	35
Blue Gain 1	31	31	31

Servicing Zenith Televisions

Power Supply

As I stated in the introduction, I decided to offer only the schematics for the CB chassis because that's where the new technology is found. So, let's begin with the circuit that usually gives more problems than the others, namely the power supply. When you look closely at figure 13-4, you see that Zenith has produced a new power supply that they refer to as a "quasi-resonant" type of switching power supply (Training Manual, page 5-42). I'll introduce you to it in as logical a way as I can.

AC enters the TV at 3R8, and goes to the bridge rectifiers via FX3401, LX3401, CX3401, and CX3402. The last three components comprise a filter that reduces noise riding on the AC voltage, but its chief function is to keep "hash" generated by the power supply from being coupled into the AC line. Note that FX3401 is rated at 4 amps, 250 volts and is a fast-blow fuse. Capacitor CX3407 completes the circuit. Its size varies from 330mfd for the 25/27-inch sets to 470mfd for the 32/36-inch ones. The 150-155 volts DC the input circuit develops is applied to ICX3412 through one of the primary windings of TX3401 (pins 18 and 11) and through RX3416 to the start-up circuit.

Since it is the heart of the CB chassis's power supply, let's take a pin-by-pin look at ICX3412. B+ (raw B+) loops through TX3401 and goes to pin 3 of ICX3412 (the drain of the MOSFET inside the chip). But it won't start until a "start-up voltage" has been developed and applied to pin 4. When the set is first plugged in, CX3410 begins to charge through start-up resistor RX3416 (47k, 3 watts). After the charge reaches about 16 volts, the regulator enables and draws about 20ma from CX3410 causing the voltage across it to fall momentarily. The IC begins switching action and outputs a series of pulses to pin 18 of the transformer. The switching action also causes a series of pulses to appear at pin 13 of TX3401. Diode DX3407 rectifies these pulses to provide a DC voltage that recharges CX3410 and becomes the run B+ for the chip. Every switching power supply has to have a control circuit to ensure that the output of the power supply remains constant over a range of input voltages. (According to specifications, this power supply works best as long as the AC input falls between 90 and 135 volts.) ICX3403, an optoisolator, and a few associated components handle the job of regulating it to provide the required outputs.

Let's take a closer look at these "associated components" by looking first at those on the "cold side" of the power supply. Neither the training material nor the service manual tells you much about what you find here, and neither piece of literature supplies you with what I consider some much-needed operating voltages. However, we can garner enough information to see how the thing works.

ICX3401 is a three-terminal voltage regulator. Pin 1 is tied to the +130-volt line, pin 3 to ground, and pin 2 to pin 2 of ICX3402 and the cathode of ICX3406. The latter is a relatively new device in the electronic world. Called a "precision shunt regulator," it is a remarkable little device. You might think of it as a "gated" zener diode or an infinite gain operational amplifier with a reference voltage tied to the negative input. The voltage at the "R" terminal controls the current flowing through it and is therefore "the control voltage." The literature doesn't tell us what the control voltage ought to be, so let's assume one for the sake of illustration. Thomson uses a precision shunt regulator in their new CTC203 chassis and lists the control voltage as a nominal 2.5 volts DC. So, I assume – for the sake of illustration – that the control voltage is 2.5 volts DC. When the voltage at the R terminal drops below 2.5 volts, ICX3406's internal impedance decreases permitting current through it to flow more freely. When the control voltage rises above 2.5 volts, its internal impedance increases "pinching" off current flow. Since the voltage drop across RX3408 develops the control voltage, the current through ICX3406 is proportional to the voltage sensed across RX3408. Because the voltage drop across RX3408 is proportional to the applied voltage, the voltage at the "R" input follows the ups and downs of the +130-volt line and causes the LED inside the optoisolator to illuminate accordingly.

The "hot side" of the optocoupler applies the sensed current (that is proportional to the output voltage error signal) through RX3405 where it is summed with the current through RX3403 that is proportional to the load. The literature points out that the bias signal across RX3405 is close to the threshold voltage of pin 1. Since the threshold voltage at pin 1 is tight, the power supply turns on a constant current sink to prevent false shutdowns when the set first fires up.

Keep in mind that the control voltages on which the power supply depends are quite small, meaning just a fraction of a volt has a significant impact on the operation of the power supply. I advise you to be careful when you probe around these small components because a slip of a probe might do more damage than you imagine. I wish I had more information to share with you, but I don't. However, I'll give you what I believe are some useful part numbers before concluding this section. I suspect that you ought to replace all of these critical components when you have a catastrophic power supply failure even if they don't appear to be damaged:

ICX3412	F-53857	mounted on a heat sink
ICX3403	162-00167-05A	optocoupler
ICX3406	221-00265-03A	shunt regulator

Now let's look at the voltages produced by our "quasi-resonant" switching power supply. When it is up and running, the power supply develops four standby voltages: +130 volts, +12 volts, +24 volts, and +15 volts. The first sources the horizontal deflection circuit. The second supplies +12 volts for the chassis as well as the +5 volts for system control via ICX3402. The third provides power for the audio amplifier. The fourth provides B+ to the video and IF sections of ICX2200. The last three sources are fuse protected. The horizontal deflection source doesn't need a fuse because it has overcurrent protection.

When the TV Turns On

The power supply switches from standby to full power when the microprocessor outputs a high at pin 32 to turn Q3402. When it turns on, Q3402 drives Q3404 and Q3405 into saturation making +15 and +9 volts available to the rest of the chassis. Have you noticed that the +12 volts switched is derived from the +15 volts standby via ICX3405?

If the TV doesn't turn on but the power supply is up and running, make these checks. I am not listing the voltages in any particular order. Let your common sense dictate where you start:

power on high at pin 32 of IC6000
power on signal coupled to the base of Q3402
power on signal coupled to the collector of Q3402
+15 volts DC at C3433
+12 volts DC at C3434
+9 volts DC at L3409

Degauss Cycle

The CB chassis employs a new degaussing method. When it receives an on command, pin 33 ("Deg-Ctl") of the microcontroller goes high. The high turns Q3403 on which actuates relay KX3401 and permitting current to flow through the degaussing coil. The circuit turns off in about a second, completing the degaussing cycle.

Servicing Zenith Televisions

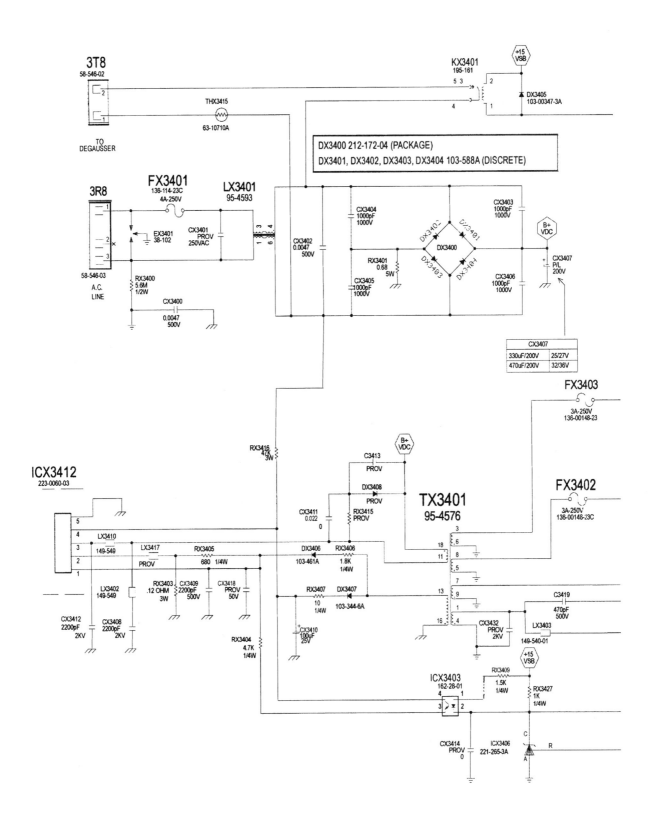

Figure 13-4 Power Supply

The B-Line Chassis

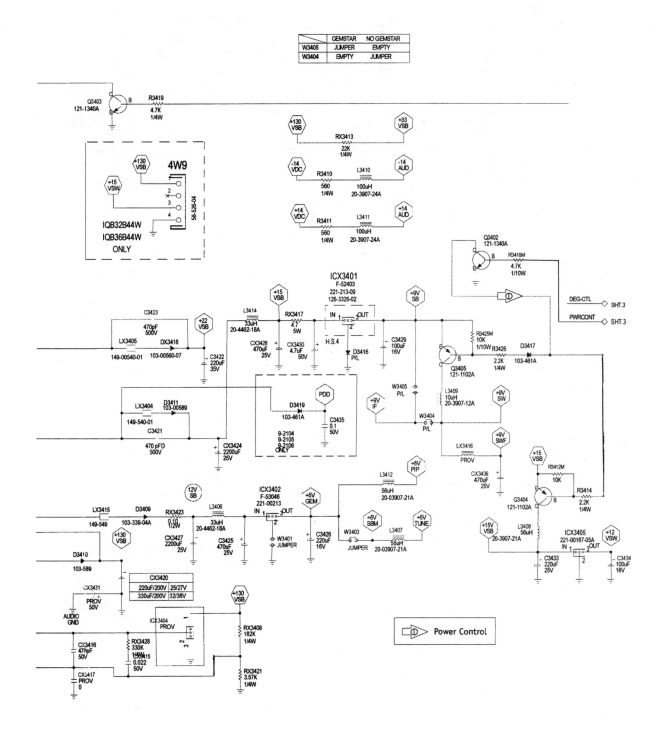

Figure 13-4 Power Supply (Continued)

Servicing Zenith Televisions

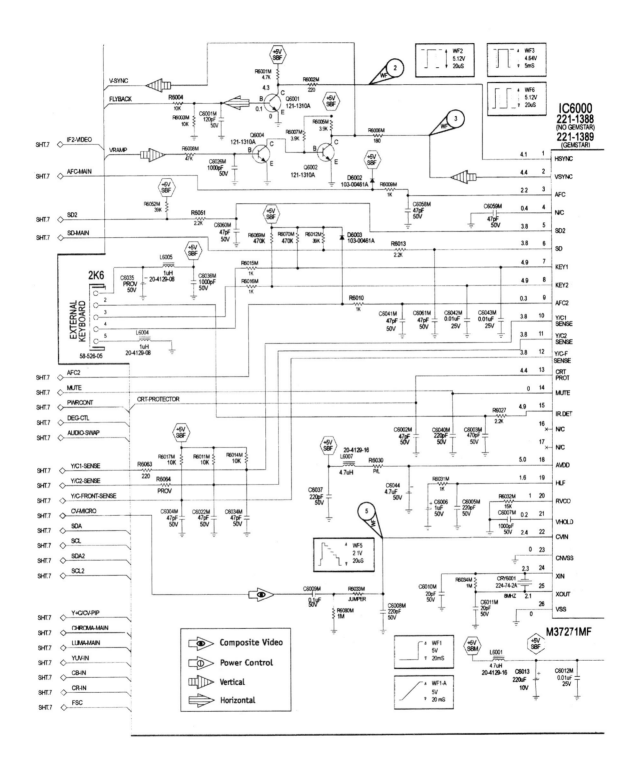

Figure 13-5 Main Microprocessor

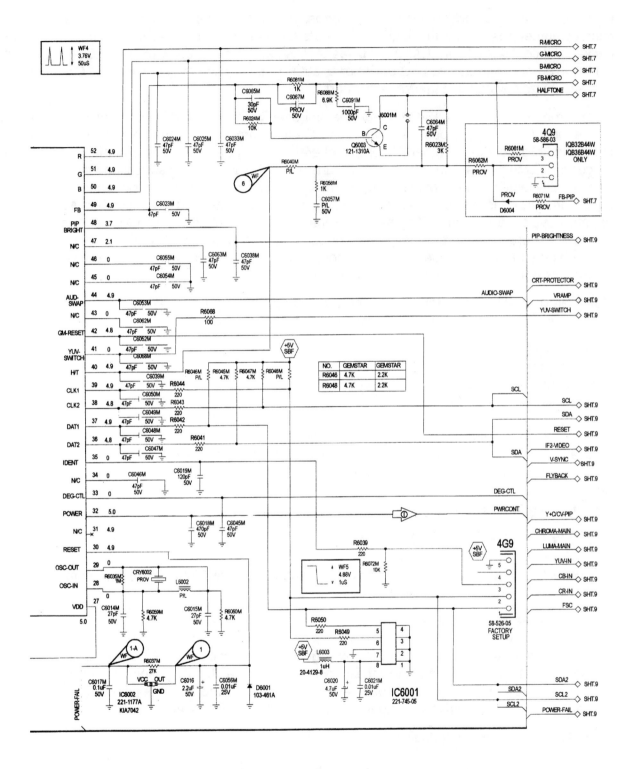

Figure 13-5 Main Microprocessor (Continued)

System Control

IC6000, the microcontroller for the CB series, is a marvelous little device. It is a single-chip microcomputer designed with CMOS silicon gate technology and comes in the familiar 52-pin DIP package. It contains 64k of ROM, 1024 bytes of RAM, and is powered by the usual 5 volts DC (within a 10% tolerance). It has two LED drive ports, a I2C bus interface, a data slicer, an OSD generator, and an eight-bit by one channel serial I/O port.

The microprocessor (figure 13-5) controls the entire television by communicating with the other chips on the module via the I2C bus. The bus allows the microprocessor to read or write data to and from the control registers of the other ICs or "peripherals." The "peripherals" include the video and audio processors, both tuners (if the set has a two-tuner PIP), the audio/video switch, and the EAROM. You are naturally familiar with all of this from your reading and experience, but the microprocessor used in the CB chassis performs some of these chores just a bit differently. You have seen how it manages the degauss cycle via software. That's one difference from the chassis we have seen in the last 12 chapters. There are others. For example, it controls vertical size and width by outputting a PWM pulse on pins 47 and 48 respectively. The more you become familiar with it, the more you see subtle differences between it and previous microprocessors.

However, it is still a microprocessor, and you troubleshoot it exactly as you would any other microprocessor. I think you will find these guidelines helpful:

IR receiver input at pin 15
keyboard input at pins 7 and 8
serial clock and serial data at pins 36 and 38 and pins 37 and 39
+5 volts at pins 18 and 27
reset at pin 30
horizontal flyback pulse at pin 1 (necessary for timing)
vertical pulse at pin 2 (if missing, the microcontroller will shut the set off)
R, G, and B output at pins 52, 51, and 50, respectively
ground at pin 29
oscillator at pins 24 and 25

Deflection Circuits

The deflection circuits contain something old and something new (figure 13-6). I will comment only on the new, most of which is found in the vertical deflection circuit.

Horizontal Deflection

Horizontal drive exits pin 19 of ICX2200 and goes to the base of Q3201, an emitter follower that provides a bit of current gain and the necessary impedance matching and isolation the circuit requires. The signal then scoots along to Q3202 (the horizontal driver) where it is amplified and applied to the primary of the driver transformer, and the rest you ought to know by now.

The Sweep-Derived Voltages

Horizontal sweep develops five voltages. Use this chart to check for them:

+14 volts at the junction of CX2101 and RX2124
-14 volts at the junction of CX2100 and RX2125

+35 volts at the junction of CX3253 and DX3252
+235 volts at the junction of CX3208 and RX3208
+33 volts at the junction of RX3216 and ZD1230

Check the CRT filament voltage at pins 1 and 2 of collector 2F5. If you use a true RMS meter, look for a voltage of about 6.1 volts AC.

Horizontal Shutdown

The HEW circuit takes a sample pulse from pin 6 of TX3204 (the flyback), rectifies and filters it, and compares it to a reference voltage fixed by ZDX3004 at transistor QX3002. When the high voltage reaches its maximum permitted value, QX3002 conducts causing QX3001 to conduct. When it turns on, QX3001 attenuates the flyback pulses arriving at pin 18 of ICX2200. If the attenuation causes the flyback pulses to drop to a volt or less, the TV enters the shutdown mode. When shutdown occurs, the microcontroller won't permit the TV to turn on until it has been reset by unplugging and plugging the set back in. The circuit is different from the XRP circuits in previous chassis, but it achieves the same results. It's also another circuit that you will have difficulty troubleshooting unless you use a scope.

Vertical Deflection

ICX2100, the vertical deflection IC, incorporates in one-chip thermal protection, a pump-up stage, and a power amplifier. Of course, none of this is new, but the manner in which it is configured is new. For example, Zenith uses two vertical signals to drive it. Both signals originate inside ICX2200 and leave at pins 13 and 14. The signal that originates at pin 13 is labeled "VDP" while the signal that leaves pin 14 is labeled "VDN." Both must be present for the vertical circuit to operate properly. May I assume that you noticed a voltage supply that consists of a positive and a negative voltage?

How do you troubleshoot the new circuit? It's not as difficult as a first glance might lead you to believe. Simply check for both vertical drive ramps at pins 4 and 5, the +14 and -14 volts at pins 6 and 1, and a vertical output ramp at pin 2. But before you change the chip, take the time to check the voltages at the other pins. A component tied to one of them might have caused the IC to stop working. Common sense, isn't it?

By the way, the part number for the vertical output IC is F-53864. When you order it by that number, you get the IC mounted on its heat sink. I don't have a part number just for the IC.

CRT Protection Circuit

Following recent trends in the industry, the CB chassis incorporates a CRT protection circuit that shuts the TV down when vertical deflection fails. The circuit is necessary to prevent, in its mildest form, a phosphor burn across the face of the tube or, in its worst form, a broken CRT neck. The protection circuit consists of C2108, D2102, D2103, R2108M, R2109M, R2112M, and R2113M. It taps a vertical pulse off pin 7 (pump up) of ICX2100, rectifies and filters it, and sends the resulting voltage to a voltage divider network the output of which is applied to pin 13 of IC6000. During normal operation, pin 13 reads between 3.6 and 5.1 volts. If the voltage drops outside the range, like 0 volts, the microprocessor responds by turning the TV off three seconds after the defect has been detected and will not permit the TV to be turned on until it has been reset. It almost goes without saying that a failure in the protection circuit also results in a set that shuts down three seconds after turnon!

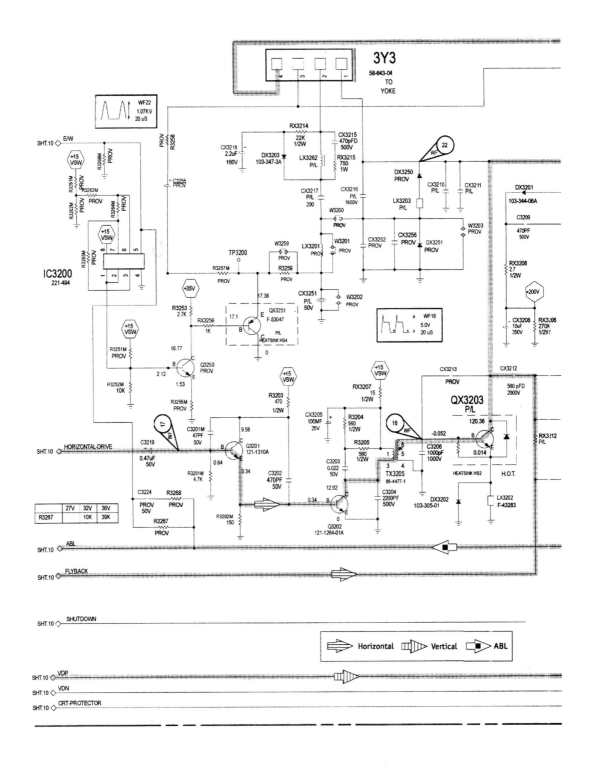

Figure 13-6 Deflection Circuits

The B-Line Chassis

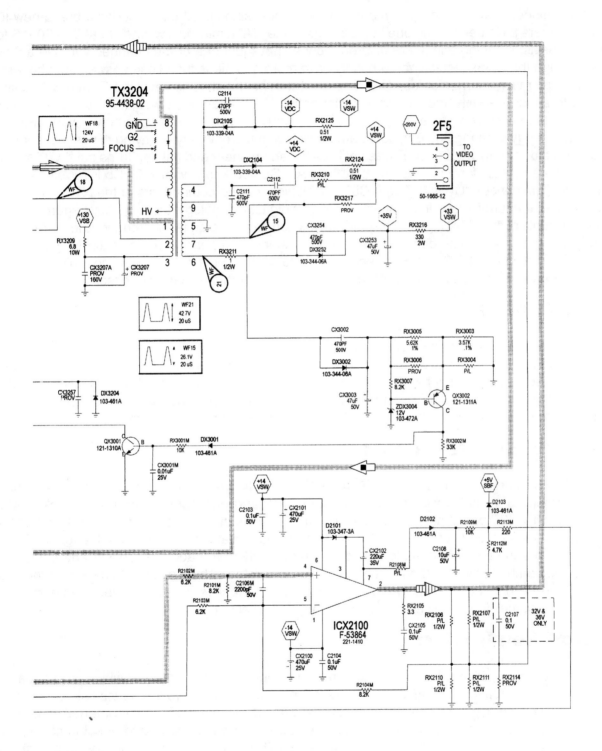

Figure 13-6 Deflection Circuits (Continued)

Video Processing System

Zenith's engineers have come up with a new video processing circuit that includes a brand-new IC. Their engineers have been using one that belongs to the "TA" family but switched to a CXA2061S for the CB chassis, meaning things are different now. That's just one of the changes you see as you move through the video circuits. You probably have more questions than I can answer, but I honestly don't have as much information as I like. I do hope that as the chassis develops, Zenith will be kind enough to supply us with "the rest of the information." For example, I would like to have a block diagram of the new video processing chip because I have difficulty visualizing how it works without one. Terminology is changing. Engineers are familiar with the new way of saying things" but some of us aren't. I didn't know what a "LOT" was until I called an engineer and asked. Turns out "LOT" stands for "line operated transformer." We call it a "flyback." Do you know what Y Cr Cb is? Well, "Y" is of course "Y" (luminance), "Cr" is "R," and "Cb" is "B" which are the components of the NTSC signal. If you deal with the terminology introduced by the new digital technology, then you have seen Y Pr Pb, the components of the ATSC signal. The digital revolution has led to quite a few changes in terminology. So has the global economy because we are now seeing European terms crop up in our service literature. Incidentally, "LOT" is a European term. I do wish manufacturers would not only document the new technology but also provide us servicers with explanations of the changes.

Main Tuner IF

One of the really big differences between the CB chassis and the circuits Zenith has been using shows up right here. Take a look at figure 13-7. IF from the tuner goes to IC1200 (a LA7577) where it is processed to obtain baseband audio and composite video and routes both signals to the A/V switch. If you have read this book from the first chapter till now, you know that using a separate chip to process the IF signal really isn't new, but it is a feature Zenith hasn't used for a while.

Jack and Switch Circuit

IC2900 (Zenith part number 221-1053) is the A/V switch (figure 13-8). It receives input from three external groups of signals, two from the rear jack pack and one from the front jack pack, and is controlled by the I2C bus at pins 26 and 27. The user selects the source he or she wants to view by using the customer menu. Let's assume the viewer wants to watch the signal from the tuner. The A/V switch routes the composite audio from the tuner directly to the audio processor and composite video to the comb filter where the luma and chroma separated. The "combed" luma and chroma go back to the A/V switch where they are routed to the video processor. As you see from figure 13-8, other signal sources follow a similar path when they are selected.

Digital Comb Filter

The next stop along the way is the digital comb filter (figure 13-9), the job of which is to separate luminance from chrominance with a minimum amount of what engineers call "cross talk." The circuit works by first converting the incoming analog signal into a digital one and using its internal memory to store one horizontal line of the signal. It uses the stored information to add to or subtract from the incoming signal to obtain the separated components. The comb filter completes its job by using a "notch filter" to eliminate vestiges of the chroma signal that still cling to the luma information. I won't saddle you with more information than this because I don't think it would be helpful when you get to the grit of troubleshooting.

Video Processor

Now let's take a look at ICX2200 (figure 13-10).

Y Signal

The Y signal (Y/CVBS) selected by the video switch goes to the Y signal processing circuit. This circuit includes a trap to eliminate any vestiges of the chroma signal that are still present, the delay line, the sharpness control, a circuit called "the clamp," the black expansion circuits, and finally the RGB processing circuit.

Chroma Signal

The chroma signal (TV/CVBS) selected by the video switch passes through the ACC (automatic color control), chroma bandpass filter, chroma amplifier, and demodulation circuits to become R-Y and B-Y signals before being input to the RGB signal processing circuit.

RGB Processing

The Y and color difference signals obtained from the Y and C signals go first to the YUV switch. The selected Y and C signals then become the RGB signals after G-Y signal has been synthesized. They go through some additional processing (dynamic color, picture, gamma compensation, brightness control, etc.) before they are output at pins 22, 23, and 24 as R,G, and B out. Pins 30 and 31, labeled B2IN and G2IN respectively, are used to produce the on screen display and should not be confused with the RGB signals.

Automatic Brightness Limiter (ABL)

The ABL circuit protects the CRT against an overcurrent condition by limiting the current flowing through it. The circuit senses current by monitoring the "low side" of the high-voltage winding inside the flyback. This winding is connected to ground via resistor R2202 which also serves as a beam-current-to-voltage converter. The ABL circuit takes the signal developed across R2202 and filters it by capacitor C2206. It then adds a bias voltage to it via a voltage divider network consisting of R2201, R2212, and R2213. Components R2204 and C2204 complete the process by smoothing the resulting ABL signal before it is applied to pin 3 of the video processor.

Take another look at figure 13-10, this time paying attention to pin 42 which is labeled "ABLFIL." Do you see capacitor C2234 tied to it? The voltage applied to pin 3 is compared to an internal reference voltage which C2234 integrates. The resulting voltage becomes the control signal used to adjust the brightness and picture controls.

It seems the ABL can be switched so that it controls either the picture or both picture and brightness via register 39 (ABLMODE) in the installer's menu (service menu). Simply set the register to 1 to activate both picture and brightness at the same time. However, picture mode only (register 39 set to 0) has a function that guarantees the brightness control works when excessive beam current begins to flow through the picture tube.

White and Black Balance

The CXA2061 has a drive control function that adjusts the gain between the RGB outputs and a cutoff control function that adjusts the DC level among the RGB outputs. The adjustments are made over

Servicing Zenith Televisions

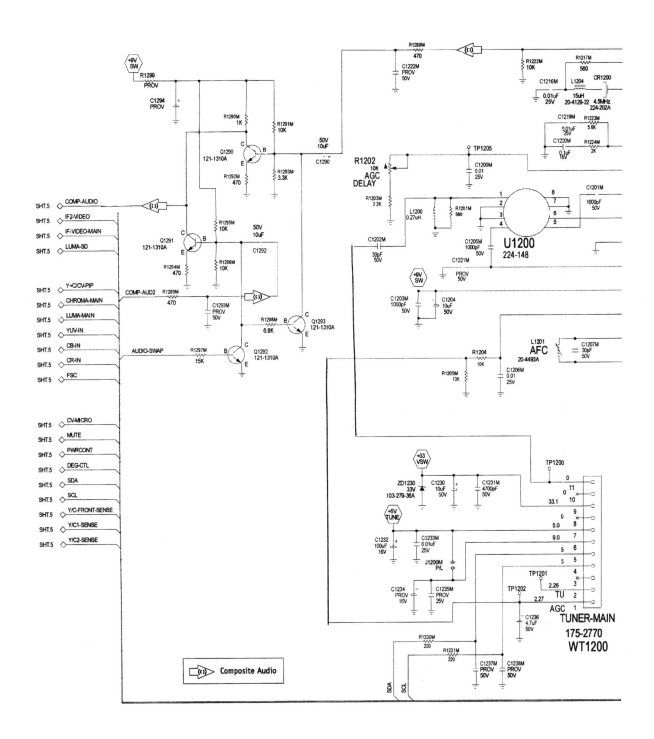

Figure 13-7 Main Tuner IF

The B-Line Chassis

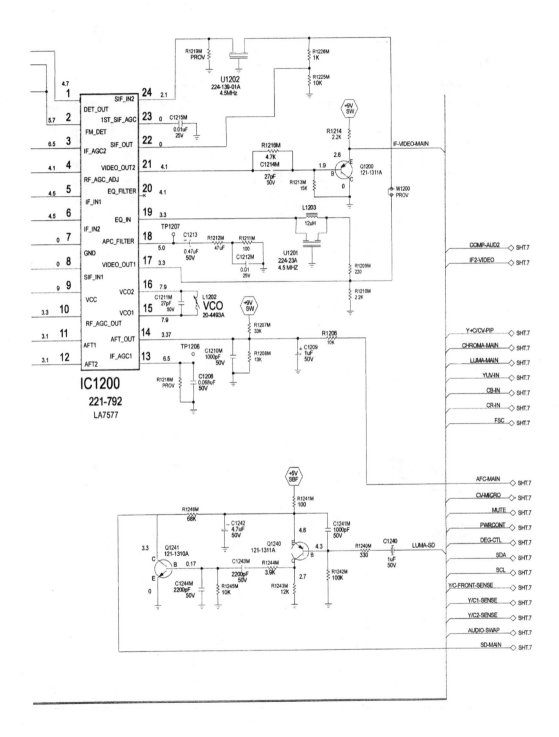

Figure 13-7 Main Tuner IF (Continued)

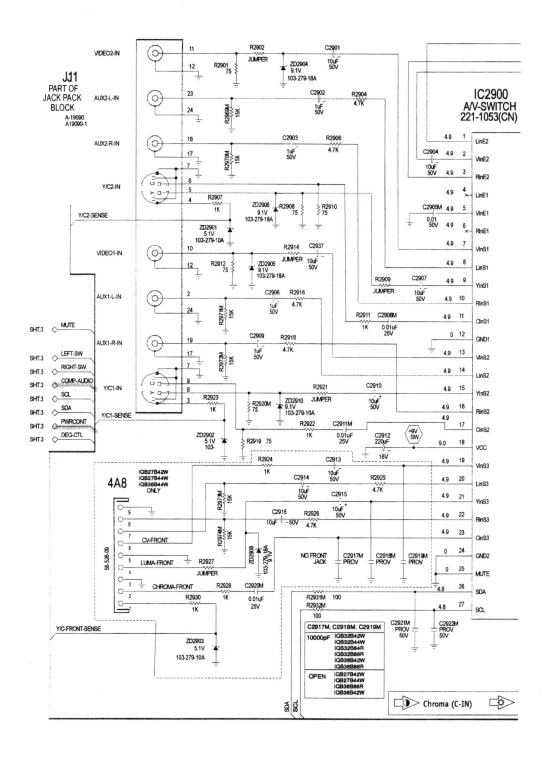

Figure 13-8 Jack and Switch Circuit

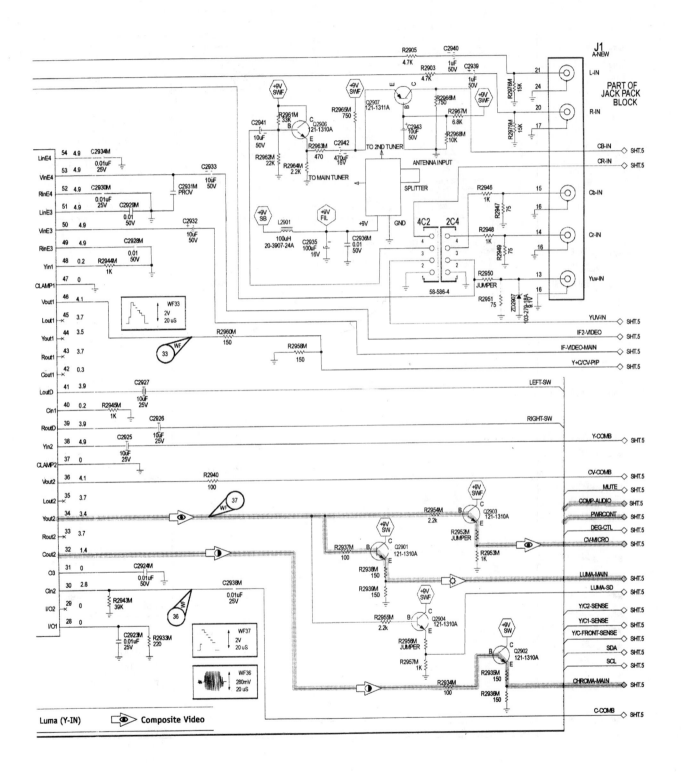

Figure 13-8 Jack and Switch Circuit (Continued)

Servicing Zenith Televisions

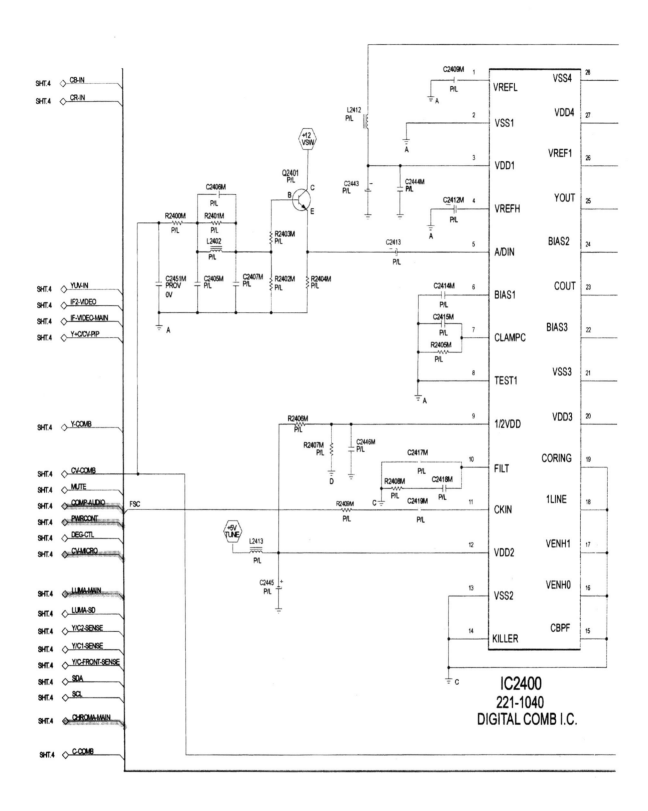

Ffigure 13-9 Digital Comb Filter

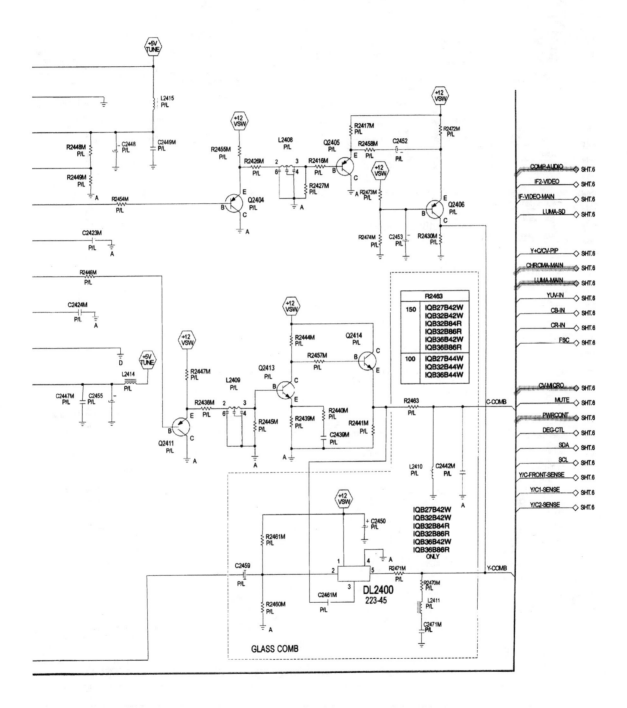

Ffigure 13-9 Digital Comb Filter (Continued)

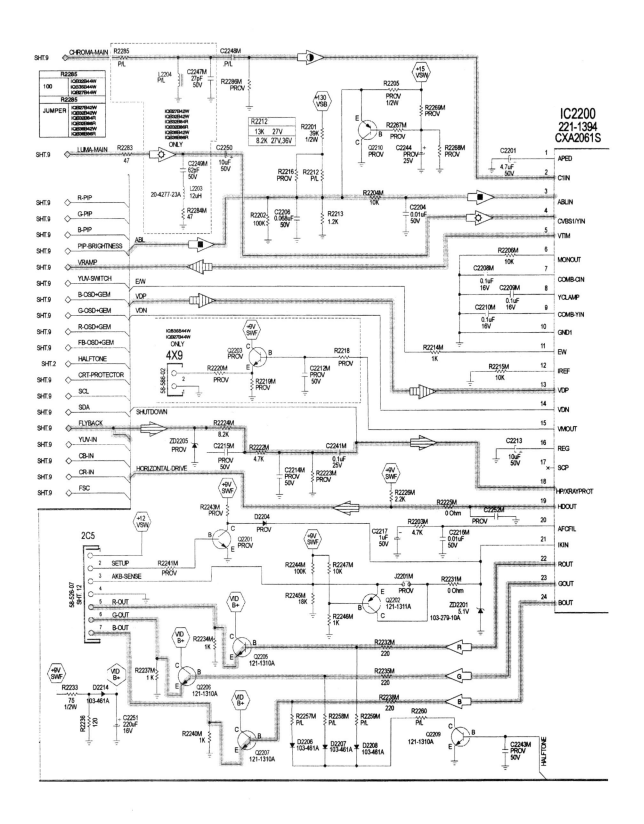

Figure 13-10 Video Processor

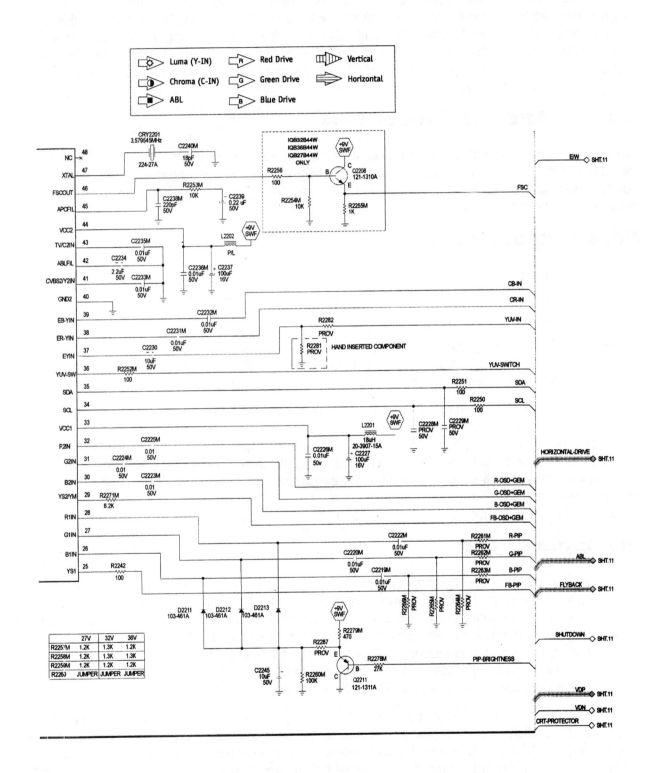

Figure 13-10 Video Processor (Continued)

the I2C bus. The circuit also has an auto cutoff function (AKB) that forms a loop between ICX2200 and the CRT. The AKB circuit has responsibility for maintaining the correct color temperature of the CRT. The cathode current at pin 21 of the CXA2061S, which is labeled "IKIN," is converted inside the IC to a voltage. The so-called "reference pulse interval" of this voltage is compared with a reference voltage generated inside the IC, and the current developed by the resulting "error voltage" charges a capacitor that is also inside the chip. The capacitor holds the charge during all intervals except the pulse interval. The circuit functions to change the DC level of the R, G, and B outputs with respect to the charge on the capacitor. Keep in mind that the reference voltage inside the IC can be adjusted for each of the R, G, and B outputs via the I2C bus.

Gemstar And PIP Connections

The Gemstar and PIP modules are very much a part of the video processing circuit. Both are module replacement circuits, but you still should know how they are wired into the circuitry of the main chassis. Therefore, the hook-up information is included in figure 13-11. Also, don't forget that you have PIP adjustments in the factory service menu.

Video Output Module

The video output module (figure 13-13) is the last stage in the video system. It is more complicated than comparable modules in other chassis because it contains circuits that sense cathode current to provide input to the AKB circuit. Since there is no need to trace each of the three drive signals arriving at connector 5C2, let's focus on the red signal. The incoming video signal goes to Q5122 that in conjunction with Q5221 forms a cascode amplifier. These transistors amplify the signal and pass it along to a coupling push-pull stage consisting of Q5123 and Q5124. After a long and involved journey, the signal finally arrives at the cathode of the picture tube where it provides drive to produce a picture.

Now a technical note. I thought I would explain just a bit what a cascode amplifier is since some of you may not know. It is a common emitter transistor in series with a common base one. Its current gain and input resistance (or impedance) is about equal to a single transistor in a common emitter configuration, and the output resistance (or impedance) is about equal to a single transistor configured as a common base amplifier. In our circuit, Q5122 is the common emitter transistor, and Q5121 the common base transistor. You often see cascade amplifiers in tuner circuits because they are quite effective dealing with signals that are 25 MHz and higher in frequency.

The Audio Circuit

The audio system is relatively straightforward (figure 13-14), consisting of an audio processor (IC1400), a line out audio processor (IC803), a power amplifier (IC804), in some instances an audio boost IC (IC1401), and a switching circuit that is a part of the jack pack (figure 13-9). The switching circuit has seven stereo audio inputs and three stereo audio outputs. Two of the three outputs depend on the video selection, but the third is independent of the selected video. The independent output can be switched between the signal selected in the other two. In other words, it permits the TV user to swap audio between the main audio source and the PIP audio source.

The audio processor IC (IC1400) is a complicated workhorse of a chip. You might think of it as a chip capable of multitasking. One section is devoted to the task of multiplexing and includes a stereo demodulator, an SAP demodulator, and a DBX decoder. Another section is devoted to sound processing and has provisions for control of three channels (left, right, and surround), a volume limiter, and bass and treble control. Yet another section oversees two auxiliary inputs and two auxiliary outputs. The

The B-Line Chassis

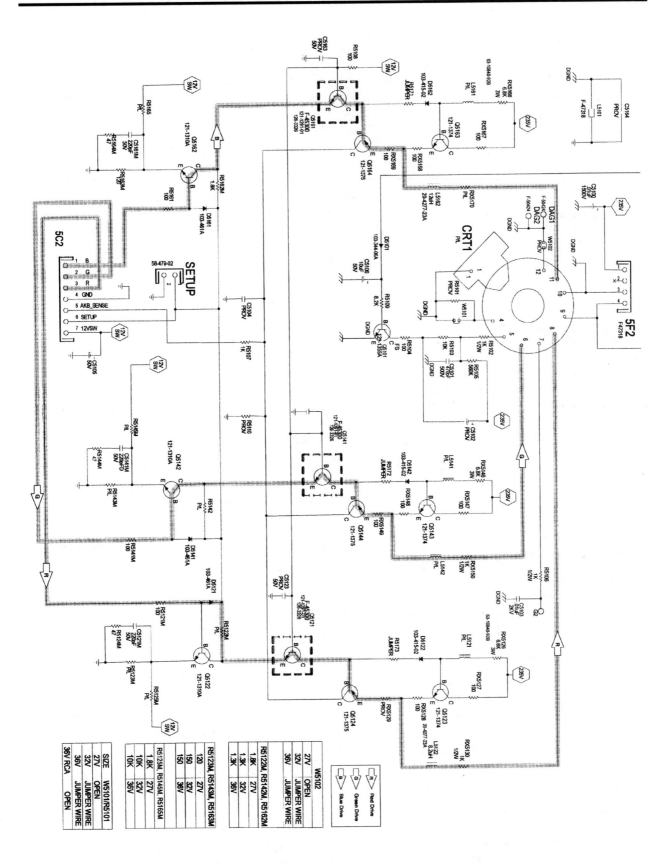

Figure 13-13 Video Output

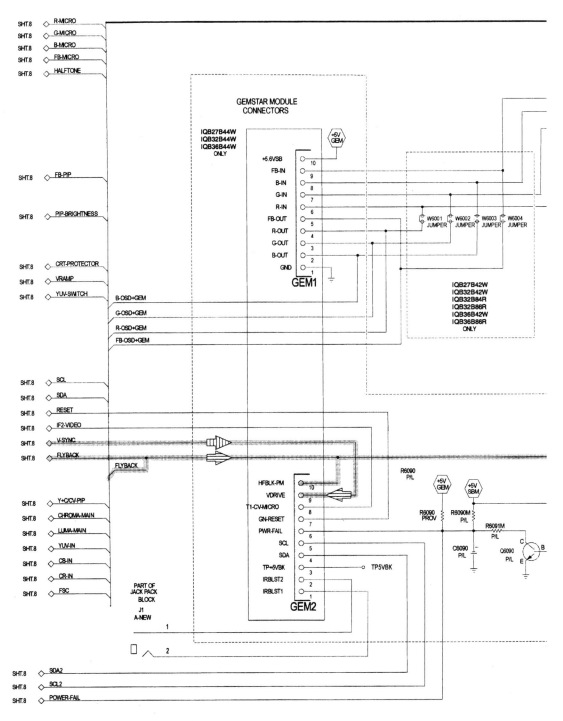

Figure 13-11 Gemstar and PIP Connections

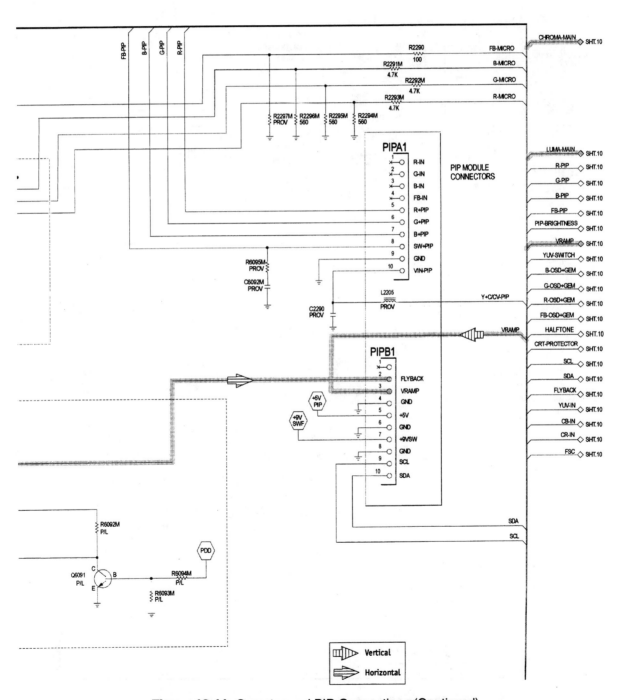

Figure 13-11 Gemstar and PIP Connections (Continued)

microprocessor oversees the audio processor's work and issues the "work orders" over I2C bus at pins 9 and 10.

Two other ICs complete the audio processing circuit. IC804 provides the power necessary to drive the speakers. It receives audio from IC1400, amplifies it, and sends it on its way to the speakers via connector 9S4. Note the capacitive coupling between it and IC1400 and it and the speakers. I regret that I don't have more information to give you about IC804, but this is all I have at the present. IC803 completes the circuit by providing variable left and right audio out via the jack pack. It, too, is capacitively coupled to the audio processor IC.

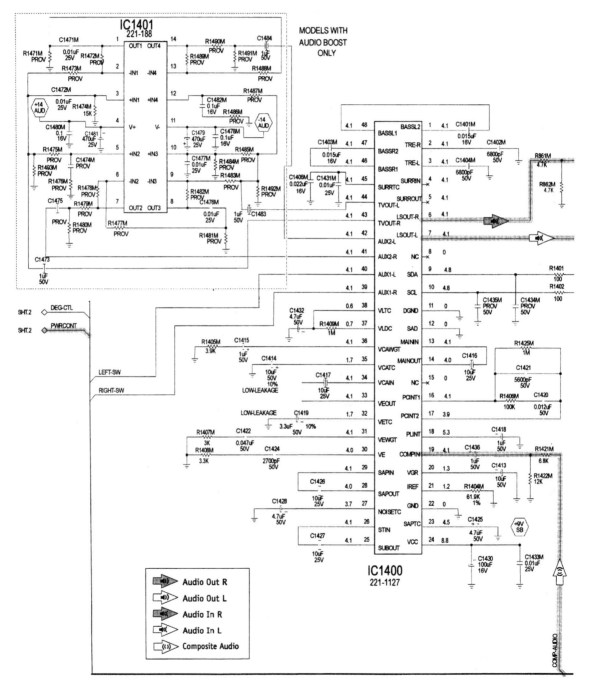

Figure 13-14 Audio Processor and Amplifier

The B-Line Chassis

Repair History

The B-Line televisions are too new to have much of a repair history, but Zenith has issued several service bulletins in which you might be interested. You probably already have the information if you are an authorized service center. If you aren't an ASC, you might never see these problems, but the information won't do you any harm.

The first bulletin calls attention to the loss of PIP on some CB-chassis and applies only to tabletop sets produced during weeks 25 through 29 and consoles produced during weeks 28 through 32 of

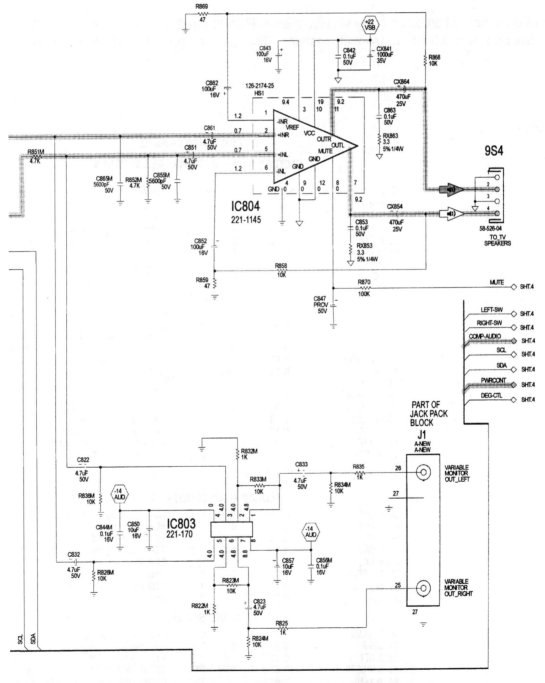

Figure 13-14 Audio Processor and Amplifier (Continued)

1999. The sixth and seventh digits of the serial number tell you when the TV was produced. For example, a set that has the serial number 921-73283419A came through the assembly line in the 28th week of 1999. PIP pictures in these sets have a high risk of blanking or distorting. The problem has been especially noted on the 9-2110 PIP module. You may either replace the PIP or repair the board by installing a jumper at locations R2545 and R2546M.

Some CB chassis have an intermittent shutdown problem. Shutdown usually occurs when the set is first turned on and just before raster appears. These sets left the factory in the 32nd week of 1999. Fix the problem by replacing the ferrite bead (Zenith calls it the "inductive bead.") labeled LX3415 with part number 149-549.

Finally, make a note of an error in the service menu. RX3404 in the power supply is a 180k, 1/2-watt resistor instead of a 3.3 ohm, 1/4-watt resistor. The resistor provides start up B+ for the power supply.

Table 13-2a

CA CONSUMER CHASSIS
FACTORY MENU RECOMMENDED SETTINGS

ITEM	RANGE	25"	27"	DESCRIPTION
00 F MODE	0-1	0	0	Factory mode - perfer to page 2-1
01 PRE PX	0-1	0	0	Used to store customer menu adjustments
02 V POS	0-24	13	13	Moves On Screen display vertically
03 H POS	0-20	6	6	Moves On Screen display horizontally
04 LEVEL	0-1	1	1	
05 BAND	0-7	0	0	Broadcast Band adjustment
06 AC ON	0-1	0	0	AC power feature

TECHNICAL SETTINGS

ITEM	RANGE	25"	27"	DESCRIPTION
07 RFB PF	0-1	1	1	Control for brightness in therefore mode
08 3.5T	0-1	0	0	3.58 MHz Trap
09 RF BRT	0-63	33	33	Control for brightness in RF mode
10 AX BRT	0-63	33	33	Control for brightness in AUX mode
11 MAX CON	0-63	63	63	Control for contrast
12 VSIZE	0-63	145	125	Vertical Size
13 HSIZE	0-254	0	0	Horizontal size
14 V PHASE	0-7	2	2	Shifts picture vertically
15 H PHASE	0-31	14	14	Shifts picture horizontally
16 AUDLVL	0-63	46	46	Sound Attenuation
17 RF AGC	0-63	31	31	Weak channel adjustment
18 H AFC	0-1	1	1	Horizontal Automatic Frquency control
19 WHCOMP	0-1	0	0	White compression
20 60HZSW	0-2	2	2	60 Hertz Switched
21 PIFVCO	1-127	31	31	PIF Voltage Controlled Oscillator

COLOR TEMP STARTING VALUE

ITEM	RANGE	25"	27"	DESCRIPTION
22 R CUT	0-254	0	0	B&W tracking adjustment
23 G CUT	0-254	0	0	B&W tracking adjustment
24 B CUT	0-254	0	0	B&W tracking adjustment
25 G GAIN	0-254	103	103	B&W tracking adjustment
26 B GAIN	0-254	75	75	B&W tracking adjustment
27 CTYPE	0-3	2	2	Chasis type
28 SCROLL	0-1	0	0	User Menu scroll type
29 KEYS	0-1	0	0	Keyboard type

AUDIO SETTINGS

ITEM	RANGE	25"	27"	DESCRIPTION
30 SPKRSW	0-1	0	0	Internal Speaker Switch
31 SURF	0-1	0	0	Surf Mode
32 A ATT	0-15	9	9	Audio Attenuator
33 ASPECTRAL	0-63	31	31	Adjustment of Stereo seperation
34 WBAND	0-63	31	31	Wide band Low Frequency Seperation

Table 13-2b

CA CONSUMER CHASSIS
FACTORY MENU RECOMMENDED SETTINGS

ITEM	RANGE	25"	27"	32"	36"	DESCRIPTION
00 F MODE	0-1	0	0	0	0	Factory mode - perfer to page 2-1
01 PRE PX	0-7	0	0	0	0	Used to store customer menu adjustments
02 V POS	0-30	15	15	15	15	Moves On Screen display vertically
03 H POS	0-45	9	9	9	9	Moves On Screen display horizontally
04 LEVEL	0-2	1	1	1	1	
05 BAND	0-7	0	0	0	0	Broadcast Band adjustment
06 AC ON	0-1	0	0	0	0	AC power feature
TECHNICAL SETTINGS						
07 RFB PF	0-1	1	1	1	1	Control for brightness in therefore mode
08 3.5T	0-1	1	1	1	1	3.58 MHz Trap
09 RF BRT	0-63	38	33	38	33	Control for brightness in RF mode
10 AX BRT	0-63	38	33	38	33	Control for brightness in AUX mode
11 MAX CON	0-63	63	63	63	63	Control for contrast
12 V SIZE	0-254	145	125	120	174	Vertical Size
13 H SIZE	0-254	0	0	174	120	Horizontal size
14 V PHASE	0-7	2	0	2	1	Shifts picture vertically
15 H PHASE	0-31	14	17	14	14	Shifts picture horizontally
16 AUDLVL	0-63	46	46	46	46	Sound Attenuation
17 RF AGC	0-63	31	31	31	31	Weak channel adjustment
18 H AFC	0-1	1	1	1	1	Horizontal Automatic Frquency control
19 WHCOMP	0-1	0	0	0	0	White compression
20 60HZSW	0-2	2	2	2	2	60 Hertz Switched
21 PIFVCO	1-127	31	31	31	31	PIF Voltage Controlled Oscillator
COLOR TEMP STARTING VALUE						
22 R CUT	0-254	0	0	0	0	B&W tracking adjustment
23 G CUT	0-254	0	0	0	0	B&W tracking adjustment
24 B CUT	0-254	0	0	0	0	B&W tracking adjustment
25 G GAIN	0-254	86	103	81	90	B&W tracking adjustment
26 B GAIN	0-254	82	75	75	90	B&W tracking adjustment
27 SCROLL	0-1	0	0	0	0	User Menu scroll type
28 6 KEYS	0-1	0	0	0	0	Keyboard type
AUDIO SETTINGS						
29 A ATT	0-15	9	9	9	9	Audio Attenuator
30 A VCO	0-63	31	31	31	31	Audio Decoder Oscillator
31 AFILTER	0-63	31	31	31	31	Audio Filter
32 ASPECTRAL	0-63	31	31	31	31	Adjustment of Stereo seperation
33 WBAND	0-63	31	31	31	31	Wide band Low Frequency Seperation
34 IN OSDC	0-1	0	0	0	0	On Screen Display Internal Oscillator
35 OSD FRAME	0-1	0	0	0	0	On Screen Display Frame
36 ASS ON	0-1	0	0	0	0	Automatic Sleep System
DIFFERENCES FOR 221-1385-01						
35 XTAL	0-1	0	0	0	0	External Crystal

Note: If you change the EPROM, change the microprocessor from 221-1385-00 to 221-1385-01

Table 13-2c

CA CONSUMER CHASSIS
FACTORY MENU RECOMMENDED SETTINGS

ITEM	RANGE	25"	27"	32"	36"	DESCRIPTION
00 F MODE	0-1	0	0	0	0	Factory mode - perfer to page 2-1
01 PRE PX	0-1	0	0	0	0	Used to store customer menu adjustments
02 V POS	0-30	15	15	15	15	Moves On Screen display vertically
03 H POS	0-45	9	9	9	9	Moves On Screen display horizontally
04 LEVEL	0-2	1	1	1	1	
05 BAND	0-7	0	0	0	0	Broadcast Band adjustment
06 AC	0-1	0	0	0	0	AC power feature

TECHNICAL SETTINGS

ITEM	RANGE	25"	27"	32"	36"	DESCRIPTION
07 RFB PF	0-1	1	1	1	1	Control for brightness in therefore mode
08 3.5T	0-1	0	0	0	0	3.58 MHz Trap
09 RF BRT	0-63	33	33	33	33	Control for brightness in RF mode
10 AX BRT	0-63	33	33	33	33	Control for brightness in AUX mode
11 MAX CON	0-63	63	63	63	63	Control for contrast
12 V SIZE	0-254	145	125	120	175	Vertical Size
13 H SIZE	0-254	0	0	174	120	Horizontal size
14 V PHASE	0-7	2	0	2	1	Shifts picture vertically
15 H PHASE	0-31	14	17	14	14	Shifts picture horizontally
16 AUDLVL	0-63	46	46	46	46	Sound Attenuation
17 RF AGC	0-63	31	31	31	31	Weak channel adjustment
18 H AFC	0-1	1	1	1	1	Horizontal Automatic Frquency control
19 WHCOMP	0-1	0	0	0	0	White Compression
20 60HZSW	0-2	2	2	2	2	60 Hertz Switched
21 PIFVCO	1-127	31	31	31	31	PIF Voltage Controlled Oscillator

COLOR TEMP STARTING VALUE

ITEM	RANGE	25"	27"	32"	36"	DESCRIPTION
22 R CUT	0-254	0	0	0	0	B&W tracking adjustment
23 G CUT	0-254	0	0	0	0	B&W tracking adjustment
24 B CUT	0-254	0	0	0	0	B&W tracking adjustment
25 G GAIN	0-254	86	103	81	90	B&W tracking adjustment
26 B GAIN	0-254	82	75	75	90	B&W tracking adjustment
27 SCROLL	0-1	0	0	0	0	User Menu scroll type
28 6 KEYS	0-1	0	1	0	0	Keyboard type

AUDIO SETTINGS

ITEM	RANGE	25"	27"	32"	36"	DESCRIPTION
29 A ATT	0-15	9	9	9	9	Audio Attenuator
30 ASPECTRAL	0-63	31	31	31	31	Adjustment of Stereo seperation
31 WBAND	0-63	31	31	31	31	Wide Band Low Frequency Seperation

PIP SETTINGS

ITEM	RANGE	25"	27"	32"	36"	DESCRIPTION
32 PIP X1 POS	0-254	17	17	17	17	PIP Position X_1
33 PIP Y1 POS	0-254	25	25	25	25	PIP Position Y_1
34 PIP X2 POS	0-254	113	113	113	113	PIP Position X_2
35 PIP Y2 POS	0-254	143	143	143	143	PIP Position Y_2
36 PIP BRIGHTNESS	0-254	205	205	205	205	PIP Brightness
37 PIPR CONTRAST	0-127	50	50	50	50	PIP Red Contrast
38 PIPG CONTRAST	0-127	50	50	50	50	PIP Green Contrast
39 PIPB CONTRAST	0-127	50	50	50	50	PIP Blue Contrast
40 PIP ADJUST	0-15	2	2	2	2	PIP Adjust
41 PIPY DELAY	0-15	4	4	4	4	PIP Y Delay Luma
42 PIP BGSTART	0-31	10	10	10	10	PIP Burst Gate Start
43 PIP ACCLEVEL	0-63	20	20	20	20	PIP Color Gain
44 PIP HX	0-63	30	30	30	30	PIP Horizontal Phase
45 PIP VXS	0-63	40	41	41	41	PIP Vertical Phase
46 PIP COMP	0-3	2	2	2	2	PIP External Syncronic

DIFFERENCES FOR 221-1386-01

ITEM	RANGE	25"	27"	32"	36"	DESCRIPTION
47 XTAL	0-1	0	0	0	0	External Crystal

Note: If you change the EPROM, change the microprocessor from 221-1386-00 to 221-1386-01

Table 13-2d

CA CONSUMER CHASSIS

FACTORY MENU RECOMMENDED SETTINGS

ITEM	RANGE	27"	32"	36"	DESCRIPTION
00 F MODE	0-1	0	0	0	Factory mode - perfer to page 2-1
01 PRE PX	0-1	0	0	0	Used to store customer menu adjustments
02 V POS	0-30	15	15	15	Moves On Screen display vertically
03 H POS	0-45	9	9	9	Moves On Screen display horizontally
04 LEVEL	0-2	1	1	1	
05 BAND	0-7	0	0	0	Broadcast Band adjustment
06 AC ON	0-1	0	0	0	AC power feature

TECHNICAL SETTINGS

ITEM	RANGE	27"	32"	36"	DESCRIPTION
07 RFB PF	0-1	1	1	1	Control for brightness in therefore mode
08 3.5T	0-1	1	1	1	3.58 MHz Trap
09 RF BRT	0-63	33	33	33	Control for brightness in RF mode
10 AX BRT	0-63	33	33	33	Control for brightness in AUX mode
11 V SIZE	0-254	170	120	175	Vertical Size
12 H SIZE	0-254		174	120	Horizontal size
13 V PHASE	0-7	0	2	1	Shifts picture vertically
14 H PHASE	0-31	13	14	14	Shifts picture horizontally
15 AUDLVL	0-63	46	46	46	Sound Attenuation
16 RF AGC	0-63	31	31	31	Weak channel adjustment
17 H AFC	0-1	1	1	1	Horizontal Automatic Frquency control
18 60 HZSW	0-2	2	2	2	60 Hertz Switched
19 PIFVCO	1-127	31	31	31	PIF Voltage Controlled Oscillator

COLOR TEMP STARTING VALUE

ITEM	RANGE	27"	32"	36"	DESCRIPTION
20 R CUT	0-254	0	0	0	B&W tracking adjustment
21 G CUT	0-254	0	0	0	B&W tracking adjustment
22 B CUT	0-254	0	0	0	B&W tracking adjustment
23 G GAIN	0-254	103	81	90	B&W tracking adjustment
24 B GAIN	0-254	75	75	90	B&W tracking adjustment
25 6 KEYS	0-1	1	0	0	Keyboard type

AUDIO SETTINGS

ITEM	RANGE	27"	32"	36"	DESCRIPTION
26 A ATT	0-15	9	9	9	Audio Attenuator
27 AVCO	0-63	31	31	31	
28 AFILTER	0-63	31	31	31	Audio Filter
29 ASPECTRAL	0-63	31	31	31	Adjustment of Stereo seperation
30 WBAND	0-63	31	31	31	Wide band Low Frequency Seperation

PIP SETTINGS

32 PIP X1 POS	0-254	17	17	17	17	PIP Position X_1
33 PIP Y1 POS	0-254	25	25	25	25	PIP Position Y_1
34 PIP X2 POS	0-254	113	113	113	113	PIP Position X_2
35 PIP Y2 POS	0-254	143	143	143	143	PIP Position Y_2
36 PIP BRIGHTNESS	0-254	205	205	205	205	PIP Brightness
37 PIPR CONTRAST	0-127	50	50	50	50	PIP Red Contrast
38 PIPG CONTRAST	0-127	50	50	50	50	PIP Green Contrast
39 PIP ADJUST	0-15	2	2	2	2	PIP Adjust
40 PIPY DELAY	0-15	10	10	10	10	PIP Y Delay Luma
41 PIP BGSTART	0-31	10	10	10	10	PIP Burst Gate Start
42 PIP ACC LEVEL	0-63	20	20	20	20	PIP Color Gain
43 PIP HX	0-63	30	30	30	30	PIP Horizontal Phase
44 PIP VXS	0-63	40	41	41	41	PIP Vertical Phase
46 PIP COMP	0-3	2	2	2	2	PIP External Syncronic

DIFFERENCES FOR 221-1387-01

39 PIPY DELAY	0-15	10	10	10	10	PIP Y Delay Luma
40 PIP BGSTART	0-31	10	10	10	10	PIP Burst Gate Start
41 PIP ACC LEVEL	0-63	20	20	20	20	PIP Color Gain
42 PIP HX	0-63	30	30	30	30	PIP Horizontal Phase
43 PIP VXS	0-63	40	41	41	41	PIP Vertical Phase
44 PIP COMP	0-3	2	2	2	2	PIP External Syncronic
45 XTAL	0-1	0	0	0	0	External crystal

Note: If you change the EPROM, change the microprocessor from 221-1387-00 to 221-1387-01

Table 13-2e

FACTORY MENU SETTINGS FOR 221-01388					
ITEM	RANGE	27"	32"	36"	DESCRIPTION
00 F Mode	0-1	0	0	0	Factory mode-perfer to page 2-1
01 Pre Px	0-1	0	0	0	Used to store customer menu adjustments
02 V Pos	0-30	15	15	15	Moves On Screen display vertically
03 H Pos	0-45	42	42	9	Moves On Screen display horizntally
04 Level	0-2	1	1	1	Level is fixed at factory
05 Band	0-7	0	0	0	Broadcast Band adjustment
06 AC On	0-1	0	0	0	AC power feature
TECHNICAL SETTINGS					
07 RFB pf	0-31	15	15	15	Control for Brightness
08 Ax Brt	0-31	15	15	15	Auxiliary Brigthness
09 Max con	0-63	63	63	63	Max Contrast
10 V Size	0-63	38	35	31	Vertical Size
11 H Size	0-63	63	35	55	Horizontal Size
12 V Phase	0-63	42	36	43	Shifts picture vertically
13 H Phase	0-63	0	24	25	Shifts picture horizontally
14 H OSC	0-31	7	7	7	H OSC
15 SCORR	0-31	7	3	0	S Correction
16 V Linea	0-15	7	4	6	V Linearity
17 Pin Amp	0-63	63	23	44	Pin Amp
18 C Pin	0-63	63	30	31	Corner Pin
19 Trapez	0-15	15	7	9	Trapezium
20 EHT Com	0-15	7	7	7	EHTComp
21 AFC Bow	0-15	7	7	7	AFCBown
22 AFC Ang	0-15	7	5	5	AFCAngle
23 UPVLIN	0-15	6	4	0	Upper VLin
24 LOVLIN	0-15	0	3	0	LowerVLin
25 AFC G	0-3	1	1	1	AFCGain
26 EWDC	0-1	0	0	0	EWDC
COLOR TEMP STARTING VALUE					
27 R Cut	0-15	0	0	0	B&W tracking adjustment
28 G Cut	0-15	0	0	0	B&W tracking adjustment
29 B Cut	0-15	0	0	0	B&W tracking adjustment
30 R Gain	0-63	31	31	31	B&W tracking adjustment
31 G Gain	0-63	31	31	31	B&W tracking adjustment
32 B Gain	0-63	31	31	31	B&W tracking adjustment
33 DynamC	0-1	0	0	0	DynamicC Video Processor flag
34 Gamma	0-3	0	0	0	Gamma Register
35 DCTRAN	0-1	1	1	1	DCTran video processor flag
36 CBPF	0-1	1	1	1	CBPF video processor flag
37 CTRAP	0-1	1	1	1	CTRAPOFF video processor flag
38 FSCSW	0-1	0	0	0	FSCSW video processor flag
39 ABLMODE	0-1	1	1	1	ABLMODE video processor flag
40 ABLVTH	0-1	1	1	1	ABLVTH video processor flag
41 6 Keys	0-1	1	0	0	Keyboard type
AUDIO SETTINGS					
42 A Att	0-15	9	9	9	Audio Attenuator
43 Avco	0-63	31	31	31	Audio Decoder Oscillator
44 Afilter	0-63	31	31	31	Audio Filter
45 Aspectral	0-63	31	31	31	Adjustment of Stereo separation
46 W Band	0-63	31	31	31	Wide band low Frequency Separation

Table 13-2e (Continued)

ITEM	RANGE	27"	32"	36"	DESCRIPTION
FACTORY MENU SETTINGS 221-01388 (continued)					
PIP SETTINGS					
47 Pip X1Pos	0-254	17	17	17	Pip position X1
48 Pip Y1Pos	0-254	25	25	25	Pip position Y1
49 Pip X2 Pos	0-254	113	113	113	Pip position X2
50 Pip Y2 Pos	0-254	143	143	143	Pip position Y2
51 Pip Adjust	0-15	2	2	2	Pip Adjustment
52 Pip Y Delay	0-15	4	4	4	Pip Y Delay
53 P YOFF	0-31	31	31	31	Pip Y Offset
54 Pip Acc	0-63	20	20	20	Pip ACC Level
55 Pip BGST	0-63	10	10	10	Pip BG Stat
56 Pip HX	0-63	22	22	22	Pip HX
57 Pip VXS	0-63	37	37	37	Pip VXS
58 PipR Contrast	0-63	50	50	50	Pip Red Contrast
59 PipG Contrast	0-63	50	50	50	Pip Green Contrast
60 PipB Contrast	0-63	50	50	50	Pip Blue Contrast
61 Pip Com	0-3	2	2	2	Pip Com
62 Yub Brt	0-31	10	10	10	YUV sub-Brightness
63 Min Cont	0-31	10	10	10	Minimum Contrast Value
64 F Jacks	0-1	1	0	0	F Front Jacks
65 Pip Brt	0-254	200	200	200	Pip Brightness
66 Xtal	0-1	0	0	0	Select crystal or RC circuit oscillator for OSD
67 Min Brt	0-63	25	25	25	Custom Minumum Brightness
68 Max Brt	0-63	35	35	35	Custom Maximum Brightness
69 Mut Brt	0-63	20	20	20	Level Brightness when mute
70 Mut Con	0-63	20	20	20	Level Contrast when mute

Table 13-2f

FACTORY MENU SETTINGS 221-01389						
ITEM	RANGE	27"	32"	36"	DESCRIPTION
00 F Mode	0-1	0	0	0	Factory mode. Refer to page 2-1
01 Pre Px	0-1	0	0	0	Used to store customer menu adjustments
02 V Pos	0-30	15	15	15	Moves On Screen display vertically
03 H Pos	0-80	42	42	42	Moves On Screen display horizintally
04 Level	0-2	1	1	1	Level is fixed at factory
05 Band	0-7	0	0	0	Broadcast Band adjustment
06 AC On	0-1	0	0	0	AC power feature
TECHNICAL SETTINGS						
07 RFB pf	0-31	15	15	15	Control for Brightness
08 Ax Brt	0-31	15	15	15	Auxiliary Brigthness
09 Max con	0-63	63	63	63	Max Contrast
10 V Size	0-63	38	35	31	Vertical Size
11 H Size	0-63	63	35	55	Horizontal Size
12 V Phase	0-63	42	36	43	Shifts picture vertically
13 H Phase	0-63	0	24	25	Shifts picture horizontally
14 H OSC	0-31	7	7	7	H OSC
15 SCORR	0-15	7	3	0	S Correction
16 V Linea	0-15	7	4	6	V Linearity
17 Pin Amp	0-63	63	23	44	Pin Amp
18 C Pin	0-63	63	30	31	Corner Pin
19 Trapez	0-15	15	7	9	Trapezium
20 EHT Com	0-15	7	7	7	EHTComp
21 AFC Bow	0-15	7	7	7	AFCBown
22 AFC Ang	0-15	7	5	5	AFCAngle
23 UPVLIN	0-15	6	4	0	Upper VLin
24 LOVLIN	0-15	0	3	0	LowerVLin
25 AFC G	0-3	1	1	1	AFCGain
26 EWDC	0-1	1	0	0	EWDC
COLOR TEMP STARTING VALUE						
27 R Cut	0-15	0	0	0	B&W tracking adjustment
28 G Cut	0-15	0	0	0	B&W tracking adjustment
29 B Cut	0-15	0	0	0	B&W tracking adjustment
30 R Gain	0-63	31	31	31	B&W tracking adjustment
31 G Gain	0-63	31	31	31	B&W tracking adjustment
32 B Gain	0-63	31	31	31	B&W tracking adjustment
33 DynamC	0-1	0	0	0	DynamicC Video Processor flag
34 Gamma	0-3	0	0	0	Gamma Register
35 DCTRAN	0-1	1	1	1	DCTran video processor flag
36 CBPF	0-1	1	1	1	CBPF video processor flag
37 CTRAP	0-1	1	1	1	CTRAPOFF video processor flag
38 FSCSW	0-1	1	1	1	FSCSW video processor flag
39 ABLMODE	0-1	1	1	1	ABLMODE video processor flag
40 ABLVTH	0-1	1	1	1	ABLVTH video processor flag
41 6 Keys	0-1	1	0	0	Keyboard type
AUDIO SETTINGS						
42 A Att	0-15	9	9	9	Audio Attenuator
43 Avco	0-63	31	31	31	Audio Decoder Oscillator
44 Afilter	0-63	31	31	31	Audio Filter
45 Aspectral	0-63	31	31	31	Adjustment of Stereo separation
46 W Band	0-63	31	31	31	Wide band low Frequency Separation

Table 13-2f (Continued)

FACTORY MENU SETTINGS 221-01389 (continued)					
ITEM	RANGE	27"	32"	36"	DESCRIPTION
PIP SETTINGS					
47 Pip X1Pos	0-254	17	17	17	Pip position X1
48 Pip Y1Pos	0-254	25	25	25	Pip position Y1
49 Pip X2 Pos	0-254	113	113	113	Pip position X2
50 Pip Y2 Pos	0-254	143	143	143	Pip position Y2
51 Pip Adjust	0-15	2	2	2	Pip Adjustment
52 Pip Y Delay	0-15	4	4	4	Pip Y Delay
53 P YOFF	0-31	31	31	31	Pip Y Offset
54 Pip Acc	0-63	20	20	20	Pip ACC Level
55 Pip BGST	0-63	10	10	10	Pip BG Stat
56 Pip HX	0-63	22	22	22	Pip HX
57 Pip VXS	0-63	37	37	37	Pip VXS
58 PipR Contrast	0-63	50	50	50	Pip Red Contrast
59 PipG Contrast	0-63	50	50	50	Pip Green Contrast
60 PipB Contrast	0-63	50	50	50	Pip Blue Contrast
61 Pip Com	0-3	2	2	2	Pip Com
62 YUV Brt	0-31	10	10	10	Yuv Brightness
63 Min Con	0-31	10	10	10	Minimum Contrast
64 F Jack	0-1	1	0	0	Front Jacks
65 Pip Brt	0-254	200	200	200	Pip Brightness
66 XTal	0-1	0	0	0	Select crystal or RC circuit oscillator for OSD
67 Min Brt	0-63	25	25	25	Minimum Brightness
68 Max Brt	0-63	35	35	35	Maximum Brightness
69 GM PIP X	0-50	15	12	11	Pip X position for gemstar
70 GM PIP Y	0-30	17	17	16	Pip Y position for gemstar
71 GM POS X	0-60	36	31	32	OSD X position for gemstar
72 GM POS Y	0-36	13	11	11	OSD Y position for gemstar
73 GM SIZ X	0-152	110	100	110	OSD X Size for gemstar
74 GM SIZ Y	0-233	100	200	200	OSD Y Size for gemstar
75 Mut Brt	0-63	20	20	20	Level Brightness when mute
76 Mut Con	0-63	20	20	20	Level Contrast when mute
77 GMVSize	0-63	30	34	35	GMVSize

Appendix

I have reserved the appendix for those miscellaneous items about which you need to know but I couldn't, for a variety of reasons, include in the body of the book.

How Zenith Classifies Their Televisions

Year	Model Year	Service Literature
1984	A-Line	902-3654
1985	B-Line	902-3655
1986	C-Line	902-3656
1987	D-Line	902-3657
1988	E-Line	902-3658
1989	F-Line	902-3659
1990	G-Line	902-3660-R1
1991	J-Line	902-3661-R1
1992	J-2 Line	902-4018
1993	L-Line	902-4022-R1
1994	M-Line	902-4053
1995	R-Line	902-4091
1996	Y-Line	See Text
1997	Z-Line	See Text
1998	A-Line	See Text
1999	B-Line	See Text

Brief Explanation of a Few Audio Terms

You can find the following material in almost any book on audio to which you turn. I am using material published in *Technical Training Program: R-Line 1995-1996*, available under the part number 923-3261 TRM which is tailored specifically to Zenith televisions.

1) Theater: Hard Room Effect – The audio system adds a 30 mS delayed L+R signal to the surround speaker and the Left speaker and subtracts it from the Right speaker. It has the effect of simulating a medium sized theater with echo from the wall of the building. Use it when watching plays, movies and concerts.

2) Night Club: Crowded Room Effect – The system adds a 20 mS delayed L-R signal to the surround speaker and the Left speaker while subtracting it from the Right. The effect simulates a small, crowded, night club. It's a good effect for listening to stand up comedy.

3) Stadium: Large Sports Arena – The effects are achieved by directing a 80 mS delayed L+R signal to the surround speaker and adding it to the Left speaker while subtracting it from the Right. It simulates a large stadium with echo from the walls of the building. It is good for watching sports events.

4) Concert Hall: Increased Stereo Separation – The system increases separation of Left and Right channels while directing passive surround sound to the surround speaker. It enhances the pleasure of listening to music.

5) Pro Logic: Dolby Pro Logic Decoding – It operates with any stereo material, but it works best with material that is encoded with Pro Logic.

6) Phantom Pro Logic: Dolby Pro Logic Decoding with Phantom Center Channel – It is the same a Pro Logic except it doesn't use external stereo. The internal Left and Right speakers serve as Left, Right, and Center. If the customer has an external stereo connected to the AVI, Zenith recommends normal Pro Logic enhancement.

Technical Training Literature

The list that follows is by no means exhaustive and doesn't include the newer manuals to which I refer in Chapters 10 through 13. However, it should give you an idea about what Zenith offers to help you keep up with the constant flow of new products they are putting on the market.

TP-32 A Line Color TV.
PV4541 Projection TV.
VR2000, 3000, 4000, 5000 Video Recorders.
VC1000 Video Color Camera.
VM6000 Video Movie Camcorder.

TP-33 PV800 Color Projection Monitor.
CV524 Stereo Adapter.
PV4539G Projection TV.

TP-34 B Line Color TV.
VM6100 Camcorders.
PV4543 Projection TV.
B Line Tuning System.
VR1800, 2100 Video Recorders.

TP-34S VHS Video Recorder Servicing.
VM6000 Camcorder.
B Line Color TV Up-Date.

TP-35 VM7000 Camcorder,
ZS3000 Zenith Satellite Systems.
VC1100 Color Camera.
VR4100, VR5100 Video Recorders.

TP-36 C Line Color TV.
VM6200 Camcorder.
C0920, C0930 9" Color TV.
C Line Tuning Systems.
C Line VHS Video Recorders.

TP-37 C Line Digital TV.
VM6150, VM7100 Camcorders.
VR1820, VR2300, VR3300 Video Recorders.
PV851 Projection Receiver/Monitor.
ZS4000 Satellite Receiver.

TP-38 D Line Color TV.
C Chassis Color TV.
VM6150 Camcorder.
D Line Video Recorders.
MTS Alignment

TP-41 E Line Color TV.
46" Digital Projection Receivers.
E Line Digital TV.
E Line VHS Video Recorders.
Servicing Digital TV.

TP-42 E Line Color TV Up-Date.
E Line Digital TV Up-Date.
PV830, PV865 Projection TV Set-Up.
E Line New VHS Video Recorders.
ZS2000, ZS6000U Satellite Receivers.

TP-43 F Line Color TV Vanguard—C3 Chassi
F Line Projection TV Surround Sound.
Digital TV Up-Date.
F Line New VHS Video Recorders.
Multi Brand Remote.

TP-44 F Line Color TV Up-Date.
PRO840, PV875 Registration And Troubleshooting.
VRS70, VCP20 Mechanical Description
VRF510HF Auto Tracking Circuit.
Servicing Satellite Systems.

TP-45 G Line Color TV Sentry 2–C2 Chassis.
F and G Line Projection TV Registration Update.
ZS7000 Satellite System.

TP45S VHS Tape Transport System.
Service Expertise.

TP-46 J Line Color TV—C5 Chassis.
J Projection TV Up-Date.
SMD Soldering Techniques
J Line Video Cassette Recorders.

TP-47 J Line Color TV Update.
C-7 Color TV Chassis.
C-3 Projection TV.
J Line VCR Update.

The Z-Line Chassis

TECHNICIAN PARTICIPATION WORKSHOP WORKBOOKS, WITHOUT TAPE

TPW-11	Z1Chassis, Color TV
TPW-13	VM7000 Camcorder
TPW-14	VM6200 Camcorder
TPW-16	E-F Line Digital Projection Training ExercisesTV
TPW-16S	Digital Projection Maintenance and Adjustment Guide
TPW-17	Digital TV Systems Training Exercises C - G Line Line
TPW-18	G-Line Video Cassette Recorders
TPW-19	J-Line video Cassette Recorders and Training Exercises with Mechanical Alignment

ADDITIONAL MANUALS

923-2353	C2 and C5 Sentry 2 Color Television Service
923-2354	C3 System 3 and Advanced System 3 Color Television Service
923-2363-01	Service Menus for C-Line through J2-LineColor Television Products
923-2375	C-3 Projection Television Service
923-2376	J Line Commercial Products
923-2377	C-6 Sentry 2 Color Television Service
923-2378	C-10 Sentry 2 Color Television Service
923-2379	C-11 Sentry 2 Color Television Service
923-2380	C-8 AVI Color Television Service
923-2381	C-8 Projection Television Service

A Sample Order Form

SERVICE LITERATURE AND TECHNICAL TRAINING PUBLICATION ORDER FORM

Part/Catalog No. *	Description	Quantity	Unit Price	Total
Tax Table: CA 7.75%, GA 6.00%, IL 7.75%, KS 4.9%, KY 6.00%, ME 6.00%, NJ 6.00%, PA 7.00%, RI 7.00%, TX 7.75%, VA 4.50%, WA 8.20%.		Subtotal		
		State/Local Sales Tax **		
		Shipping & Handling ***		
		Total		

(ALL PRICES ARE SUBJECT TO CHANGE WITHOUT NOTICE.)

*   If Part Number is not listed, use Catalog Number.
**   Please include State and Local Tax with all orders as listed above.
***   Shipping and Handling charges are $10.00 per order on test equipment (All Technical Manuals and Video tapes include Shipping and Handling).

PLEASE ALLOW 2 WEEKS FOR DELIVERY - NO RETURNS!

First Name   Middle Initial   Last Name

( )
Company   Phone # with area code

Street Address   Apt. #

City   State   Zip

Check One:
Check   ☐ Money Order ☐
MasterCard ☐ Visa   ☐

Credit Card #   Exp. Date

Signature

Mail your order to:
Zenith Electronics Corporation
Service Literature Distribution
1000 Milwaukee Ave.
Glenview, IL 60025

Order by Phone:
(847) 391-8941 or (847) 391-8738

Order by Fax:
(847) 391-8726

Servicing RCA/GE Televisions

Bob Rose

Let's assume you're a competent technician, with a good work ethic, knowledgeable, having the proper tools to do your job efficiently. You're hungry to learn, to find a way to get an edge on the competition-better yet, to increase your productivity, and hence, your compensation.

In Servicing RCA/GE Televisions, author Bob Rose has compiled years of personal experience to share his knowledge about the unique CTC chassis. From the early CTC130 through the CTC195/197 series, Bob reveals the most common faults and quickest ways to find them, as well as some not-so-common problems, quirks and oddities he's experienced along the way. From the RCA component numbering system to the infamous "tuner wrap" problem, Bob gives you all you need to make faster diagnoses and efficient repairs, with fewer call-backs-and that's money in the bank!

Troubleshooting & Repair
352 pages • paperback • 8-3/8" x 10-7/8"
ISBN 0-7906-1171-6 • Sams 61171 • $34.95

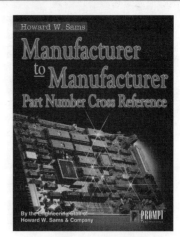

Howard W. Sams Manufacturer to Manufacturer Part Number Cross Reference

Engineering Staff of Howard W. Sams & Company

Desperate to get that replacement part? Have a workshop full of parts you aren't sure are the right replacements? With the Manufacturer-to-Manufacturer Part Number Cross Reference you will have the source for finding all the possible alternative replacement parts. The engineers at Howard W. Sams, using years of hands-on experience, have put together this handy cross reference guide.

Created from the Howard W. Sams database that was developed through the production of PHOTOFACT® service documentation, this guide gives you the knowledge to use electrically compatible parts already in stock. It saves time and money – an investment for anyone servicing consumer electronics devices.

Manufacturers Include:
==> Thompson
==> Sharp
==> Sears
==> Hitachi
==> Sony
==> Sanyo
==> Zenith

Professional Reference
320 pages • paperback • 8-1/2"x 11"
ISBN 0-7906-1207-0 • Sams 61207 • $29.95

**To order or locate your nearest Prompt® Publications distributor :
1-800-428-7267 or www.samswebsite.com**

Prices subject to change.

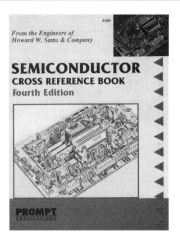

Semiconductor Cross Reference Book, *Fourth Edition*

The Engineers of Howard W. Sams &Co.

This newly revised and updated reference book is the most comprehensive guide to replacement data available for engineers, technicians, and those who work with semiconductors. With more than 490,000 part numbers, type numbers, and other identifying numbers listed, technicians will have no problem locating replacement or substitution information.

The *Semiconductor Cross Reference Book* covers all major types of semiconductors, including bipolar transistors, FETs, diodes, rectifiers, ICs, SCRs, LEDs, modules, and thermal devices. It also features replacements for NTE, ECG, Radio Shack, and TCE making this book four cross references in one. And lastly, this reference includes up-to-date listing of all original equipment manufacturers, making it easy to find all necessary parts.

This title is also available on CD-ROM!

Professional Reference
671 pages • paperback • 8-1/2" x 11"
ISBN: 0-7906-1080-9 • Sams: 61080 • $29.95

Semiconductor Cross Reference CD-ROM

The Engineers of Howard W. Sams & Co.

Now get all the benefits of the *Howard W. Sams Cross Reference Book* in an easy-to-use CD-ROM version. There are no block numbers to look up, and the search is completed in an instant, with all the information you need right there on the screen before you.

In addition to the wealth of information included in the print version, this CD-ROM version includes an **additional cross reference** of similar type numbers that have the same semiconductor replacements. In essence, this CD will replace two or three reference books in one!

Professional Reference
674 pages • paperback • 8-1/2" x 11"
ISBN: 0-7906-1140-6 • Sams 61140 • $29.95

**To order or locate your nearest Prompt® Publications distributor :
1-800-428-7267 or www.samswebsite.com**

Prices subject to change.

Guide to HDTV Systems

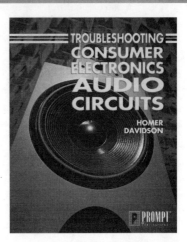

Troubleshooting Consumer Electronics Audio Circuits

Conrad Persson

As HDTV is developed, refined, and becomes more available to the masses, technicians will be required to service them. Until now, precious little information has been available on the subject. This book provides a detailed background on what HDTV is, the technical standards involved, how HDTV signals are generated and transmitted, and a generalized description of the circuitry an HDTV set consists of. Some of the topics include the ATSC digital TV standard, receiver characteristics, NTSC/HDTV compatibility, scanning methods, test equipment, servicing considerations, and more.

Homer L. Davidson

Everything from the basic amplifier of the small phono player to the high-powered auto receiver-amplifier is covered in this book. This book can provide critical electronic information for the homeowner, hobbyist, electronic student, and technician.

Troubleshooting Consumer Electronics Audio Circuits has solutions for audio electronics circuit problems, including:
- How to take critical voltages
- How to use and locate defective audio components with the multitester
- How to identify weak, distorted, intermittent, and noisy circuits
- How to service simple and inexpensive consumer electronics audio circuits
- How to troubleshoot the special audio circuits of the erase head, tape head, phono, remote control, mute, speaker relay, and tube bias circuits.

Video Technology
256 pages • paperback • 7-3/8" x 9-1/4"
ISBN 0-7906-1166-X • Sams 61166 • $34.95

Troubleshooting & Repair
368 pages • paperback • 7-3/8" x 9-1/4"
ISBN 0-7906-1165-1 • Sams 61165 • $34.95

**To order or locate your nearest Prompt® Publications distributor :
1-800-428-7267 or www.samswebsite.com**

Prices subject to change.

1999 Computer Monitor Troubleshooting Tips

M.I. Technologies, Inc.

The information in this book will save hours of troubleshooting time, the cost of searching for schematics or tips, and the irritation of slow repairs.

1999 Computer Monitor Tips includes over 3500 troubleshooting and repair tips listed by manufacturer and model number, featuring major names such as Apple, Acer, Compaq, CTX, Dell, Gateway, Goldstar, IBM, NEC, Packard Bell, Sony, Viewsonic, Zenith and many more. Repair procedures for the KD1700 series monitor, including FREE schematics and parts list is also included.

It also contains FREE FCC number identification software to cross-reference model information to the original manufacturer and contact information, as well as an overview of how VGA monitors work, an overview of monitor EEPROM repair, an overview of ESR in-circuit capacitor testers and international monitor manufacturer contact listings.

Troubleshooting and Repair
312 pages • paperback • 8-1/2" x 11"
ISBN: 0-7906-1179-1 • Sams: 61179 • $49.95

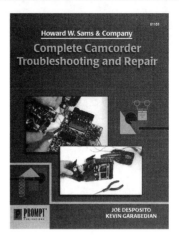

Complete Camcorder Troubleshooting & Repair

Joe Desposito & Kevin Garabedian

A video camcorder's circuits perform many important tasks such as processing video and audio signals, controlling motors, and supplying power to the machine. Though camcorders are complex, you don't need complex tools or test equipment to repair or maintain them, and this book will show the technician or hobbyists how to care for their video camcorder.

Complete Camcorder Troubleshooting & Repair contains sound troubleshooting procedures, beginning with an examination of the external parts of the camcorder then narrowing the view to gears, springs, pulleys, lenses, and other mechanical parts. *Complete Camcorder Troubleshooting & Repair* also features numerous case studies of particular camcorder models, in addition to illustrating how to troubleshoot audio and video circuits, special effect circuits, and more.

Troubleshooting & Repair
336 pages • paperback • 8-1/2" x 11"
ISBN: 0-7906-1105-8 • Sams: 61105 • $34.95

**To order or locate your nearest Prompt® Publications distributor :
1-800-428-7267 or www.samswebsite.com**

Prices subject to change.

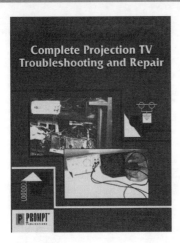

Complete Projection TV Troubleshooting and Repair

Joe Desposito and Kevin Garabedian

Complete Projection TV Troubleshooting & Repair is just the tool every professional technician, enthusiastic hobbyist or do-it-yourselfer needs to successfully repair projection TVs.

This book covers everything from the basics of projection circuits, tools and test equipment, TV types and measurements, to the finely detailed repair techniques. It also shows how to troubleshoot tuner/demodulator circuits, audio and video circuits, horizontal and vertical circuits, high-voltage circuits, microprocessor circuits, power supplies, CRTs, projection TV controls and remote control circuits.

This book also contains a case study for the Mitsubishi Model VS-458RS – a step-by-step repair with photographs and detailed schematics.

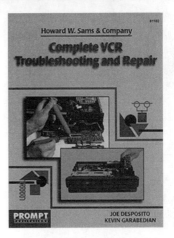

Complete VCR Troubleshooting & Repair

Joe Desposito & Kevin Garabedian

Though VCRs are complex, you don't need complex tools or test equipment to repair them. *Complete VCR Troubleshooting and Repair* contains sound troubleshooting procedures beginning with an examination of the external parts of the VCR, then narrowing the view to gears, springs, pulleys, belts and other mechanical parts. This book also shows how to troubleshoot tuner/demodulator circuits, audio and video circuits, special effect circuits, sensors and switches, video heads, servo systems and more.

Complete VCR Troubleshooting & Repair also contains nine detailed VCR case studies, each focusing on a particular model of VCR with a very specific and common problem. The case studies guide you through the repair from start to finish, using written instruction, helpful photographs and Howard W. Sams' own *VCRfacts*® schematics. Some of the problems covered include failure to rewind, tape loading problems and intermittent clock display.

Troubleshooting and Repair
184 pages • paperback • 8-1/2" x 11"
ISBN: 0-7906-1134-1 • Sams: 61134 • $34.95

Troubleshooting & Repair
184 pages • paperback • 8-1/2" x 11"
ISBN: 0-7906-1102-3 • Sams: 61102 • $34.95

**To order or locate your nearest Prompt® Publications distributor :
1-800-428-7267 or www.samswebsite.com**

Prices subject to change.

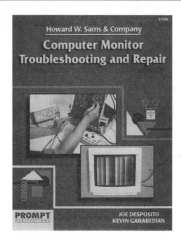

Computer Monitor Troubleshooting & Repair

Joe Desposito & Kevin Garabedian

The explosion of computer systems for homes, offices, and schools has resulted in a subsequent demand for computer monitor repair information. *Computer Monitor Troubleshooting & Repair* makes it easier for any technician, hobbyist or computer owner to successfully repair dysfunctional monitors. Learn the basics of computer monitors with chapters on tools and test equipment, monitor types, special procedures, how to find a problem and how to repair faults in the CRT. Other chapters show how to troubleshoot circuits such as power supply, high voltage, vertical, sync and video.

This book also contains six case studies which focus on a specific model of computer monitor. The problems addressed include a completely dead monitor, dysfunctional horizontal width control, bad resistors, dim display and more.

Troubleshooting & Repair
308 pages • paperback • 8-1/2" x 11"
ISBN: 0-7906-1100-7 • Sams: 61100 • $34.95

Troubleshooting & Repair Guide to TV, *Second Edition*

The Engineers of Howard W. Sams & Co.

The Howard W. Sams Troubleshooting & Repair Guide to TV is the most complete and up-to-date television repair book available, with tips on how to troubleshoot the newest circuits in today's TVs. Written for novice and professionals alike, this guide contains comprehensive and easy-to-follow basic electronics information, coverage of television basics, and extensive coverage of common TV symptoms and how to correct them. Also included are tips on how to find problems, and a question-and-answer section at the end of each chapter.

This repair guide is illustrated with useful photos, schematics, graphs, and flowcharts. It covers audio, video, technician safety, test equipment, power supplies, picture-in-picture, and much more. *The Troubleshooting & Repair Guide to TV* was written, illustrated, and assembled by the engineers and technicians of Howard W. Sams & Company, who have a combined 200 years of troubleshooting experience.

Troubleshooting & Repair
263 pages • paperback • 8-1/2" x 11"
ISBN: 0-7906-1146-7 • Sams: 61146 • $34.95

**To order or locate your nearest Prompt® Publications distributor :
1-800-428-7267 or www.samswebsite.com**

Prices subject to change.

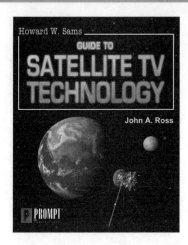

Howard W. Sams Guide to Satellite TV Technology

John A. Ross

This book covers all aspects of satellite television technology in a style that breaks "tech-talk" down into easily understood reading. It is intended to assist consumers with the installation, maintenance, and repair of their satellite systems. It also contains sufficient technical content to appeal to technicians as a reference.

Coverage includes C, Ku, and DBS signals. Chapters include How Satellite Television Technology Works, Parts of a Satellite Television Reception System, Installing the Hardware Portion of Your System, Installing the Electronics of Your System, Installing Your DSS, DBS, or Primestar System, Setting Up a Multi-Receiver Installation, Maintaining the System, Repairing Your System, and more.

Communications
464 pages • paperback • 7-3/8" x 9-1/4"
ISBN 0-7906-1176-7 • Sams 61176 • $39.95

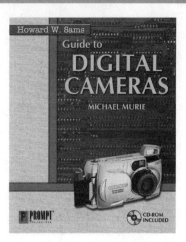

Complete Guide to Digital Cameras

Michael Murie

The *Complete Guide to Digital Cameras* will appeal to anyone who has recently purchased or is considering an investment in a digital camera. The first section introduces the reader to digital cameras, how they work, uses and features, and how to buy one. The second section gives tips on use, available options, and how to transfer images from camera to computer. The third section focuses on manipulating the images on computer in varying file formats, and looks at some color printers presently available. Along with model comparisons and index of currently available cameras, a CD-ROM contains sample images, trial software, and utilities.

The *Complete Guide to Digital Cameras* is the answer to all your questions. Author Michael Murie, avid photographer and multimedia developer, has compiled a comprehensive volume, covering all aspects of the digital camera world. As a bonus, we include a CD-ROM with sample images and software, as well as a comprehensive table of the latest digital cameras, comparing features and technical specifications.

Video Technology
536 pages • paperback • 7-3/8" x 9-1/4"
ISBN 0-7906-1175-9 • Sams 61175 • $39.95

**To order or locate your nearest Prompt® Publications distributor :
1-800-428-7267 or www.samswebsite.com**

Prices subject to change.

In-Home VCR Mechanical Repair & Cleaning Guide

Curt Reeder

Like any machine that is used in the home or office, a VCR requires minimal service to keep it functioning well and for a long time. However, a technical or electrical engineering degree is not required to begin regular maintenance on a VCR. *The In-Home VCR Mechanical Repair & Cleaning Guide* shows readers the tricks and secrets of VCR maintenance using just a few small hand tools, such as tweezers and a power screwdriver.

This book is also geared toward entrepreneurs who may consider starting a new VCR service business of their own. The vast information contained in this guide gives a firm foundation on which to create a personal niche in this unique service business.

This book is compiled from the most frequent VCR malfunctions author Curt Reeder has encountered in the six years he has operated his in-home VCR repair and cleaning service.

Troubleshooting & Repair
222 pages • paperback • 8-3/8" x 10-7/8"
ISBN: 0-7906-1076-0 • Sams: 61076 • $24.95

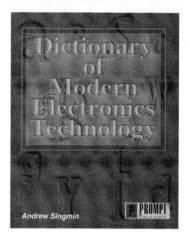

Dictionary of Modern Electronics Technology

Andrew Singmin

New technology overpowers the old everyday. One minute you're working with the quickest and most sophisticated electronic equipment, and the next minute you're working with a museum piece. The words that support your equipment change just as fast.

If you're looking for a dictionary that thoroughly defines the ever-changing and advancing world of electronic terminology, look no further than the Modern Dictionary of Electronics Technology. With up-to-date definitions and explanations, this dictionary sheds insightful light on words and terms used at the forefront of today's integrated circuit industry and surrounding electronic sectors.

Whether you're a device engineer, a specialist working in the semiconductor industry, or simply an electronics enthusiast, this dictionary is a necessary guide for your electronic endeavors.

Electronics Technology
220 pages • paperback • 7 3/8" x 9 1/4"
ISBN: 0-7906-1164-4 • Sams 61164 • $34.95

**To order or locate your nearest Prompt® Publications distributor :
1-800-428-7267 or www.samswebsite.com**

Prices subject to change.